More information about this series at https://link.springer.com/bookseries/558

Naoki Kobayashi · James Worrell
Editors

Foundations of Software Science and Computation Structures

27th International Conference, FoSSaCS 2024
Held as Part of the European Joint Conferences on Theory and Practice of Software, ETAPS 2024
Luxembourg City, Luxembourg, April 6–11, 2024
Proceedings, Part II

Springer

Editors
Naoki Kobayashi
The University of Tokyo
Tokyo, Japan

James Worrell
University of Oxford
Oxford, UK

ISSN 0302-9743 ISSN 1611-3349 (electronic)
Lecture Notes in Computer Science
ISBN 978-3-031-57230-2 ISBN 978-3-031-57231-9 (eBook)
https://doi.org/10.1007/978-3-031-57231-9

This Springer imprint is published by the registered company Springer Nature Switzerland AG
The registered company address is: Gewerbestrasse 11, 6330 Cham, Switzerland

Paper in this product is recyclable.

ETAPS Foreword

Welcome to the 27th ETAPS! ETAPS 2024 took place in Luxembourg City, the beautiful capital of Luxembourg.

ETAPS 2024 is the 27th instance of the European Joint Conferences on Theory and Practice of Software. ETAPS is an annual federated conference established in 1998, and consists of four conferences: ESOP, FASE, FoSSaCS, and TACAS. Each conference has its own Program Committee (PC) and its own Steering Committee (SC). The conferences cover various aspects of software systems, ranging from theoretical computer science to foundations of programming languages, analysis tools, and formal approaches to software engineering. Organising these conferences in a coherent, highly synchronized conference programme enables researchers to participate in an exciting event, having the possibility to meet many colleagues working in different directions in the field, and to easily attend talks of different conferences. On the weekend before the main conference, numerous satellite workshops took place that attracted many researchers from all over the globe.

ETAPS 2024 received 352 submissions in total, 117 of which were accepted, yielding an overall acceptance rate of 33%. I thank all the authors for their interest in ETAPS, all the reviewers for their reviewing efforts, the PC members for their contributions, and in particular the PC (co-)chairs for their hard work in running this entire intensive process. Last but not least, my congratulations to all authors of the accepted papers!

ETAPS 2024 featured the unifying invited speakers Sandrine Blazy (University of Rennes, France) and Lars Birkedal (Aarhus University, Denmark), and the invited speakers Ruzica Piskac (Yale University, USA) for TACAS and Jérôme Leroux (Laboratoire Bordelais de Recherche en Informatique, France) for FoSSaCS. Invited tutorials were provided by Tamar Sharon (Radboud University, the Netherlands) on computer ethics and David Monniaux (Verimag, France) on abstract interpretation.

As part of the programme we had the first ETAPS industry day. The goal of this day was to bring industrial practitioners into the heart of the research community and to catalyze the interaction between industry and academia. The day was organized by Nikolai Kosmatov (Thales Research and Technology, France) and Andrzej Wąsowski (IT University of Copenhagen, Denmark).

ETAPS 2024 was organized by the SnT - Interdisciplinary Centre for Security, Reliability and Trust, University of Luxembourg. The University of Luxembourg was founded in 2003. The university is one of the best and most international young universities with 6,000 students from 130 countries and 1,500 academics from all over the globe. The local organisation team consisted of Peter Y.A. Ryan (general chair), Peter B. Roenne (organisation chair), Maxime Cordy and Renzo Gaston Degiovanni (workshop chairs), Magali Martin and Isana Nascimento (event manager), Marjan Skrobot (publicity chair), and Afonso Arriaga (local proceedings chair). This team also organised the online edition of ETAPS 2021, and now we are happy that they agreed to also organise a physical edition of ETAPS.

ETAPS 2024 is further supported by the following associations and societies: ETAPS e.V., EATCS (European Association for Theoretical Computer Science), EAPLS (European Association for Programming Languages and Systems), and EASST (European Association of Software Science and Technology).

The ETAPS Steering Committee consists of an Executive Board, and representatives of the individual ETAPS conferences, as well as representatives of EATCS, EAPLS, and EASST. The Executive Board consists of Marieke Huisman (Twente, chair), Andrzej Wąsowski (Copenhagen), Thomas Noll (Aachen), Jan Kofroň (Prague), Barbara König (Duisburg), Arnd Hartmanns (Twente), Caterina Urban (Inria), Jan Křetínský (Munich), Elizabeth Polgreen (Edinburgh), and Lenore Zuck (Chicago).

Other members of the steering committee are: Maurice ter Beek (Pisa), Dirk Beyer (Munich), Artur Boronat (Leicester), Luís Caires (Lisboa), Ana Cavalcanti (York), Ferruccio Damiani (Torino), Bernd Finkbeiner (Saarland), Gordon Fraser (Passau), Arie Gurfinkel (Waterloo), Reiner Hähnle (Darmstadt), Reiko Heckel (Leicester), Marijn Heule (Pittsburgh), Joost-Pieter Katoen (Aachen and Twente), Delia Kesner (Paris), Naoki Kobayashi (Tokyo), Fabrice Kordon (Paris), Laura Kovács (Vienna), Mark Lawford (Hamilton), Tiziana Margaria (Limerick), Claudio Menghi (Hamilton and Bergamo), Andrzej Murawski (Oxford), Laure Petrucci (Paris), Peter Y.A. Ryan (Luxembourg), Don Sannella (Edinburgh), Viktor Vafeiadis (Kaiserslautern), Stephanie Weirich (Pennsylvania), Anton Wijs (Eindhoven), and James Worrell (Oxford).

I would like to take this opportunity to thank all authors, keynote speakers, attendees, organizers of the satellite workshops, and Springer Nature for their support. ETAPS 2024 was also generously supported by a RESCOM grant from the Luxembourg National Research Foundation (project 18015543). I hope you all enjoyed ETAPS 2024.

Finally, a big thanks to both Peters, Magali and Isana and their local organization team for all their enormous efforts to make ETAPS a fantastic event.

April 2024

Marieke Huisman
ETAPS SC Chair
ETAPS e.V. President

Preface

This volume contains the papers presented at the 27th International Conference on Foundations of Software Science and Computation Structures (FoSSaCS 2024), which was held during April 8–11, 2024 in Luxembourg City, Luxembourg. The conference is dedicated to foundational research with a clear significance for software science and brings together research on theories and methods to support the analysis, integration, synthesis, transformation, and verification of programs and software systems.

In addition to an invited talk by Jérôme Leroux (Laboratoire Bordelais de Recherche en Informatique, France) on "Ackermannian Completion of Separators", the program consisted of 24 talks on contributed papers, selected from 79 submissions. Each submission was assessed by three or more Program Committee members, with the help of external reviewers. The conference management system EasyChair was used to handle the submissions, to conduct the electronic Program Committee discussions, and to assist with the assembly of the proceedings.

We wish to thank all the authors who submitted papers for consideration, the members of the Program Committee for their conscientious work, and all additional reviewers who assisted the Program Committee in the evaluation process. We would also like to thank Andrzej Murawski, the FoSSaCS Steering Committee Chair for various pieces of advice, and the members of the ESOP/FASE/FoSSaCS joint Artifact Evaluation Committee for the artifact evaluation. Finally, we would like to thank the ETAPS organization for providing an excellent environment for FoSSaCS, the other conferences and the workshops.

February 2024

Naoki Kobayashi
James Worrell

Organization

Program Committee Chairs

Naoki Kobayashi	The University of Tokyo, Japan
James Worrell	University of Oxford, UK

Program Committee

Sandra Alves	University of Porto, Portugal
Mauricio Ayala-Rincón	Universidade de Brasília, Brazil
Stephanie Balzer	CMU, USA
Udi Boker	Reichman University, Israel
James Brotherston	University College London, UK
Corina Cirstea	University of Southampton, UK
Yuxin Deng	East China Normal University, China
Claudia Faggian	CNRS - Université Paris Cité, France
Pierre Ganty	IMDEA Software Institute, Spain
Ichiro Hasuo	National Institute of Informatics, Japan
Naoki Kobayashi	The University of Tokyo, Japan
Robbert Krebbers	Radboud University, the Netherlands
Antonin Kucera	Masaryk University, the Czech Republic
Karoliina Lehtinen	CNRS - Université Aix-Marseille, France
Bas Luttik	Eindhoven University of Technology, the Netherlands
Rasmus Ejlers Møgelberg	IT University of Copenhagen, Denmark
Luca Padovani	Università di Camerino, Italy
Catuscia Palamidessi	Inria, France
Paritosh Pandya	IIT Bombay, India
Elaine Pimentel	University College London, UK
Damien Pous	CNRS - ENS Lyon, France
Ana Sokolova	University of Salzburg, Austria
Lidia Tendera	University of Opole, Poland
Nikos Tzevelekos	Queen Mary University of London, UK
Tarmo Uustalu	Reykjavik University, Iceland
Franck van Breugel	York University, Canada
James Worrell	University of Oxford, UK

ESOP/FASE/FoSSaCS Joint Artifact Evaluation Committee

AEC Co-chairs

Tobias Kappé	Open Universiteit and ILLC, University of Amsterdam, The Netherlands
Ryosuke Sato	University of Tokyo, Japan
Stefan Winter	LMU Munich, Germany

AEC Members

Arwa Hameed Alsubhi	University of Glasgow, UK
Levente Bajczi	Budapest University of Technology and Economics, Hungary
James Baxter	University of York, UK
Matthew Alan Le Brun	University of Glasgow, UK
Laura Bussi	University of Pisa, Italy
Gustavo Carvalho	Universidade Federal de Pernambuco, Brazil
Chanhee Cho	Carnegie Mellon University, USA
Ryan Doenges	Northeastern University, USA
Zainab Fatmi	University of Oxford, UK
Luke Geeson	University College London, UK
Hans-Dieter Hiep	Leiden University, Belgium
Philipp Joram	Tallinn University of Technology, Estonia
Ulf Kargén	Linköping University, Sweden
Hiroyuki Katsura	University of Tokyo, Japan
Calvin Santiago Lee	Reykjavík University, Iceland
Livia Lestingi	Politecnico di Milano, Italy
Nuno Macedo	University of Porto and INESC TEC, Portugal
Kristóf Marussy	Budapest University of Technology and Economics, Hungary
Ivan Nikitin	University of Glasgow, UK
Hugo Pacheco	University of Porto, Portugal
Lucas Sakizloglou	Brandenburgische Technische Universität Cottbus-Senftenberg, Germany
Michael Schröder	TU Wien, Austria
Michael Schwarz	TU Munich, Germany
Wenjia Ye	University of Hong Kong, China

Additional Reviewers

Abraham, Erika
Ajdarow, Michal
An, Jie
Asada, Kazuyuki
Avanzini, Martin
Balasubramanian, A. R.
Barbosa, João
Basold, Henning
Batz, Kevin
Beohar, Harsh
Bertrand, Nathalie
Beyersdorff, Olaf
Bohn, León
Bonelli, Eduardo
Bonsangue, Marcello
Breuvart, Flavien
Bruyère, Véronique
Carette, Titouan
Chadha, Rohit
Clemente, Lorenzo
Cockett, Robin
Czerwiński, Wojciech
D'Osualdo, Emanuele
Dagnino, Francesco
De Moura, Flavio L. C.
De, Abhishek
Di Stasio, Antonio
Espírito Santo, José
Fahrenberg, Uli
Feng, Yuan
Fijalkow, Nathanaël
Filiot, Emmanuel
Fokkink, Wan
Frumin, Daniil
Galal, Zeinab
Geatti, Luca
Geuvers, Herman
van Glabbeek, Rob
van Gool, Sam
Goy, Alexandre
Guha, Shibashis
Guttenberg, Roland
Hague, Matthew
Hainry, Emmanuel
Harper, Robert
Hausmann, Daniel
Hedges, Jules
Hinrichsen, Jonas Kastberg
Ho, Hsi-Ming
Jaber, Guilhem
Jafarrahmani, Farzad
Jakl, Tomas
Jancar, Petr
Kanazawa, Makoto
Kaposi, Ambrus
Katsumata, Shin-Ya
Kavvos, Alex
Keiren, Jeroen J. A.
Kelmendi, Edon
Klaška, David
Klock Ii, Felix S.
Knight, Sophia
Koutavas, Vasileios
Krivine, Jean
König, Barbara
Laurent, Olivier
Leroux, Jérôme
Lhote, Nathan
Li, Yong
Long, Huan
Lopez, Aliaume
Loreti, Michele
Maarand, Hendrik
Madnani, Khushraj
Mallik, Kaushik
Martens, Jan
Marti, Johannes
Mascle, Corto
Mazzocchi, Nicolas
McDermott, Dylan
Melliès, Paul-André
Mery, Daniel
Michaliszyn, Jakub
Michielini, Vincent
Miculan, Marino
Moot, Richard

Morawska, Barbara
Mulder, Ike
Nguyễn, Lê Thành Dũng
Novotný, Petr
Paquet, Hugo
Piedeleu, Robin
Pinto, Luís
Proença, José
Pérez, Jorge A.
Rehak, Vojtech
Riba, Colin
Rivas, Exequiel
Rogalewicz, Adam
Rot, Jurriaan
Rowe, Reuben
Sakayori, Ken
Sarkis, Ralph
Schmid, Todd
Schmitz, Sylvain
Schröder, Lutz
Sin'Ya, Ryoma
Skrzypczak, Michał
Sobociński, Paweł
Staton, Sam
Stein, Dario
Takagi, Tsubasa
Tini, Simone
Totzke, Patrick
Urbat, Henning
Valencia, Frank
Vandenhove, Pierre
Varacca, Daniele
Veltri, Niccolò
Ventura, Daniel
Waga, Masaki
Wagemaker, Jana
Wan, Cheng-Syuan
Weil-Kennedy, Chana
Winskel, Glynn
Witkowski, Piotr
Wißmann, Thorsten
Wolter, Frank

Contents – Part II

Types and Programming Languages

Logic and Proofs

Infinite-State Systems

Contents – Part I

Invited Talk

Infinite Games

Categorical Semantics

Automata and Synthesis

Types and Programming Languages

From Rewrite Rules to Axioms in the $\lambda\Pi$-Calculus Modulo Theory

Valentin Blot[1], Gilles Dowek[1], Thomas Traversié[1,2](✉), and Théo Winterhalter[1]

[1] Université Paris-Saclay, Inria, ENS Paris-Saclay, CNRS, LMF, Gif-sur-Yvette, France
{valentin.blot,gilles.dowek,thomas.traversie,theo.winterhalter}@inria.fr
[2] Université Paris-Saclay, CentraleSupélec, MICS, Gif-sur-Yvette, France

Abstract. The $\lambda\Pi$-calculus modulo theory is an extension of simply typed λ-calculus with dependent types and user-defined rewrite rules. We show that it is possible to replace the rewrite rules of a theory of the $\lambda\Pi$-calculus modulo theory by equational axioms, when this theory features the notions of proposition and proof, while maintaining the same expressiveness. To do so, we introduce in the target theory a heterogeneous equality, and we build a translation that replaces each use of the conversion rule by the insertion of a transport. At the end, the theory with rewrite rules is a conservative extension of the theory with axioms.

Keywords: Rewrite rules · Equality · Logical Framework.

1 Introduction

For Poincaré, the reasoning by which we deduce that $2+2 = 4$ is not a meaningful proof, but a simple verification. He concludes that the goal of exact sciences is to "dispense with these direct verifications" [20]. Far from being solely a philosophical issue, this principle impacts the foundations of logical systems and in particular the choice between *axioms* and *rewrite rules*. For instance, in systems with axioms $x + \mathsf{succ}\ y = \mathsf{succ}\ (x+y)$ and $x+0 = x$, we can *prove* that $2+2 = 4$. On the other hand, in systems with rewrite rules $x + \mathsf{succ}\ y \hookrightarrow \mathsf{succ}\ (x+y)$ and $x+0 \hookrightarrow x$, we just need to prove $4 = 4$ as we can *compute* that $(2+2=4) \equiv (4=4)$. In that respect, logical systems with computation rules are convenient tools for making proofs. That is why rewrite rules have been added to systems such as AGDA [5] or COQ [12] and why Dowek [9,10] developed Deduction modulo theory, an extension of first-order logic that mixes computation and proof. Since logical systems with rewrite rules are more user-friendly, one may ask whether or not the results are the same as in axiomatic logical systems.

Rewrite rules are at the core of the $\lambda\Pi$-calculus modulo theory, an extension of simply typed λ-calculus with dependent types and user-definable rewrite rules [6]. The combination of β-reduction and of the rewrite rules of a signature Σ forms the conversion $\equiv_{\beta\Sigma}$. If we know that $t : A$ with conversion $A \equiv_{\beta\Sigma} B$,

N. Kobayashi and J. Worrell (Eds.): FoSSaCS 2024, LNCS 14575, pp. 3–23, 2024.
https://doi.org/10.1007/978-3-031-57231-9_1

then we can derive that $t : B$. In this system, a theory is a set of rewrite rules, together with a set of axioms (that are typed constants). The $\lambda\Pi$-calculus modulo theory is a powerful logical framework in which many theories can be expressed, such as Predicate logic, Simple type theory or the Calculus of constructions [3]. It is the theory behind the DEDUKTI language [2,16] and the LAMBDAPI proof assistant.

In this paper, we choose to study the replacement of rewrite rules by axioms in the $\lambda\Pi$-calculus modulo theory. Since it is a logical framework, the result applies to many theories. Moreover, as DEDUKTI is geared towards the interoperability between proof systems, if we want to exchange proofs between a system with rewrite rules and a system without rewrite rules *via* DEDUKTI, we need to replace rewrite rules by axioms in the $\lambda\Pi$-calculus modulo theory. Working in this logical framework rather than in an extension of Martin-Löf type theory [17] is therefore relevant on both theoretical and practical levels, but complicates the task as the $\lambda\Pi$-calculus modulo theory does not feature identity types or an infinite hierarchy of sorts.

One method to replace rewrite rules by axioms is to mimic the behavior of the conversion rule using transports: if we have $t : A$ and $A \equiv_{\beta\Sigma} B$ with p an equality between A and B, then we can deduce that $\mathsf{transp}\ p\ t : B$, but we do not directly have $t : B$. However trivial this seems, we face several challenges when trying to demonstrate it fully: the insertion of transports in terms and types is difficult due to the presence of dependent types, and the building of transports is involved as we cannot have inside the $\lambda\Pi$-calculus modulo theory an equality between types.

A similar problem is the elimination of equality reflection from extensional systems. Equality reflection states that $\ell = r$ implies $\ell \equiv r$, just like $\ell \hookrightarrow r$ implies $\ell \equiv r$ in systems with rewrite rules. In extensional systems, typing is eased by a more powerful conversion. Hofmann [14,15] investigated categorically the problem. Oury [19] developed a translation of proofs from an extensional version of the Calculus of Constructions to the Calculus of Inductive Constructions with equality axioms. Winterhalter, Sozeau and Tabareau [23,24] built upon this result to reduce the number of axioms needed.

The replacement of rewrite rules by axioms paves the way for the interpretation of a theory into another inside the $\lambda\Pi$-calculus modulo theory. Indeed, when interpreting a theory into another, we represent each constant of the source theory by a term in the target theory, but we cannot generally do the same for rewrite rules. We can however pre-process the source theory to replace its rewrite rules by axioms, and then interpret it. The interpretation of theories allows to prove relative consistency and relative normalization theorems [8].

Contribution. The main contribution of this paper is the translation of a theory with rewrite rules to a theory with equational axioms. To do so, we restrict the theories considered to theories with an encoding of the notions of proposition and proof inside the $\lambda\Pi$-calculus modulo theory. So as to compare objects that possibly do not have the same type, we define a heterogeneous equality—following the one defined by McBride [18]. The restriction considered allows us to build an equality between particular types—called small types. We define a type system with typed conversion for the $\lambda\Pi$-calculus modulo theory, so that the proofs are done by induction on the derivation trees more easily.

Outline of the paper. In Section 2, we present the $\lambda\Pi$-calculus modulo theory, we detail a prelude encoding of the notions of proposition and proof in it, and we identify the assumptions made on the considered theories. The heterogeneous equality and the equality between small types are presented in Section 3. The replacement of rewrite rules by axioms and the translation of terms, judgments and theories are presented in Section 4.

2 Theories in the $\lambda\Pi$-Calculus Modulo Theory

In this section, we give a more detailed overview of the $\lambda\Pi$-calculus modulo theory [6] and its type system. In particular, we present an encoding of the notions of proposition and proof in the $\lambda\Pi$-calculus modulo theory [3]. We characterize small types—a subclass of types for which we can define an equality.

2.1 The $\lambda\Pi$-Calculus Modulo Theory

The $\lambda\Pi$-calculus, also known as the Edinburgh Logical Framework [13], is an extension of simply typed λ-calculus with dependent types. The $\lambda\Pi$-calculus modulo theory ($\lambda\Pi/\equiv$) [6] is an extension of the $\lambda\Pi$-calculus, in which user-definable rewrite rules have been added [7]. Its syntax is given by:

Sorts	$s ::= \mathtt{TYPE} \mid \mathtt{KIND}$
Terms	$t, u, A, B ::= c \mid x \mid s \mid \Pi x : A.\ B \mid \lambda x : A.\ t \mid t\ u$
Contexts	$\Gamma ::= \langle\rangle \mid \Gamma, x : C$
Signatures	$\Sigma ::= \langle\rangle \mid \Sigma, c : D \mid \Sigma, \ell \hookrightarrow r$

where c is a constant and x is a variable (ranging over disjoint sets), C and r are terms, D is a closed term (*i.e.* a term with no free variables) and ℓ is a term such that $\ell = c\, t_1 \dots t_k$ with c a constant. TYPE and KIND are two sorts: terms of type TYPE are called types, and terms of type KIND are called kinds. $\Pi x : A.\ B$ is a dependent product, $\lambda x : A.\ t$ is an abstraction and $t\ u$ is an application. $\Pi x : A.\ B$ is simply written $A \to B$ if x does not appear in B. Signatures and contexts are finite sequences, and are written $\langle\rangle$ when empty. Signatures contain both typed constants and rewrite rules (written $\ell \hookrightarrow r$). $\lambda\Pi/\equiv$ is a logical framework, in which Σ is fixed by the user depending on the logic they are working in.

The relation $\hookrightarrow_{\beta\Sigma}$ is generated by β-reduction and by the rules of Σ. More explicitly, $\hookrightarrow_{\beta\Sigma}$ is the smallest relation, closed by context, such that if t rewrites to u for some rule in Σ or by β-reduction then $t \hookrightarrow_{\beta\Sigma} u$. Conversion $\equiv_{\beta\Sigma}$ is the reflexive, symmetric, and transitive closure of $\hookrightarrow_{\beta\Sigma}$.

2.2 The Type System of the $\lambda\Pi$-Calculus Modulo Theory

We introduce in Figs. 1 and 2 typing rules for $\lambda\Pi/\equiv$. Fig. 1 presents the usual typing rules while Fig. 2 focuses on the conversion rules. We write $\vdash \Gamma$ when the context Γ is well formed and $\Gamma \vdash t : A$ when t is of type A in the context Γ. $\langle\rangle \vdash t : A$ is simply written $\vdash t : A$. The notation $(\vdash \Gamma_1) \equiv (\vdash \Gamma_2)$ means that Γ_1 and Γ_2 are both well formed, have the same length and have the same variables with convertible types. We write $(\Gamma_1 \vdash t_1 : A_1) \equiv (\Gamma_2 \vdash t_2 : A_2)$ when t_1 and t_2 are convertible with $\Gamma_1 \vdash t_1 : A_1$ and $\Gamma_2 \vdash t_2 : A_2$. In particular, convertible terms $t_1 \equiv t_2$ are authorized to have different types—provided that both types are convertible—and to be typed in different contexts—provided that both contexts are convertible. In ConvRule, $\boldsymbol{x}$ is a vector representing the free variables of ℓ. The standard weakening rule and substitution lemma can be derived from this type system.

$$\frac{}{\vdash \langle\rangle}\ [\textsc{Empty}] \qquad \frac{\vdash \Gamma \qquad \Gamma \vdash A : s}{\vdash \Gamma, x : A}\ [\textsc{Decl}]\ x \notin \Gamma \qquad \frac{\vdash \Gamma}{\Gamma \vdash \mathtt{TYPE} : \mathtt{KIND}}\ [\textsc{Sort}]$$

$$\frac{\vdash \Gamma \qquad \vdash A : s}{\Gamma \vdash c : A}\ [\textsc{Const}]\ c : A \in \Sigma \qquad \frac{\vdash \Gamma}{\Gamma \vdash x : A}\ [\textsc{Var}]\ x : A \in \Gamma$$

$$\frac{\Gamma \vdash A : \mathtt{TYPE} \qquad \Gamma, x : A \vdash B : s}{\Gamma \vdash \Pi x : A.\ B : s}\ [\textsc{Prod}]$$

$$\frac{\Gamma \vdash A : \mathtt{TYPE} \qquad \Gamma, x : A \vdash B : s \qquad \Gamma, x : A \vdash t : B}{\Gamma \vdash \lambda x : A.\ t : \Pi x : A.\ B}\ [\textsc{Abs}]$$

$$\frac{\Gamma \vdash t : \Pi x : A.\ B \qquad \Gamma \vdash u : A}{\Gamma \vdash t\ u : B[x \mapsto u]}\ [\textsc{App}]$$

$$\frac{\Gamma \vdash t : A \qquad (\Gamma \vdash A : s) \equiv (\Gamma \vdash B : s)}{\Gamma \vdash t : B}\ [\textsc{Conv}]$$

Fig. 1. Typing rules of the $\lambda\Pi$-calculus modulo theory

Lemma 1 (Substitution).

- *If we have* $\vdash \Gamma, x : A, \Delta$ *and* $\Gamma \vdash u : A$, *then* $\vdash \Gamma, \Delta[x \mapsto u]$.
- *If we have* $\Gamma, x : A, \Delta \vdash t : B$ *and* $\Gamma \vdash u : A$, *then* $\Gamma, \Delta[x \mapsto u] \vdash t[x \mapsto u] : B[x \mapsto u]$.
- *If we have* $(\vdash \Gamma_1, x : A_1, \Delta_1) \equiv (\vdash \Gamma_2, x : A_2, \Delta_2)$ *and* $\Gamma_1 \vdash u : A_1$, *then* $(\vdash \Gamma_1, \Delta_1[x \mapsto u]) \equiv (\vdash \Gamma_2, \Delta_2[x \mapsto u])$.

$$\frac{\Gamma \vdash u : A}{(\Gamma \vdash u : A) \equiv (\Gamma \vdash u : A)}\ [\textsc{ConvRefl}] \qquad \frac{(\Gamma \vdash u : A) \equiv (\Gamma \vdash v : B)}{(\Gamma \vdash v : B) \equiv (\Gamma \vdash u : A)}\ [\textsc{ConvSym}]$$

$$\frac{(\Gamma \vdash u : A) \equiv (\Gamma \vdash v : B) \qquad (\Gamma \vdash v : B) \equiv (\Gamma \vdash w : C)}{(\Gamma \vdash u : A) \equiv (\Gamma \vdash w : C)}\ [\textsc{ConvTrans}]$$

$$\frac{(\vdash \Gamma_1) \equiv (\vdash \Gamma_2) \qquad (\Gamma_1 \vdash A_1 : s) \equiv (\Gamma_2 \vdash A_2 : s)}{(\vdash \Gamma_1, x : A_1) \equiv (\vdash \Gamma_2, x : A_2)}\ [\textsc{ConvDecl}]\ x \notin \Gamma_1, \Gamma_2$$

$$\frac{(\vdash \Gamma_1) \equiv (\vdash \Gamma_2) \qquad \vdash A : s}{(\Gamma_1 \vdash c : A) \equiv (\Gamma_2 \vdash c : A)}\ [\textsc{ConvConst}]\ c : A \in \Sigma$$

$$\frac{(\vdash \Gamma_1) \equiv (\vdash \Gamma_2)}{(\Gamma_1 \vdash x : A_1) \equiv (\Gamma_2 \vdash x : A_2)}\ [\textsc{ConvVar}]\ x : A_1 \in \Gamma_1, x : A_2 \in \Gamma_2$$

$$\frac{\begin{array}{c}(\Gamma_1 \vdash A_1 : \mathtt{TYPE}) \equiv (\Gamma_2 \vdash A_2 : \mathtt{TYPE}) \\ (\Gamma_1, x : A_1 \vdash B_1 : s) \equiv (\Gamma_2, x : A_2 \vdash B_2 : s)\end{array}}{(\Gamma_1 \vdash \Pi x : A_1.\ B_1 : s) \equiv (\Gamma_2 \vdash \Pi x : A_2.\ B_2 : s)}\ [\textsc{ConvProd}]$$

$$\frac{\begin{array}{c}(\Gamma_1 \vdash A_1 : \mathtt{TYPE}) \equiv (\Gamma_2 \vdash A_2 : \mathtt{TYPE}) \\ (\Gamma_1, x : A_1 \vdash B_1 : s) \equiv (\Gamma_2, x : A_2 \vdash B_2 : s) \\ (\Gamma_1, x : A_1 \vdash t_1 : B_1) \equiv (\Gamma_2, x : A_2 \vdash t_2 : B_2)\end{array}}{(\Gamma_1 \vdash \lambda x : A_1.\ t_1 : \Pi x : A_1.\ B_1) \equiv (\Gamma_2 \vdash \lambda x : A_2.\ t_2 : \Pi x : A_2.\ B_2)}\ [\textsc{ConvAbs}]$$

$$\frac{\begin{array}{c}(\Gamma_1 \vdash t_1 : \Pi x : A_1.\ B_1) \equiv (\Gamma_2 \vdash t_2 : \Pi x : A_2.\ B_2) \\ (\Gamma_1 \vdash u_1 : A_1) \equiv (\Gamma_2 \vdash u_2 : A_2)\end{array}}{(\Gamma_1 \vdash t_1\ u_1 : B_1[x \mapsto u_1]) \equiv (\Gamma_2 \vdash t_2\ u_2 : B_2[x \mapsto u_2])}\ [\textsc{ConvApp}]$$

$$\frac{\Gamma \vdash A : \mathtt{TYPE} \qquad \Gamma, x : A \vdash t : B \qquad \Gamma, x : A \vdash B : s \qquad \Gamma \vdash u : A}{(\Gamma \vdash (\lambda x : A.\ t)\ u : B[x \mapsto u]) \equiv (\Gamma \vdash t[x \mapsto u] : B[x \mapsto u])}\ [\textsc{ConvBeta}]$$

$$\frac{\boldsymbol{x} : \boldsymbol{B} \vdash \ell : A \qquad \boldsymbol{x} : \boldsymbol{B} \vdash r : A \qquad \Gamma \vdash \boldsymbol{t} : \boldsymbol{B}}{(\Gamma \vdash \ell[\boldsymbol{x} \mapsto \boldsymbol{t}] : A[\boldsymbol{x} \mapsto \boldsymbol{t}]) \equiv (\Gamma \vdash r[\boldsymbol{x} \mapsto \boldsymbol{t}] : A[\boldsymbol{x} \mapsto \boldsymbol{t}])}\ [\textsc{ConvRule}]\ \ell \hookrightarrow r \in \Sigma$$

$$\frac{\Gamma \vdash u : A \qquad (\Gamma \vdash A : s) \equiv (\Gamma \vdash B : s)}{(\Gamma \vdash u : A) \equiv (\Gamma \vdash u : B)}\ [\textsc{ConvConv}]$$

Fig. 2. Convertibility rules of the λΠ-calculus modulo theory

– *If we have* $(\Gamma_1, x : A_1, \Delta_1 \vdash t_1 : B_1) \equiv (\Gamma_2, x : A_2, \Delta_2 \vdash t_2 : B_2)$ *and* $\Gamma_1 \vdash u : A_1$*, then* $(\Gamma_1, \Delta_1[x \mapsto u] \vdash t_1[x \mapsto u] : B_1[x \mapsto u]) \equiv (\Gamma_2, \Delta_2[x \mapsto u] \vdash t_2[x \mapsto u] : B_2[x \mapsto u])$.

Proof. We proceed by induction on the typing derivation.

We chose to present a type system with *typed* conversion (written $\equiv$)—so as to easily do proofs on the derivations—while the usual type system for $\lambda\Pi/\equiv$ features *untyped* conversion (written $\equiv_{\beta\Sigma}$). The equivalence between type systems with typed conversion and type systems with untyped conversion has been a longstanding question: Geuvers and Werner [11] investigated the case of Pure Type Systems with $\beta\eta$-convertibility, Adams [1] proved the equivalence in the case of functional Pure Type Systems, and Siles [21,22] later proved the equivalence in the general case of the Pure Type Systems. The case of $\lambda\Pi/\equiv$, in which we have β-convertibility but also user-defined rewrite rules, remains to be investigated.

We write $|\Sigma|$ for the set of constants of Σ, and $\Lambda(\Sigma)$ for the set of terms t whose constants belong to $|\Sigma|$. We say that $\mathcal{T} = \Sigma$ is a theory when for each rule $\ell \hookrightarrow r \in \Sigma$ we have ℓ and r in $\Lambda(\Sigma)$, when $\hookrightarrow_{\beta\Sigma}$ is confluent on $\Lambda(\Sigma)$, and when every rule of Σ preserves typing in Σ (that is when for all context Γ and for all term $A \in \Lambda(\Sigma)$, if $\Gamma \vdash \ell : A$ then $\Gamma \vdash r : A$).

Example 1 (Natural numbers and lists). We can define in $\lambda\Pi/\equiv$ a partial theory of natural numbers and indexed lists of natural numbers. nat represents the type of natural numbers and list represents the dependent type of indexed lists of natural numbers. cons adds a new element to a list, concat concatenates two lists, and isRev checks if the first given list is the reverse of the second.

$$\mathsf{nat} : \mathsf{TYPE} \qquad 0 : \mathsf{nat} \qquad \mathsf{succ} : \mathsf{nat} \to \mathsf{nat} \qquad + : \mathsf{nat} \to \mathsf{nat} \to \mathsf{nat}$$

$$x + 0 \hookrightarrow x \qquad x + \mathsf{succ}\ y \hookrightarrow \mathsf{succ}\ (x + y) \qquad \mathsf{list} : \mathsf{nat} \to \mathsf{TYPE} \qquad \mathsf{nil} : \mathsf{list}\ 0$$

$$\mathsf{cons} : \Pi x : \mathsf{nat}.\ \mathsf{list}\ x \to \mathsf{nat} \to \mathsf{list}\ (\mathsf{succ}\ x)$$

$$\mathsf{isRev} : \Pi x : \mathsf{nat}.\ \mathsf{list}\ x \to \mathsf{list}\ x \to \mathsf{TYPE}$$

$$\mathsf{concat} : \Pi x, y : \mathsf{nat}.\ \mathsf{list}\ x \to \mathsf{list}\ y \to \mathsf{list}\ (x + y)$$

In the context $\ell : \mathsf{list}\ (\mathsf{succ}\ 0)$, we have $\mathsf{concat}\ (\mathsf{succ}\ 0)\ 0\ \ell\ \mathsf{nil}$ of type $\mathsf{list}\ (\mathsf{succ}\ 0 + 0)$. If we want to compare ℓ and this new list with isRev, we cannot directly do it because they do not have the same type. However, we can use the conversion rule with $\mathsf{list}\ (\mathsf{succ}\ 0 + 0) \equiv_{\beta\Sigma} \mathsf{list}\ (\mathsf{succ}\ 0)$. This conversion derives from the rewrite rule $x + 0 \hookrightarrow x$ instantiated with $x := \mathsf{succ}\ 0$.

2.3 A Prelude Encoding for the $\lambda\Pi$-Calculus Modulo Theory

It is possible to introduce in $\lambda\Pi/\equiv$ the notions of proposition and proof [3]. In particular, this encoding—called prelude encoding—gives the possibility to quantify on certain propositions through codes, which is not possible inside the standard $\lambda\Pi/\equiv$. This encoding is defined by following signature.

Definition 1. *The signature Σ_{pre} contains the following constants and rewrite rules:*

$$
\begin{array}{ll}
Set : \mathtt{TYPE} & o : Set \\
El : Set \to \mathtt{TYPE} & Prf : El\ o \to \mathtt{TYPE} \\
\leadsto_d : \Pi x : Set.\ (El\ x \to Set) \to Set & \Rightarrow_d : \Pi x : El\ o.\ (Prf\ x \to El\ o) \to El\ o \\
\pi : \Pi x : El\ o.\ (Prf\ x \to Set) \to Set & \forall : \Pi x : Set.\ (El\ x \to El\ o) \to El\ o \\
El\ (x \leadsto_d y) \hookrightarrow \Pi z : El\ x.\ El\ (y\ z) & Prf\ (x \Rightarrow_d y) \hookrightarrow \Pi z : Prf\ x.\ Prf\ (y\ z) \\
El\ (\pi\ x\ y) \hookrightarrow \Pi z : Prf\ x.\ El\ (y\ z) & Prf\ (\forall\ x\ y) \hookrightarrow \Pi z : El\ x.\ Prf\ (y\ z)
\end{array}
$$

We declare the constant *Set*, which represents the universe of types, along with the injection *El* that maps terms of type *Set* into TYPE. o is a term of type *Set* such that *El* o defines the universe of propositions. The injection *Prf* maps propositions into TYPE. $\leadsto_d$ (respectively $\Rightarrow_d$) is written infix and is used to represent dependent function types between terms of type *Set* (respectively *El* o). The symbol π (respectively $\forall$) is used to represent dependent function types between elements of type *El* o and *Set* (respectively *Set* and *El* o).

The main advantage of this encoding is that it allows us to quantify on propositions. Indeed, in $\lambda\Pi/\equiv$, we cannot quantify on TYPE. Instead, we can quantify on objects of type *El* o, and then inject them into TYPE using *Prf*.

2.4 Small Types and Small Derivations

As we work in $\lambda\Pi/\equiv$ rather than in an extension of Martin-Löf type theory, we do not have a pre-defined equality. Moreover, we cannot define an equality between types since such object would have type TYPE $\to$ TYPE $\to$ TYPE, which is not allowed in $\lambda\Pi/\equiv$.

If we want to compare types *Prf* a and *Prf* b, we cannot do it directly, but we can compare a and b (that are of type *El* o). We can proceed similarly to compare types *El* a and *El* b (with a and b of type *Set*). In that respect, we want types to be into a special form—called small type—that takes advantages of the prelude encoding, so as to compare them if necessary. To put types of the prelude encoding into this special form, we use the reverse of the rewrite rules of Σ_{pre} to represent dependent types with the symbols $\leadsto_d$, $\Rightarrow_d$, π and $\forall$ whenever it is possible. This is achieved by the partial function ν, defined by:

$$\nu(Set) = Set \qquad \nu(Prf\ a) = Prf\ a \qquad \nu(El\ a) = El\ a$$

$$
\nu(\Pi x : A.\ B) = \begin{array}{ll}
Prf\ (a \Rightarrow_d (\lambda x : Prf\ a.\ b)) & \text{if } \nu(A) = Prf\ a \text{ and } \nu(B) = Prf\ b \\
El\ (a \leadsto_d (\lambda x : El\ a.\ b)) & \text{if } \nu(A) = El\ a \text{ and } \nu(B) = El\ b \\
Prf\ (\forall\ a\ (\lambda x : El\ a.\ b)) & \text{if } \nu(A) = El\ a \text{ and } \nu(B) = Prf\ b \\
El\ (\pi\ a\ (\lambda x : Prf\ a.\ b)) & \text{if } \nu(A) = Prf\ a \text{ and } \nu(B) = El\ b \\
\Pi x : \nu(A).\ \nu(B) & \text{otherwise}
\end{array}
$$

Therefore, when $\nu(A)$ is defined, we have $A \equiv_{\beta\Sigma_{pre}} \nu(A)$. Note that ν is partial because we do not handle the case where a type is a β-reducible expression, as in practice we will not have types under λ-abstraction form.

To continue to characterize a particular form of types, we define the three following grammars:

$$\mathcal{S} ::= Set \mid \mathcal{S} \to \mathcal{S} \qquad\qquad \mathcal{P} ::= Prf\ a \mid \mathcal{P} \to \mathcal{S} \mid \Pi z : \mathcal{S}.\ \mathcal{P}$$

$$\mathcal{E} ::= El\ b \mid \mathcal{E} \to \mathcal{S} \mid \Pi z : \mathcal{S}.\ \mathcal{E}$$

with $a : El\ o$ and $b : Set$. The notation $A \in \mathcal{S}$ means that A is generated by the grammar $\mathcal{S}$. The grammar $\mathcal{S}$ generates types that only contain *Set*. Therefore, if $\nu(A) \in \mathcal{S}$ then $\nu(A) = A$. The grammars $\mathcal{P}$ and $\mathcal{E}$ generate types that contain a central symbol *Prf* or *El*.

Definition 2 (Small type, Small context). *A type A is small when $\nu(A)$ is defined and $\nu(A) \in \mathcal{S} \cup \mathcal{P} \cup \mathcal{E}$. In that case, $\nu(A)$ is called the small form of A. A context Γ is small when for every $x : A \in \Gamma$ we have that A is a small type.*

Example 2. $Prf\ a \to Prf\ b$, with $a, b : El\ o$, is a small type since its small form $Prf\ (a \Rightarrow_d (\lambda z.\ b))$ is generated by the grammar $\mathcal{P}$. The type $\Pi x : Prf\ b.\ El\ c$, with $c : Set$ depending on x, is a small type since its small form $El\ (\pi\ b\ (\lambda x : Prf\ b.\ c))$ is generated by the grammar $\mathcal{E}$. The type $Prf\ a \to Set \to Prf\ b$ is not small, since $\nu(Prf\ a \to Set \to Prf\ b) = Prf\ a \to Set \to Prf\ b \notin \mathcal{S} \cup \mathcal{P} \cup \mathcal{E}$.

We would ideally like all the types to be small, so that we can compare them if necessary. Therefore, if $\Gamma \vdash t : A$, we want A to be a small type, or t to be a small type and $A = \mathsf{TYPE}$. However, small types are built using the constants of Σ_{pre}. In particular, the type of the constants o, $\leadsto_d$, $\Rightarrow_d$ and $\forall$ are small, but the types of π, *Prf* and *El* are not. Note that the type of an application of π, *Prf* or *El* is small. We thus come up with the following notion.

Definition 3 (Small judgment). *$\vdash \Gamma$ is a small judgment when Γ is a small context. $\Gamma \vdash t : A$ is a small judgment when Γ is a small context and when*

- *$t : A \in \Sigma_{pre}$,*
- *or t is the type of a constant of Σ_{pre},*
- *or A is a small type,*
- *or t is a small type.*

$(\Gamma_1 \vdash t_1 : A_1) \equiv (\Gamma_2 \vdash t_2 : A_2)$ is a small judgment when $\Gamma_1 \vdash t_1 : A_1$ and $\Gamma_2 \vdash t_2 : A_2$ are small.

Definition 4 (Small derivation). *A small derivation is a derivation in which all the judgments are small.*

2.5 Theories with Prelude Encoding

We define the theories we will consider in the rest of the paper: theories that features the prelude encoding inside $\lambda\Pi/\equiv$.

Definition 5 (Theory with prelude encoding). *We say that a theory $\mathcal{T} = \Sigma$ in the $\lambda\Pi/\equiv$ is a theory with prelude encoding when:*

– *there exists* $\Sigma_{\mathcal{T}}$ *such that* $\Sigma = \Sigma_{pre} \cup \Sigma_{\mathcal{T}}$ *and* $\Sigma_{pre} \cap \Sigma_{\mathcal{T}} = \emptyset$,
– *for every* $c : A \in \Sigma_{\mathcal{T}}$, A *is small and admits a small derivation* $\vdash A : \mathsf{TYPE}$,
– *for every* $\ell \hookrightarrow r \in \Sigma_{\mathcal{T}}$, *we have small derivations* $\boldsymbol{x} : \boldsymbol{B} \vdash \ell : A$ *and* $\boldsymbol{x} : \boldsymbol{B} \vdash r : A$ *with* A *a small type, where* $\boldsymbol{x}$ *represents the free variables of* ℓ.

A theory with prelude encoding is a theory with the constants and rewrite rules Σ_{pre}, and additional user-defined constants and rewrite rules. To ensure that $\Sigma_{\mathcal{T}}$ is encoded *inside* the prelude encoding, we can only define new constants whose types are small. We do not allow the use of rewrite rules $\ell \hookrightarrow r$ when ℓ has TYPE in its type. In particular, we cannot define new rewrite rules on *Prf* or *El* and change the behavior of these constants. It follows that the three grammars $\mathcal{S}$, $\mathcal{P}$ and $\mathcal{E}$ generate disjoint types.

In the following examples, we present three theories with prelude encoding in $\lambda\Pi/\equiv$. The examples of predicate logic and set theory illustrate that the restrictions considered are generally respected, even for expressive theories.

Example 3 (Predicate logic). Predicate logic can be encoded in a theory with prelude encoding. We declare constants for tautology and contradiction $\top, \bot : El\ o$, for negation $\neg : El\ o \to El\ o$, for conjunction and disjunction $\wedge, \vee : El\ o \to El\ o \to El\ o$, and for existential quantification $\exists : \Pi z : Set.\ (El\ z \to El\ o) \to El\ o$. The semantics of tautology is defined by the rewrite rule $\top \hookrightarrow \forall\ o\ (\lambda x : El\ o.\ x \Rightarrow x)$, which is equivalent to the more common form $Prf\ \top \hookrightarrow \Pi z : El\ o.\ Prf\ z \to Prf\ z$. The rewrite rule $Prf\ (A \wedge B) \hookrightarrow \Pi P : El\ o.\ (Prf\ A \to Prf\ B \to Prf\ P) \to Prf\ P$ can be encoded by $A \wedge B \hookrightarrow \forall\ o\ (\lambda P.\ (A \to B \to P) \to P)$. The rule $Prf\ (\neg A) \hookrightarrow Prf\ A \to Prf\ \bot$ is forbidden, but $\neg A \hookrightarrow A \Rightarrow \bot$ is allowed. We proceed similarly the other rewrite rules.

Example 4 (Natural numbers and lists). We can define our small theory of natural numbers and lists in the prelude encoding, by replacing TYPE by *Set* (in the universe of types) or *El* o (in the universe of propositions), and by adding *El* and *Prf* at the necessary positions.

$$\mathsf{nat} : Set \qquad 0 : El\ \mathsf{nat} \qquad \mathsf{succ} : El\ \mathsf{nat} \to El\ \mathsf{nat} \qquad + : El\ \mathsf{nat} \to El\ \mathsf{nat} \to El\ \mathsf{nat}$$

$$\mathsf{list} : El\ \mathsf{nat} \to Set \qquad x + 0 \hookrightarrow x \qquad x + \mathsf{succ}\ y \hookrightarrow \mathsf{succ}\ (x + y)$$

$$\mathsf{nil} : El\ (\mathsf{list}\ 0) \qquad \mathsf{cons} : \Pi x : El\ \mathsf{nat}.\ El\ \mathsf{list}\ x \to El\ \mathsf{nat} \to El\ (\mathsf{list}\ (\mathsf{succ}\ x))$$

$$\mathsf{isRev} : \Pi x : El\ \mathsf{nat}.\ El\ (\mathsf{list}\ x) \to El\ (\mathsf{list}\ x) \to El\ o$$

$$\mathsf{concat} : \Pi x, y : El\ \mathsf{nat}.\ El\ (\mathsf{list}\ x) \to El\ (\mathsf{list}\ y) \to El\ (\mathsf{list}\ (x + y))$$

Example 5 (Set theory). The implementation in DEDUKTI of set theory [4] is a theory with prelude encoding. In this implementation, sets are represented by a more primitive notion of pointed graphs: we have graph and node of type *Set*. The predicate $\eta : El\ \mathsf{graph} \to El\ \mathsf{node} \to El\ \mathsf{node} \to El\ o$ is such that $\eta\ a\ x\ y$ is the proposition asserting that there is an edge in a from y to x. The operator $\mathsf{root} : El\ \mathsf{graph} \to El\ \mathsf{node}$ returns the root of a, which is a node.

In practice, the derivations of small judgments are small derivations. As we consider theories with prelude encoding, the only way of introducing a judgment that is not small is through λ-abstractions. For instance in Example 4 the judgment $\vdash El\ (\mathsf{list}\ ((\lambda x : El\ \mathsf{nat}.\ \lambda y : Set.\ x)\ 0\ \mathsf{nat})) : \mathtt{TYPE}$ is small, but in its derivation we have $\vdash \lambda x : El\ \mathsf{nat}.\ \lambda y : Set.\ x : El\ \mathsf{nat} \rightarrow Set \rightarrow El\ \mathsf{nat}$ which is not a small judgment. However, $\vdash El\ (\mathsf{list}\ 0) : \mathtt{TYPE}$ admits a small derivation. If the derivation is not small, we can in practice apply β-reduction on the fragments of the derivation that are not small to obtain a small derivation.

3 Equalities

Since we want to replace rewrite rules $\ell \hookrightarrow r$ by equational axioms $\ell = r$, we need to define an equality in the target theory. In this section, we present a heterogeneous equality and a method to compare small types. The heterogeneous equality is necessary to compare objects that do not have the same type. Although we cannot define an equality between types in $\lambda\Pi/\equiv$, it is possible to develop an equality between small types, taking advantage of their structure.

3.1 Heterogeneous Equality

In our development, we need to have an equality between two translations of the same term. However, the two translations do not necessarily have the same type, as we may have introduced transports over the course of the translation. To that end, we define a heterogeneous equality inspired by the one of McBride [18]. Our heterogeneous equality is defined by the constant schemas $\mathsf{heq}_{A,B} : A \rightarrow B \rightarrow El\ o$ where A and B are of type $\mathtt{TYPE}$. We write $u\ {}_A{\approx}_B\ v$ for $Prf\ (\mathsf{heq}_{A,B}\ u\ v)$. Heterogeneous equality is reflexive, symmetric, and transitive.

$$\begin{aligned}
&\mathsf{refl}_A : \Pi u : A.\ u\ {}_A{\approx}_A\ u \\
&\mathsf{sym}_{A,B} : \Pi u : A.\ \Pi v : B.\ u\ {}_A{\approx}_B\ v \rightarrow v\ {}_B{\approx}_A\ u \\
&\mathsf{trans}_{A,B,C} : \Pi u : A.\ \Pi v : B.\ \Pi w : C.\ u\ {}_A{\approx}_B\ v \rightarrow v\ {}_B{\approx}_C\ w \rightarrow u\ {}_A{\approx}_C\ w
\end{aligned}$$

When two objects have the same type, heterogeneous equality acts as Leibniz equality. In particular, we can replace u by v in the universes of propositions and types. The result of a Leibniz substitution on t remains equal to t.

$$\begin{aligned}
&\mathsf{leib}_A^{\mathsf{Prf}} && : \Pi u, v : A.\ \Pi p : u\ {}_A{\approx}_A\ v.\ \Pi P : A \rightarrow El\ o.\ Prf\ (P\ u) \rightarrow Prf\ (P\ v) \\
&\mathsf{eqLeib}_A^{\mathsf{Prf}} && : \Pi u, v : A.\ \Pi p : u\ {}_A{\approx}_A\ v.\ \Pi P : A \rightarrow El\ o.\ \Pi t : Prf\ (P\ u). \\
& && \quad \mathsf{leib}_A^{\mathsf{Prf}}\ u\ v\ p\ P\ t\ {}_{Prf\ (P\ v)}{\approx}_{Prf\ (P\ u)}\ t
\end{aligned}$$

The same axiom schemas exist for the universe of types, with superscript El instead of Prf, El instead of Prf, and Set instead of $El\ o$.

Finally, we add axioms for the congruence of each constructor of $\lambda\Pi/\equiv$.

Application constructor. For the application, we take:

$$\begin{aligned}\mathsf{app}_{A_1,A_2,B_1,B_2} : \Pi t_1 : (\Pi x : A_1.\ B_1).\ \Pi t_2 : (\Pi x : A_2.\ B_2).\\ \Pi u_1 : A_1.\ \Pi u_2 : A_2.\ t_1 \approx t_2 \to u_1 \approx u_2\\ \to t_1\ u_1\ {}_{B_1[x\mapsto u_1]}{\approx}_{B_2[x\mapsto u_2]}\ t_2\ u_2\end{aligned}$$

For the λ-abstraction and Π-type constructors, we cannot directly build equality axioms. Indeed, if we want to define an equality between functional terms t_1 of type $\Pi x : A_1.\ B_1$ and t_2 of type $\Pi x : A_2.\ B_2$, we need to ensure that types A_1 and A_2 are equal. Therefore, we would like to have

$$\begin{aligned}\mathsf{fun}_{A_1,A_2,B_1,B_2} : \Pi t_1 : (\Pi x : A_1.\ B_1).\ \Pi t_2 : (\Pi y : A_2.\ B_2).\ A_1 \approx A_2\\ \to (\Pi x : A_1.\ \Pi y : A_2.\ x \approx y \to t_1\ x \approx t_2\ y)\\ \to t_1 \approx t_2\end{aligned}$$

but we cannot take such an axiom, since the heterogeneous equality is not defined to compare objects that have type TYPE, and $A_1 \approx A_2$ is therefore ill typed. This shortcoming is addressed by developing an equality between small types.

3.2 Equality between Small Types

We cannot build an equality between types, since such an equality would have type $\mathsf{TYPE} \to \mathsf{TYPE} \to \mathsf{TYPE}$, which is impossible in $\lambda\Pi/\equiv$. An option would be to take axiom schemas $A \approx B$ for every equality between types A and B. Such an equality would be too far from standard and would require additional axioms to build transports. An alternative is to define an equality between small types. By construction, if $\nu(A) \in \mathcal{P}$, then $\nu(A)$ is generated from $\mathit{Prf}\ a$ for some $a : \mathit{El}\ o$, and if $\nu(A) \in \mathcal{E}$, then $\nu(A)$ is generated from $\mathit{El}\ a$ for some $a : \mathit{Set}$. If the small form of A contains $\mathit{Prf}\ a$ and the small form of B contains $\mathit{Prf}\ b$, then we want an equality between a and b. We define the partial function κ on small forms by

$$\kappa(\mathit{Prf}\ a_1, \mathit{Prf}\ a_2) = a_1 \approx a_2 \qquad \kappa(\mathit{El}\ a_1, \mathit{El}\ a_2) = a_1 \approx a_2$$

$$\kappa(S, S) = \mathsf{True} \text{ if } S \in \mathcal{S} \qquad \kappa(T_1 \to S, T_2 \to S) = \kappa(T_1, T_2) \text{ if } S \in \mathcal{S}$$

$$\kappa(\Pi z : S.\ T_1, \Pi z : S.\ T_2) = \Pi z : S.\ \kappa(T_1, T_2) \text{ if } S \in \mathcal{S}$$

where $\mathsf{True} := \Pi P : \mathit{El}\ o.\ \mathit{Prf}\ P \to \mathit{Prf}\ P$, so we can always give a witness of $\kappa(S, S)$ if $S \in \mathcal{S}$. By convention, we simply write $\kappa(A, B)$ for the result of $\kappa(\nu(A), \nu(B))$.

Example 6. $\kappa(\Pi x : \mathit{Set}.\ \mathit{Prf}\ P \to \mathit{Prf}\ Q, \Pi x : \mathit{Set}.\ \mathit{Prf}\ R) = \Pi x : \mathit{Set}.\ (P \Rightarrow_d \lambda z : P.\ Q) \approx R$ since $\nu(\Pi x : \mathit{Set}.\ \mathit{Prf}\ P \to \mathit{Prf}\ Q) = \Pi x : \mathit{Set}.\ \mathit{Prf}\ (P \Rightarrow_d (\lambda z : P.\ Q))$.

We can now go back to the definition of equality axioms for the constructors of $\lambda\Pi/\equiv$.

Function constructor. If A_1 and A_2 are small types, we can take $\kappa(A_1, A_2)$. We do not compare objects of type TYPE anymore, but objects that have either type $El\ o$ or type Set. The axiom schema for the function constructor is thus:

$$\begin{aligned}\mathsf{fun}_{A_1,A_2,B_1,B_2} : \Pi t_1 : (\Pi x : A_1.\ B_1).\ \Pi t_2 : (\Pi y : A_2.\ B_2).\ \kappa(A_1, A_2)\\ \to (\Pi x : A_1.\ \Pi y : A_2.\ x \approx y \to t_1\ x \approx t_2\ y)\\ \to t_1 \approx t_2\end{aligned}$$

This axiom schema is a generalization of the *functional extensionality* principle with distinct domains A_1 and A_2 in the case of heterogeneous equality. Functional extensionality states that two pointwise-equal functions are equal. If the domains A_1 and A_2 are generated by $\mathcal{S}$, then they are syntactically equal and we can derive a simpler axiom schema:

$$\begin{aligned}\mathsf{fun}_{A,B_1,B_2} : \Pi t_1 : (\Pi x : A.\ B_1).\ \Pi t_2 : (\Pi x : A.\ B_2).\ (\Pi x : A.\ t_1\ x \approx t_2\ x)\\ \to t_1 \approx t_2\end{aligned}$$

Π-type constructor. The congruence axiom for dependent types aims at building $\kappa(\Pi x : A_1.\ B_1, \Pi x : A_2\ B_2)$. There are different cases depending on the grammars generating $\nu(A_1)$, $\nu(A_2)$, $\nu(B_1)$ and $\nu(B_2)$. If $\nu(A_1)$, $\nu(A_2)$, $\nu(B_1)$, $\nu(B_2) \in \mathcal{S}$, then $\Pi x : A_1.\ B_1$ and $\Pi x : A_2.\ B_2$ are syntactically equal and we can build an object of type True. If $\nu(A_1)$, $\nu(A_2) \in \mathcal{S}$ and $\nu(B_1)$, $\nu(B_2) \in \mathcal{P} \cup \mathcal{E}$, then $A_1 = A_2$ and $\kappa(\Pi x : A_1.\ B_1, \Pi x : A_2\ B_2) = \Pi x : A_1.\ \kappa(B_1, B_2)$. If $\nu(A_1)$, $\nu(A_2) \in \mathcal{P} \cup \mathcal{E}$ and $\nu(B_1)$, $\nu(B_2) \in \mathcal{S}$, then $B_1 = B_2$ and $\kappa(\Pi x : A_1.\ B_1, \Pi x : A_2\ B_2) = \kappa(A_1, A_2)$. If $\nu(A_1)$, $\nu(A_2)$, $\nu(B_1)$, $\nu(B_2) \in \mathcal{P} \cup \mathcal{E}$, then there are four cases, corresponding to $\leadsto_d$, $\Rightarrow_d$, π and $\forall$. For instance, if $\nu(A_1)$, $\nu(A_2)$, $\nu(B_1)$ and $\nu(B_2)$ are all generated by $\mathcal{E}$, then necessarily we have $\nu(A_1) = El\ a_1$, $\nu(A_2) = El\ a_2$, $\nu(B_1) = El\ b_1$ and $\nu(B_2) = El\ b_2$. Therefore $\kappa(\Pi x : A_1.\ B_1, \Pi x : A_2.\ B_2) := (a_1 {\leadsto_d} (\lambda x : El\ a_1.\ b_1)) \approx (a_2 {\leadsto_d} (\lambda y : El\ a_2.\ b_2))$. The axiom is:

$$\begin{aligned}\mathsf{prod}_{\leadsto_d} : \Pi a_1, a_2 : Set.\ \Pi b_1 : (El\ a_1 \to Set).\ \Pi b_2 : (El\ a_2 \to Set).\ a_1 \approx a_2\\ \to (\Pi x : El\ a_1.\ \Pi y : El\ a_2.\ x \approx y \to b_1\ x \approx b_2\ y)\\ \to (a_1 \leadsto_d b_1) \approx (a_2 \leadsto_d b_2)\end{aligned}$$

Note that this axiom is derivable from the previous axioms. We proceed similarly for the cases $\Rightarrow_d$, π and $\forall$.

We write Σ_{eq} for the signature formed by the axiom schemas defining the heterogeneous equality. Reflexivity, symmetry, and transitivity are standard axioms of equality. We have also added axioms stating that a heterogeneous equality comparing two objects of the same type acts like Leibniz equality. Finally, we have an axiom for the application constructor and one axiom for the abstraction constructor—that is functional extensionality. Both axioms are used by Oury [19], who also assumes the uniqueness of identity proofs principle that entails the Leibniz principle we use.

4 Replacing Rewrite Rules

When working in theories with prelude encoding, rewriting originates from the rewrite rules of Σ_{pre} (which are generic rewrite rules), from the rewrite rules $\Sigma_{\mathcal{T}}$ (which are defined by the user) and from β-reduction. The goal of this work is to replace the user-defined rewrite rules $\Sigma_{\mathcal{T}}$ by equational axioms. In the rest of the paper, we write $\vdash_{\mathcal{R}}$ for a derivation inside the source theory—the theory with user-defined rewrite rules—and $\vdash$ for a derivation inside the target theory—the theory with axioms instead of user-defined rewrite rules.

We now have all the tools to replace rewrite rules by equational axioms. To do so, we build suitable transports, such that if $\Gamma \vdash t : A$ and $\Gamma \vdash p : \kappa(A, B)$, then $\Gamma \vdash \mathsf{transp}\ p\ t : B$. The goal is to insert such transports into the terms instead of using conversion with the rules of $\Sigma_{\mathcal{T}}$. In the signature, each rewrite rule $\ell \hookrightarrow r$ is replaced by the equational axiom $\bar{\ell} \approx \bar{r}$.

4.1 Transports

If we have $\Gamma \vdash t : A$ and $\Gamma \vdash p : \kappa(A, B)$, we want to transport t from A to B, that is to build a term $\mathsf{transp}\ p\ t$ such that $\Gamma \vdash \mathsf{transp}\ p\ t : B$. A paramount result is that t and $\mathsf{transp}\ p\ t$ are heterogeneously equal.

Lemma 2 (Transport). *Given $\Gamma \vdash t : A$ and $\Gamma \vdash p : \kappa(A, B)$ with A and B small types, there exists* $\mathsf{transp}\ p\ t$*, called transport of t along p, such that:*

- *$\Gamma \vdash \mathsf{transp}\ p\ t : B$,*
- *there exists* $\mathsf{eqTransp}$ *such that* $\Gamma \vdash \mathsf{eqTransp}\ p\ t : \mathsf{transp}\ p\ t\ {}_{B}{\approx}_{A}\ t$.

Proof. A and B are small types and we have an equality $\kappa(A, B)$. If $A, B \in \mathcal{S}$ then $\nu(A) = \nu(B) = A = B$ and we take $\mathsf{transp}\ p\ t := t$ and $\mathsf{eqTransp}\ p\ t := \mathsf{refl}_A\ t$. Otherwise, by construction of κ, we know that $\nu(A), \nu(B) \in \mathcal{P}$, or $\nu(A), \nu(B) \in \mathcal{E}$, and that $\nu(A)$ and $\nu(B)$ have the same structure. Moreover, using $A \equiv_{\beta\Sigma_{pre}} \nu(A)$, we have $\Gamma \vdash t : \nu(A)$. We proceed by induction on the grammar $\mathcal{P}$ (we proceed similarly for the grammar $\mathcal{E}$).

- If $\nu(A) = Prf\ a$ and $\nu(B) = Prf\ b$, then we have $\Gamma \vdash p : a \approx b$. We take $\mathsf{transp}\ p\ t := \mathsf{leib}^{\mathsf{Prf}}_{El\ o}\ a\ b\ p\ (\lambda w : El\ o.\ w)\ t$. We conclude using $\mathsf{eqLeib}^{\mathsf{Prf}}_{El\ o}$.
- If $\nu(A) = A' \to S$ and $\nu(B) = B' \to S$, with $A', B' \in \mathcal{P}$ and $S \in \mathcal{S}$, then we have $\kappa(A', B') = \kappa(A, B)$. From $\Gamma \vdash p : \kappa(A', B')$ we can build some p' such that $\Gamma \vdash p' : \kappa(B', A')$ (using sym). By weakening, we also have $p' : \kappa(B', A')$ in the context $\Gamma, m_b : B'$. By induction, we have $\mathsf{transp}\ p'\ m_b : A'$ and $\mathsf{eqTransp}\ p'\ m_b : \mathsf{transp}\ p'\ m_b \approx m_b$ in the context $\Gamma, m_b : B'$. We take $\mathsf{transp}\ p\ t := \lambda m_b : B'.\ t\ (\mathsf{transp}\ p'\ m_b)$. Using trans and app we obtain an equality $t\ (\mathsf{transp}\ p'\ m_b) \approx t\ m_a$ in the context $\Gamma, m_a : A', m_b : B', p_m : m_a \approx m_b$. Using fun and $\equiv_{\beta\Sigma_{pre}}$, we have $\lambda m_b : B'.\ t\ (\mathsf{transp}\ p'\ m_b) \approx t$ in the context Γ.

– If $\nu(A) = \Pi z : S.\ A'$ and $\nu(B) = \Pi z : S.\ B'$ with $A', B' \in \mathcal{P}$ and $S \in \mathcal{S}$, then we have $\kappa(A, B) = \Pi z : S.\ \kappa(A', B')$. By weakening and application, we have $\Gamma, z : S \vdash p\ z : \kappa(A', B')$. By induction we have $\mathsf{transp}\ (p\ z)\ (t\ z) : B'$ and $\mathsf{eqTransp}\ (p\ z)\ (t\ z) : \mathsf{transp}\ (p\ z)\ (t\ z) \approx t\ z$ in the context $\Gamma, z : S$. We take $\mathsf{transp}\ p\ t := \lambda z : S.\ \mathsf{transp}\ (p\ z)\ (t\ z)$. We obtain $\lambda z : S.\ \mathsf{transp}\ (p\ z)\ (t\ z) \approx t$ using fun and $\equiv_{\beta\Sigma_{pre}}$. □

The transport of t from A to B depends on the small form of A and B. In that respect, there exists a different transport for each possible family of small form, and such transport is indexed over an equality of a small type.

4.2 Translation of Terms

To translate a theory with rewrite rules into a theory with equational axioms, we add transports at the proper locations in the terms and types. If we have $\Gamma \vdash_{\mathcal{R}} t : A$ in the source theory, we want to find $\overline{\Gamma}$, $\bar{t}$ and $\overline{A}$ that are translations of Γ, t and A, and such that $\overline{\Gamma} \vdash \bar{t} : \overline{A}$ in the target theory.

We add transports in a term by induction on a typing derivation—which is not unique—so we may have different translations for a same term. As such, we define a relation $\triangleleft$ where $\bar{t} \triangleleft t$ states that $\bar{t}$ is a translation of t. The relation is defined by induction on the terms of $\lambda\Pi/\equiv$. Variables, constants, TYPE and KIND are translations of themselves. The translations of λ-abstractions $\lambda x : A.\ t$, dependent types $\Pi x : A.\ B$ and applications $t\ u$ rely on the translations of t, u, A and B. The most important part of the definition is that the translation is stable by transports: if $\bar{t}$ is a translation of t, then $\mathsf{transp}\ p\ \bar{t}$ is also a translation of t, with p typically an equality. This relation captures all possible translations, but some are not correct as they may not be well typed. For instance, $\lambda x : \overline{A}.\ \bar{t}$ is not a valid translation of $\lambda x : A.\ t$ when the variable x used in $\bar{t}$ does not expect type $\overline{A}$ but another translation $\overline{A}'$.

Definition 6. *The translation relation* $\triangleleft$ *is defined by:*

$$\frac{}{x \triangleleft x} \qquad \frac{}{c \triangleleft c} \qquad \frac{}{\mathtt{TYPE} \triangleleft \mathtt{TYPE}} \qquad \frac{}{\mathtt{KIND} \triangleleft \mathtt{KIND}}$$

$$\frac{\overline{A} \triangleleft A \qquad \bar{t} \triangleleft t}{(\lambda x : \overline{A}.\ \bar{t}) \triangleleft (\lambda x : A.\ t)} \qquad \frac{\overline{A} \triangleleft A \qquad \overline{B} \triangleleft B}{(\Pi x : \overline{A}.\ \overline{B}) \triangleleft (\Pi x : A.\ B)}$$

$$\frac{\bar{t} \triangleleft t \qquad \bar{u} \triangleleft u}{(\bar{t}\ \bar{u}) \triangleleft (t\ u)} \qquad \frac{\bar{t} \triangleleft t}{(\mathsf{transp}\ p\ \bar{t}) \triangleleft t}$$

where p *is an arbitrary term.*

Due to the typing rules of $\lambda\Pi/\equiv$, transports for objects that have TYPE in their type do not exist. Therefore, the only well-typed translations of TYPE, KIND, *Set*, *Prf* and *El* are themselves, and the well-typed translations of $\Pi x : A.\ B$ are of the form $\Pi x : \overline{A}.\ \overline{B}$ with $\overline{A} \triangleleft A$ and $\overline{B} \triangleleft B$. It follows that a well-typed translation of a small type is still a small type. In particular, if $A \in \mathcal{S}$ then for any $\overline{A}$ we have $\overline{A} := A$; if $\nu(A) \in \mathcal{P}$ then $\nu(\overline{A}) \in \mathcal{P}$; and if $\nu(A) \in \mathcal{E}$ then $\nu(\overline{A}) \in \mathcal{E}$.

We extend the relation to contexts and signatures. For each rewrite rule $\ell \hookrightarrow r$ of a signature, we have $\boldsymbol{x} : \boldsymbol{B} \vdash_{\mathcal{R}} \ell : A$ and $\boldsymbol{x} : \boldsymbol{B} \vdash_{\mathcal{R}} r : A$, for some $\boldsymbol{B}$ and A, and some $\boldsymbol{x}$ representing the free variables of ℓ. The translation of the rewrite rule $\ell \hookrightarrow r$ is given by the equational axiom $\mathsf{eq}_{\ell r} : \Pi \boldsymbol{x} : \overline{\boldsymbol{B}}.\ \bar{\ell}\ {}_{\overline{A}}{\approx}_{\overline{A}}\ \bar{r}$. Since the type of a term is not unique in $\lambda\Pi/\equiv$, we have made a choice of $\boldsymbol{B}$ and A, which is not a problem as we will see in the proof of Theorem 1.

Definition 7. *$\triangleleft$ is defined on contexts and signatures by:*

$$\frac{}{\langle\rangle \triangleleft \langle\rangle} \qquad \frac{\overline{\Gamma} \triangleleft \Gamma \qquad \overline{A} \triangleleft A}{(\overline{\Gamma}, x : \overline{A}) \triangleleft (\Gamma, x : A)} \qquad \frac{\overline{\Sigma} \triangleleft \Sigma \qquad \overline{A} \triangleleft A}{(\overline{\Sigma}, c : \overline{A}) \triangleleft (\Sigma, c : A)}$$

$$\frac{\overline{\Sigma} \triangleleft \Sigma \qquad \bar{\ell} \triangleleft \ell \qquad \bar{r} \triangleleft r \qquad \overline{\boldsymbol{B}} \triangleleft \boldsymbol{B} \qquad \overline{A} \triangleleft A}{(\overline{\Sigma}, \mathsf{eq}_{\ell r} : \Pi \boldsymbol{x} : \overline{\boldsymbol{B}}.\ \bar{\ell}\ {}_{\overline{A}}{\approx}_{\overline{A}}\ \bar{r}) \triangleleft (\Sigma, \ell \hookrightarrow r)}$$

Lemma 3. *If $\bar{t} \triangleleft t$ and $\bar{u} \triangleleft u$ then $\bar{t}[x \mapsto \bar{u}] \triangleleft t[x \mapsto u]$.*

Proof. By induction on the derivation of $\bar{t} \triangleleft t$. For the case with the transport, we can prove that $(\mathsf{transp}\ p\ t)[x \mapsto u] = \mathsf{transp}\ p[x \mapsto u]\ t[x \mapsto u]$. □

Definition 8 (Relation $\sim$). *We say that $t_1 \sim t_2$ when there exists some t such that $t_1 \triangleleft t$ and $t_2 \triangleleft t$.*

Lemma 4. *$\sim$ is an equivalence relation.*

Proof. $\sim$ is reflexive, symmetric and transitive. When proving transitivity we exploit the fact that whenever $t \triangleleft u_1$ and $t \triangleleft u_2$, we have $u_1 = u_2$. Reflexivity is proved by induction on the term. □

An important result we need to prove is that two well-typed translations t_1 and t_2 of the same term t are heterogeneously equal. By construction, both terms do not necessarily have the same type or the same context. We will always consider $\Gamma_1 \vdash t_1 : A_1$ and $\Gamma_2 \vdash t_2 : A_2$, where Γ_1 and Γ_2 have the same length and the same variables (with possibly different types). The equality between t_1 and t_2 must be typed in some context, but Γ_1 and Γ_2 are not sufficient. That is why we define a common context $\Gamma_1 \star \Gamma_2$ (written $\mathsf{Pack}\ \Gamma_1\ \Gamma_2$ in the work of Winterhalter *et al.* [23]) by duplicating each variable and by assuming a witness of heterogeneous equality between these two duplicates. More precisely, we partially define $\star$ by induction on small contexts:

$$\langle\rangle \star \langle\rangle := \langle\rangle$$

$$(\Gamma_1, x : A_1) \star (\Gamma_2, x : A_2) := \Gamma_1 \star \Gamma_2, x_1 : A_1[\gamma_1], x_2 : A_2[\gamma_2], p_x : x_1 \approx x_2$$

where γ_1 substitutes variables z by z_1 and γ_2 substitutes variables z by z_2. We write γ_{12} for the substitution that replaces the variables z_1 and z_2 by z and the variable p_z by $\mathsf{refl}\ z$.

Lemma 5. *If $\Gamma \star \Gamma \vdash t : A$, then we can derive $\Gamma \vdash t[\gamma_{12}] : A[\gamma_{12}]$.*

Proof. We proceed by induction on the length of Γ. If we have $\langle\rangle \star \langle\rangle \vdash t : A$ then by definition we have $\langle\rangle \vdash t : A$. Suppose that we have $(\Gamma, x : B) \star (\Gamma, x : B) \vdash t : A$. We apply successively Lemma 1 to replace x_2 and x_1 by x and then p_x by $\mathsf{refl}\ x$. □

The following lemma states that two translations of a same term are heterogeneously equal.

Lemma 6 (Equal translations). *Let $t_1 \sim t_2$ such that $\Gamma_1 \vdash t_1 : A_1$ and $\Gamma_2 \vdash t_2 : A_2$ with Γ_1 and Γ_2 small contexts.*

1. *If $\Gamma_1 \vdash A_1 : \mathtt{TYPE}$ and $\Gamma_2 \vdash A_2 : \mathtt{TYPE}$, then there exists some p such that $\Gamma_1 \star \Gamma_2 \vdash p : t_1[\gamma_1]\ {}_{A_1[\gamma_1]}{\approx}_{A_2[\gamma_2]}\ t_2[\gamma_2]$.*
2. *If t_1 and t_2 are small types, then there exists some p such that $\Gamma_1 \star \Gamma_2 \vdash p : \kappa(t_1[\gamma_1], t_2[\gamma_2])$.*

Proof. We proceed by induction on the derivation of $t_1 \sim t_2$. We show two interesting cases.

- Transport $(\mathsf{transp}\ p\ t_1) \sim t_2$
  We have $\Gamma_1 \vdash \mathsf{transp}\ p\ t_1 : A_1$ and $\Gamma_2 \vdash t_2 : A_2$. By inversion of typing, we have $\Gamma_1 \vdash t_1 : A_1'$ and $\Gamma_1 \vdash p : \kappa(A_1', A_1)$. By induction there exists some p_t such that $\Gamma_1 \star \Gamma_2 \vdash p_t : t_1[\gamma_1] \approx t_2[\gamma_2]$. We also have $\Gamma_1 \vdash \mathsf{eqTransp}\ p\ t_1 : \mathsf{transp}\ p\ t_1 \approx t_1$. We derive that $\Gamma_1 \star \Gamma_2 \vdash (\mathsf{eqTransp}\ p\ t_1)[\gamma_1] : (\mathsf{transp}\ p\ t_1)[\gamma_1] \approx t_1[\gamma_1]$. We conclude using transitivity.
- Application $(t_1\ u_1) \sim (t_2\ u_2)$
  Suppose that $t_1\ u_1$ and $t_2\ u_2$ are small types. Then the only possible cases are $t_1 = t_2 = \mathit{Prf}$ or $t_1 = t_2 = \mathit{El}$. If $t_1 = t_2 = \mathit{Prf}$, then we have $\Gamma_1 \vdash \mathit{Prf}\ u_1 : \mathtt{TYPE}$ and $\Gamma_2 \vdash \mathit{Prf}\ u_2 : \mathtt{TYPE}$. Since $\kappa(\mathit{Prf}\ u_1, \mathit{Prf}\ u_2) = u_1 \approx u_2$, the result is simply the induction hypothesis $\Gamma_1 \star \Gamma_2 \vdash p : u_1[\gamma_1] \approx u_2[\gamma_2]$. We proceed similarly for $\mathit{El}\ u_1 \sim \mathit{El}\ u_2$.
  Suppose that we have $\Gamma_1 \vdash t_1\ u_1 : T_1$ and $\Gamma_2 \vdash t_2\ u_2 : T_2$ with $\Gamma \vdash T_1 : \mathtt{TYPE}$ and $\Gamma \vdash T_2 : \mathtt{TYPE}$. Then by inversion of typing we have $\Gamma_1 \vdash u_1 : B_1$ and $\Gamma_2 \vdash u_2 : B_2$ and $\Gamma_1 \vdash t_1 : \Pi x : A_1.\ B_1$ and $\Gamma_2 \vdash t_2 : \Pi x : A_2.\ B_2$, with $T_1 \equiv_{\beta\Sigma_{pre}} B_1[x \mapsto u_1]$ and $T_2 \equiv_{\beta\Sigma_{pre}} B_2[x \mapsto u_2]$. By induction hypotheses, we have $\Gamma_1 \star \Gamma_2 \vdash p_t : t_1[\gamma_1] \approx t_2[\gamma_2]$ and $\Gamma_1 \star \Gamma_2 \vdash p_u : u_1[\gamma_1] \approx u_2[\gamma_2]$. We conclude using app. □

4.3 Translation of Judgments

In Section 4.2 we have seen all the possible translations for *terms*. However, the only translations that matter are the translations of *judgments*: context formation judgments and typing judgments.

Definition 9. *For any $\vdash_{\mathcal{R}} \Gamma$ we define a set $[\![\vdash_{\mathcal{R}} \Gamma]\!]$ of valid judgments such that $\vdash \overline{\Gamma} \in [\![\vdash_{\mathcal{R}} \Gamma]\!]$ if and only if $\overline{\Gamma} \lhd \Gamma$. For any $\Gamma \vdash_{\mathcal{R}} t : A$ we define a set $[\![\Gamma \vdash_{\mathcal{R}} t : A]\!]$ of valid judgments such that $\overline{\Gamma} \vdash \bar{t} : \overline{A} \in [\![\Gamma \vdash_{\mathcal{R}} t : A]\!]$ if and only if $\vdash \overline{\Gamma} \in [\![\vdash_{\mathcal{R}} \Gamma]\!]$, $\bar{t} \lhd t$ and $\overline{A} \lhd A$.*

We are now able to prove that it is possible to switch between two translations of a small type.

Lemma 7 (Switching translations). *Suppose that we have A a small type, $\overline{\Gamma} \vdash \bar{t} : \overline{A} \in [\![\Gamma \vdash_{\mathcal{R}} t : A]\!]$ and $\overline{\Gamma} \vdash \overline{A}' : \mathsf{TYPE} \in [\![\Gamma \vdash_{\mathcal{R}} A : \mathsf{TYPE}]\!]$ with $\overline{\Gamma}$ a small context. Then there exists $\bar{t}'$ such that $\overline{\Gamma} \vdash \bar{t}' : \overline{A}' \in [\![\Gamma \vdash_{\mathcal{R}} t : A]\!]$.*

Proof. If $\nu(A) \in \mathcal{S}$, then $\overline{A} := A$ and $\overline{A}' := A$, and we take $\bar{t}' := \bar{t}$. If $\nu(A) \in \mathcal{P}$, then $\nu(\overline{A}), \nu(\overline{A}') \in \mathcal{P}$ (this is similar for $\mathcal{E}$). As $\overline{A}$ and $\overline{A}'$ are two translations of A, we have $\overline{A} \sim \overline{A}'$. From Lemma 6, we have $\overline{\Gamma} \star \overline{\Gamma} \vdash p : \kappa(\overline{A}[\gamma_1], \overline{A}'[\gamma_2])$. Using Lemma 5 we obtain $\overline{\Gamma} \vdash p[\gamma_{12}] : \kappa(\overline{A}, \overline{A}')$. Using Lemma 2, there exists some $\mathsf{transp}\ p[\gamma_{12}]\ \bar{t} \lhd t$ (since $\bar{t} \lhd t$) such that $\overline{\Gamma} \vdash \mathsf{transp}\ p[\gamma_{12}]\ \bar{t} : \overline{A}'$. □

4.4 Translation of Theories

Now that we have translated terms and judgments, we want to translate theories, so that the translation of every provable judgment in the source theory is provable in the target theory. The target theory $\mathcal{T}^{ax} = \Sigma_{pre} \cup \Sigma_{eq} \cup \overline{\Sigma}_{\mathcal{T}}$ is obtained by adding the axioms of equality to the signature, and by translating $\Sigma_{\mathcal{T}}$. To do so, we translate each typed constant and rewrite rule one by one. At the end, the rewrite rules of $\Sigma_{\mathcal{T}}$ have been replaced by equational axioms.

The paramount result of this paper is the following theorem. The first item concerns context formation. The second item is about the translation of typing judgments. The third item focuses on convertible contexts. The fourth and fifth items are about the conversion rules. It is worth noting that in the second item we use the universal quantifier on $\overline{\Gamma}$ instead of using the existential quantifier. We have opted for the universal quantifier so we can obtain the induction hypotheses for a common context.

Theorem 1 (Elimination of the rewrite rules). *Let a theory $\mathcal{T} = \Sigma$ in $\lambda\Pi/\!\equiv$ such that $\mathcal{T}$ is a theory with prelude encoding and such that all the derivations considered are small derivations. There exists a signature $\overline{\Sigma}_{\mathcal{T}} \lhd \Sigma_{\mathcal{T}}$ such that the theory $\mathcal{T}^{ax} = \Sigma_{pre} \cup \Sigma_{eq} \cup \overline{\Sigma}_{\mathcal{T}}$ satisfies:*

1. *If $\vdash_{\mathcal{R}} \Gamma$, then there exists $\vdash \overline{\Gamma} \in [\![\vdash_{\mathcal{R}} \Gamma]\!]$.*
2. *If $\Gamma \vdash_{\mathcal{R}} t : A$, then for every $\vdash \overline{\Gamma} \in [\![\vdash_{\mathcal{R}} \Gamma]\!]$ there exist $\bar{t}$ and $\overline{A}$ such that $\overline{\Gamma} \vdash \bar{t} : \overline{A} \in [\![\Gamma \vdash_{\mathcal{R}} t : A]\!]$.*
3. *If $(\vdash_{\mathcal{R}} \Gamma_1) \equiv (\vdash_{\mathcal{R}} \Gamma_2)$, then for every $\vdash \overline{\Gamma}_1 \in [\![\vdash_{\mathcal{R}} \Gamma_1]\!]$ and $\vdash \overline{\Gamma}_2 \in [\![\vdash_{\mathcal{R}} \Gamma_2]\!]$, we have $\vdash \overline{\Gamma}_1 \star \overline{\Gamma}_2$.*
4. *If $(\Gamma_1 \vdash_{\mathcal{R}} u_1 : A_1) \equiv (\Gamma_2 \vdash_{\mathcal{R}} u_2 : A_2)$ with $\Gamma_1 \vdash_{\mathcal{R}} A_1 : \mathsf{TYPE}$ and $\Gamma_2 \vdash_{\mathcal{R}} A_2 : \mathsf{TYPE}$, then for every $\vdash \overline{\Gamma}_1 \in [\![\vdash_{\mathcal{R}} \Gamma_1]\!]$ and $\vdash \overline{\Gamma}_2 \in [\![\vdash_{\mathcal{R}} \Gamma_2]\!]$, we have $\overline{\Gamma}_1 \vdash \bar{u}_1 : \overline{A}_1 \in [\![\Gamma_1 \vdash_{\mathcal{R}} u_1 : A_1]\!]$ and $\overline{\Gamma}_2 \vdash \bar{u}_2 : \overline{A}_2 \in [\![\Gamma_2 \vdash_{\mathcal{R}} u_2 : A_2]\!]$ and there exists some p such that $\overline{\Gamma}_1 \star \overline{\Gamma}_2 \vdash p : \bar{u}_1[\gamma_1]\ {}_{\overline{A}_1[\gamma_1]}\!\approx_{\overline{A}_2[\gamma_2]}\ \bar{u}_2[\gamma_2]$.*

5. *If* $(\Gamma_1 \vdash_{\mathcal{R}} u_1 : \mathsf{TYPE}) \equiv (\Gamma_2 \vdash_{\mathcal{R}} u_2 : \mathsf{TYPE})$, *then for every* $\vdash \overline{\Gamma}_1 \in [\![\vdash_{\mathcal{R}} \Gamma_1]\!]$ *and* $\vdash \overline{\Gamma}_2 \in [\![\vdash_{\mathcal{R}} \Gamma_2]\!]$, *we have* $\overline{\Gamma}_1 \vdash \bar{u}_1 : \mathsf{TYPE} \in [\![\Gamma_1 \vdash_{\mathcal{R}} u_1 : \mathsf{TYPE}]\!]$ *and* $\overline{\Gamma}_2 \vdash \bar{u}_2 : \mathsf{TYPE} \in [\![\Gamma_2 \vdash_{\mathcal{R}} u_2 : \mathsf{TYPE}]\!]$ *and there exists some* p *such that* $\overline{\Gamma}_1 \star \overline{\Gamma}_2 \vdash p : \kappa(\bar{u}_1[\gamma_1], \bar{u}_2[\gamma_2])$.

Proof. The proof of the five items is done by induction on the typing derivations, assuming the existence of $\overline{\Sigma}_{\mathcal{T}}$. We show three relevant cases.

– PROD:

$$\frac{\Gamma \vdash_{\mathcal{R}} A : \mathsf{TYPE} \qquad \Gamma, x : A \vdash_{\mathcal{R}} B : s}{\Gamma \vdash_{\mathcal{R}} \Pi x : A.\ B : s}$$

Take $\vdash \overline{\Gamma} \in [\![\vdash_{\mathcal{R}} \Gamma]\!]$. By induction hypothesis, we have $\overline{\Gamma} \vdash \overline{A} : \mathsf{TYPE} \in [\![\Gamma \vdash_{\mathcal{R}} A : \mathsf{TYPE}]\!]$. We have $(\overline{\Gamma}, x : \overline{A}) \lhd (\Gamma, x : A)$ and we know that the only translation of sort s is itself, therefore by induction hypothesis we have $\overline{\Gamma}, x : \overline{A} \vdash \overline{B} : s \in [\![\Gamma, x : A \vdash_{\mathcal{R}} B : s]\!]$. We conclude that $\overline{\Gamma} \vdash \Pi x : \overline{A}.\ \overline{B} : s$ using the PROD rule.

– CONV:

$$\frac{\Gamma \vdash_{\mathcal{R}} t : A \qquad (\Gamma \vdash_{\mathcal{R}} A : s) \equiv (\Gamma \vdash_{\mathcal{R}} B : s)}{\Gamma \vdash_{\mathcal{R}} t : B}$$

Take $\vdash \overline{\Gamma} \in [\![\vdash_{\mathcal{R}} \Gamma]\!]$. As we consider small derivations, either A is a small type or A and B are the same type.
If A is a small type, then by induction hypothesis we have $\overline{\Gamma} \star \overline{\Gamma} \vdash p : \kappa(\overline{A}[\gamma_1], \overline{B}[\gamma_2])$. By Lemma 5 we obtain $\overline{\Gamma} \vdash p[\gamma_{12}] : \kappa(\overline{A}, \overline{B})$. By Lemma 7 and induction hypothesis we have $\overline{\Gamma} \vdash \bar{t} : \overline{A} \in [\![\Gamma \vdash_{\mathcal{R}} t : A]\!]$. Thanks to Lemma 2, there exists some $\bar{t}'$ such that $\overline{\Gamma} \vdash \bar{t}' : \overline{B} \in [\![\Gamma \vdash_{\mathcal{R}} t : B]\!]$.
If A and B are the same type, then no conversion is needed and the result is simply given the induction hypothesis $\overline{\Gamma} \vdash \bar{t} : \overline{A}$.

– CONVREFL:

$$\frac{\Gamma \vdash_{\mathcal{R}} u : A}{(\Gamma \vdash_{\mathcal{R}} u : A) \equiv (\Gamma \vdash_{\mathcal{R}} u : A)}$$

Take $\vdash \overline{\Gamma} \in [\![\vdash_{\mathcal{R}} \Gamma]\!]$. By induction hypothesis, we have $\overline{\Gamma} \vdash \bar{u} : \overline{A} \in [\![\Gamma \vdash_{\mathcal{R}} u : A]\!]$.
If $\Gamma \vdash_{\mathcal{R}} A : \mathsf{TYPE}$, then we build $\overline{\Gamma} \star \overline{\Gamma} \vdash p : \bar{u}[\gamma_1] \approx \bar{u}[\gamma_2]$ using all the congruence rules of $\approx$.
We proceed similarly for the case $A = \mathsf{TYPE}$.

The existence of $\overline{\Sigma}_{\mathcal{T}}$ is proved by induction on the length of $\Sigma_{\mathcal{T}}$, using the previous five items and $\langle\rangle \lhd \langle\rangle$. □

Corollary 1 (Preservation). *If* $\vdash_{\mathcal{R}} t : A$ *and* $\vdash A : s \in [\![\vdash_{\mathcal{R}} A : s]\!]$, *then there exists* $\bar{t}$ *such that* $\vdash \bar{t} : A$.

Proof. By Theorem 1 we have $\vdash \bar{t}' : \overline{A}' \in [\![\vdash_{\mathcal{R}} t : A]\!]$. Using Lemma 7 with $\overline{A} := A$, we have some $\bar{t}$ such that $\vdash \bar{t} : A \in [\![\vdash_{\mathcal{R}} t : A]\!]$. □

We directly derive the two following conservativity and consistency results. We say that a theory $\mathcal{T}_2$ is conservative over a theory $\mathcal{T}_1$ when every formula in the common language of $\mathcal{T}_1$ and $\mathcal{T}_2$ that is provable in $\mathcal{T}_2$ is also provable in $\mathcal{T}_1$.

Corollary 2 (Conservativity). *$\mathcal{T}$ is a conservative extension of $\mathcal{T}^{ax}$.*

Corollary 3 (Relative consistency). *If $\mathcal{T}^{ax}$ is consistent then $\mathcal{T}$ is also consistent.*

5 Conclusion

Discussion. In this paper, we showed that it is possible to replace user-defined rewrite rules by equational axioms, in the case of the $\lambda\Pi$-calculus modulo theory. This result works for theories with prelude encoding—which is satisfied by expressive theories such as predicate logic and set theory—and for small derivations—which is in practice the case. So as to replace rewrite rules by equational axioms, we have defined a heterogeneous equality with standard axioms—reflexivity, symmetry, transitivity, Leibniz principle—and congruences for each constructor. At the end, the theory with rewrite rules is a conservative extension of the theory with axioms.

Related work. The similar problem of the translation from an extensional system to an intensional system has been investigated by Oury [19]. He proposed a translation from the Extensional Calculus of Constructions to the Calculus of Inductive Constructions with additional axioms that define a heterogeneous equality. Winterhalter, Sozeau and Tabareau provided a translation from extensional type theory to intensional type theory [23,24]. They took advantage of the presence of dependent pairs to encode a heterogeneous equality, unlike Oury who defined it with axioms.

In this paper, we have shown the existence of a translation from a theory with rewrite rules to a theory with equational axioms. Technical challenges appear as we are not in an extensional type system. In particular, Oury and Winterhalter *et al.* had a homogeneous equality in their source theory and introduce a heterogeneous equality in the target theory. In this work, the source theory does not contain a homogeneous equality, and the target theory only contains a heterogeneous equality.

The major difference with previous works is that we are in a logical framework without an infinite hierarchy of sorts $s_i : s_{i+1}$ for $i \in \mathbb{N}$. In $\lambda\Pi/\equiv$, we only have $\texttt{TYPE} : \texttt{KIND}$, which is the reason why we cannot define an equality between types. As such an equality is of paramount importance in the transports, we have considered a subclass of types—called small types—for which we can define an equality. However, it is worth noting that the sorts of $\lambda\Pi/\equiv$ allowed a simplification: by construction, there is no transports on types, so the translation of a dependent function type is directly a dependent function type.

References

1. Adams, R.: Pure type systems with judgemental equality. Journal of Functional Programming **16**(2), 219–246 (2006). https://doi.org/10.1017/S0956796805005770
2. Assaf, A., Burel, G., Cauderlier, R., Delahaye, D., Dowek, G., Dubois, C., Gilbert, F., Halmagrand, P., Hermant, O., Saillard, R.: Dedukti: a Logical Framework based on the $\lambda\Pi$-Calculus Modulo Theory (2016), manuscript
3. Blanqui, F., Dowek, G., Grienenberger, E., Hondet, G., Thiré, F.: A modular construction of type theories. Logical Methods in Computer Science **Volume 19, Issue 1** (Feb 2023). https://doi.org/10.46298/lmcs-19(1:12)2023, https://lmcs.episciences.org/10959
4. Blot, V., Dowek, G., Traversié, T.: An Implementation of Set Theory with Pointed Graphs in Dedukti. In: LFMTP 2022 - International Workshop on Logical Frameworks and Meta-Languages : Theory and Practice. Haïfa, Israel (Aug 2022), https://inria.hal.science/hal-03740004
5. Cockx, J., Abel, A.: Sprinkles of extensionality for your vanilla type theory (2016)
6. Cousineau, D., Dowek, G.: Embedding Pure Type Systems in the Lambda-Pi-Calculus Modulo. In: Della Rocca, S.R. (ed.) Typed Lambda Calculi and Applications. pp. 102–117. Springer Berlin Heidelberg, Berlin, Heidelberg (2007)
7. Dershowitz, N., Jouannaud, J.P.: Rewrite Systems. In: Handbook of Theoretical Computer Science, Volume B: Formal Models and Sematics (1991)
8. Dowek, G., Miquel, A.: Relative normalization (2007), manuscript
9. Dowek, G.: La part du calcul. Habilitation à diriger des recherches, Université de Paris 7 (Jun 1999), https://inria.hal.science/tel-04114581
10. Dowek, G., Werner, B.: Proof Normalization Modulo. Research Report RR-3542, INRIA (1998), https://inria.hal.science/inria-00073143, projet COQ
11. Geuvers, H., Werner, B.: On the Church-Rosser property for expressive type systems and its consequences for their metatheoretic study. In: Proceedings Ninth Annual IEEE Symposium on Logic in Computer Science. pp. 320–329 (1994). https://doi.org/10.1109/LICS.1994.316057
12. Gilbert, G., Leray, Y., Tabareau, N., Winterhalter, T.: The Rewster: The Coq Proof Assistant with Rewrite Rules (2023), https://types2023.webs.upv.es/TYPES2023.pdf
13. Harper, R., Honsell, F., Plotkin, G.: A Framework for Defining Logics. Journal of the ACM **40**(1), 143–184 (January 1993). https://doi.org/10.1145/138027.138060, https://doi.org/10.1145/138027.138060
14. Hofmann, M.: Conservativity of equality reflection over intensional type theory. In: Berardi, S., Coppo, M. (eds.) Types for Proofs and Programs. pp. 153–164. Springer Berlin Heidelberg, Berlin, Heidelberg (1996)
15. Hofmann, M.: Extensional Constructs in Intensional Type Theory. Springer London (1997). https://doi.org/10.1007/978-1-4471-0963-1
16. Hondet, G., Blanqui, F.: The New Rewriting Engine of Dedukti. In: FSCD 2020 - 5th International Conference on Formal Structures for Computation and Deduction. p. 16. No. 167, Paris, France (Jun 2020). https://doi.org/10.4230/LIPIcs.FSCD.2020.35, https://inria.hal.science/hal-02981561
17. Martin-Löf, P.: Constructive mathematics and computer programming. Studies in logic and the foundations of mathematics **104**, 167–184 (1984), https://api.semanticscholar.org/CorpusID:61930968
18. McBride, C.: Dependently Typed Functional Programs and their Proofs. Ph.D. thesis, University of Edinburgh (1999)

19. Oury, N.: Extensionality in the Calculus of Constructions. In: Hurd, J., Melham, T. (eds.) Theorem Proving in Higher Order Logics. pp. 278–293. Springer Berlin Heidelberg, Berlin, Heidelberg (2005)
20. Poincaré, H.: La Science et l'Hypothèse. Flammarion (1902)
21. Siles, V.: Investigation on the typing of equality in type systems. Ph.D. thesis, Ecole Polytechnique (Nov 2010), https://pastel.archives-ouvertes.fr/pastel-00556578
22. Siles, V., Herbelin, H.: Pure Type System conversion is always typable. Journal of Functional Programming **22**(2), 153 – 180 (May 2012). https://doi.org/10.1017/S0956796812000044, https://inria.hal.science/inria-00497177
23. Winterhalter, T., Sozeau, M., Tabareau, N.: Eliminating Reflection from Type Theory. In: CPP 2019 - 8th ACM SIGPLAN International Conference on Certified Programs and Proofs. pp. 91–103. ACM, Lisbonne, Portugal (Jan 2019). https://doi.org/10.1145/3293880.3294095, https://hal.science/hal-01849166
24. Winterhalter, T.: Formalisation and meta-theory of type theory. Ph.D. thesis, Université de Nantes (2020)

Light Genericity

Beniamino Accattoli[1] and Adrienne Lancelot[1,2](✉)

[1] Inria & LIX, Ecole Polytechnique, UMR 7161, Palaiseau, France
{beniamino.accattoli,adrienne.lancelot}@inria.fr
[2] Université Paris Cité, CNRS, IRIF, F-75013, Paris, France

Abstract. To better understand Barendregt's genericity for the untyped call-by-value λ-calculus, we start by first revisiting it in call-by-name, adopting a lighter statement and establishing a connection with contextual equivalence. Then, we use it to give a new, lighter proof of maximality of head contextual equivalence, *i.e.* that $\mathcal{H}^*$ is a maximal consistent equational theory. We move on to call-by-value, where we adapt these results and also introduce a new notion dual to light genericity, that we dub *co-genericity*. Lastly, we give alternative proofs of (co-)genericity based on applicative bisimilarity.

Keywords: lambda-calculus · semantics · call-by-value.

1 Introduction

Barendregt's genericity lemma [14, Prop. 14.3.24] is a classic result in the theory of the untyped λ-calculus. It expresses the fact that *meaningless terms*—also called *unsolvable terms*, a notion generalizing the bad behaviour of the paradigmatic looping term $\Omega := (\lambda x.xx)(\lambda x.xx)$—are sort of *black holes* for evaluation: if evaluation should ever enter them, it would never get out. This is specified somewhat dually, saying that if a term t containing a meaningless term u evaluates to a normal form, that is, if t is *observable*, then replacing u with any other term in t gives a term t' that is also observable. Roughly, if one can observe a term containing a black hole then evaluation never enters the black hole.

Genericity is arguably more than a lemma, but it is so labeled because its main use is as a tool in Barendregt's proofs of *collapsibility of meaningless terms*, that is, the fact that the equational theory $\mathcal{H}$ equating all meaningless terms is *consistent*, i.e. it does not equate all terms. Such collapsibility is one of the cornerstones of the semantics of the untyped λ-calculus.

Recap about Meaningless Terms. Meaningless terms were first studied in the 1970s, by Wadsworth [45,46] and Barendregt [12,13], while working on denotational models and the representation of partial recursive functions (PRFs). The starting point is that the natural choice of representing the *being undefined* of PRFs—considered as the paradigmatic meaningless computation—with *terms not having a normal form* leads to a problematic representation of PRFs. The issue is visible also at the equational level, as all theories collapsing all diverging terms are

N. Kobayashi and J. Worrell (Eds.): FoSSaCS 2024, LNCS 14575, pp. 24–46, 2024.
https://doi.org/10.1007/978-3-031-57231-9_2

inconsistent. Wadsworth and Barendregt then identify the class of *unsolvable terms* as a better notion of meaningless terms: the representation of PRFs using them as undefined terms is better behaved, they are collapsible, and in particular they are identified in Scott's first D_∞ model of the untyped λ-calculus.

Unsolvable terms are defined via a contextual property, but they are also characterized as being diverging for *head* β-reduction $\to_{\mathtt{h}}$, rather than plain β-reduction $\to_\beta$. The dual notion of *solvable terms*, which are terminating for head reduction, are taken as the right notion of *defined term*, replacing the natural but misleading idea that β-normal forms are the right notion of defined term.

Barendregt classic book from the 1980s [14] is built around the concept of (un)solvability. Visser and Statman noted that (un)solvability is not the only partition of terms providing good representations of PRFs and being collapsible, as summarized by Barendregt [15]. Typically, *(in)scrutable terms*, first studied by Paolini and Ronchi della Rocca [38,36,41] (under the name *(non-)potentially valuable terms*), provide an alternative good partition. In call-by-name (CbN), (in)scrutable terms correspond to *weak* head normalizing/diverging terms.

This Paper. The work presented here stems from the desire to obtain genericity for the untyped *call-by-value* λ-calculus. Perhaps surprisingly, the call-by-value (shortened to CbV) λ-calculus behaves quite differently with respect to meaningless terms. Accattoli and Guerrieri's recent study of meaningless terms in CbV [6] indeed stresses two key differences: *genericity fails in CbV*, and *collapsibility fails as well*, as any equational theory equating CbV meaningless terms is *inconsistent*, if one considers as meaningless the CbV analogous of unsolvable terms. Accattoli and Guerrieri also show that collapsibility can be recovered by adopting a different notion of meaningless terms, namely *CbV inscrutable terms*, but they do not prove genericity for them.

In this paper, we do prove a genericity result for inscrutable terms, and also provide a new proof of their collapsibility. These results, however, are only a small part of the contributions of this paper.

Contribution 1: the Very Statement of Genericity. We start by focussing on the statement of genericity. The literature contains various versions. The one used by Barendregt for proving collapsibility is the following (where unsolvable terms are identified with $\to_{\mathtt{h}}$-diverging terms), here dubbed as *heavy*:

> ***Heavy genericity:*** *let u be $\to_{\mathtt{h}}$-diverging and C be a context such that $C\langle u\rangle \to_\beta^* n$ with n β-normal. Then, $C\langle t\rangle \to_\beta^* n$ for all t.*

In Takahashi's elegant proof of heavy genericity [44]—which is an inspiration for our work—the following statement is called *fundamental property of unsolvable λ-terms*, which we here consider as an alternative, *lighter* statement for genericity:

> ***Light genericity:*** *let u be $\to_{\mathtt{h}}$-diverging and C be a context such that $C\langle u\rangle$ is $\to_{\mathtt{h}}$-normalizing. Then, $C\langle t\rangle$ is $\to_{\mathtt{h}}$-normalizing for all t.*

We adopt the lighter statement as the proper one for genericity for three reasons:

1. *Powerful enough.* We show that the collapsibility of unsolvable terms follows already from the light notion: there is no need to consider reductions to β-normal form, nor the fact that the normal forms of $C\langle u\rangle$ and $C\langle t\rangle$ coincide.
2. *Economical and natural.* The light version involves less concepts and it is more in line with the motivations behind (un)solvability: if the right notion of *defined terms* is head normalizable terms, it is somewhat odd to state genericity with respect to β-normal forms.
3. *Modularity.* In CbV, it is less clear what notion of normal form to use for the heavy statement, as shall be explained below. The light version, instead, adapts naturally. It is also impossible to have a heavy form of the *co-genericity* property given below, since the involved terms have no (full) normal form.

We then adapt Takahashi's proof of heavy genericity to the light case.

Contribution 2: (Open) Contextual Equivalence/Pre-Order. Once one adopts the light statement, a connection with contextual equivalence becomes evident. Precisely, consider the contextual *pre-order* (that is, the asymmetric variant of contextual equivalence) induced by head reduction:

Head contextual pre-order: *$t \precsim_C^{\mathtt{h}} u$ if $C\langle t\rangle$ $\rightarrow_{\mathtt{h}}$-normalizing implies $C\langle u\rangle$ $\rightarrow_{\mathtt{h}}$-normalizing, for all closing contexts C.*

Light genericity seems to rephrase that $\rightarrow_{\mathtt{h}}$-diverging terms are minimum terms with respect to $\precsim_C^{\mathtt{h}}$. There is however a small yet non-trivial glitch: contextual pre-orders/equivalences are defined using *closing* contexts, while genericity is defined using arbitrary, *possibly open* contexts. Is the closing requirement essential in the definition of contextual notions? To our knowledge, this question has not been addressed in the literature. In fact, there is no absolute answer for all cases, as it depends on the notion of observation and on the underlying calculus.

We show that, for head reduction, open and closed contextual notions do coincide, what we refer to as the fact that head reduction is *openable.* As it is often the case with behavioral notions, proving head reduction openable cannot be done by simply unfolding the definitions, and requires some work.

The proof that we provide is—we believe—particularly elegant. It is obtained as the corollary of a further contribution, the revisitation of another classic result from the theory of the untyped λ-calculus, described next.

Contribution 3: Maximality. Barendregt proves that open head contextual equivalence—what he denotes as the equational theory $\mathcal{H}^*$—is maximal among consistent equational theories, i.e. any extension of $\mathcal{H}^*$ is inconsistent (moreover, $\mathcal{H}^*$ is the *unique maximum* theory among those collapsing unsolvable terms). His proof uses Böhm theorem, an important and yet non-trivial result. We give a new proof based only on light genericity, which is an arguably simpler result than Böhm theorem, obtained adapting a similar result for CbV by Egidi et al. [19].

Contribution 4: Call-by-Value. Finally, we study the CbV case, adopting inscrutable terms as notion of meaningless terms. In Plotkin's original CbV calculus

[40], however, these terms cannot be characterized as diverging for some strategy. Moreover, in Plotkin's calculus evaluation is not openable, that is, open and closed contextual notions do *not* coincide. In both cases, the issue is connected to the management of open terms.

We then adopt Accattoli and Paolini's *value substitution calculus* (VSC) [9], which is an extension of Plotkin's calculus solving its well-known issues with open terms and having the same (closed) contextual equivalence. Therein, inscrutable terms are characterized as those diverging for weak evaluation $\to_{\mathtt{w}}$.

For the VSC, we prove light genericity for $\to_{\mathtt{w}}$-diverging terms. We use a different technique with respect to the CbN case, namely we rely on Ehrhard's CbV multi types [20] (multi types are also known as *non-idempotent intersection types*), because Takahashi's technique does not easily adapt to the CbV case. We also give a proof of maximality (essentially Egidi et al. [19]'s argument used as blueprint for the CbN case) from which it follows that evaluation in the VSC is openable, in contrast with evaluation in Plotkin's calculus.

As hinted at above, it is relevant that in CbV we study light genericity rather than the heavy variant because the notion of full normal form in the CbV case is less standard. Firstly, it differs between Plotkin's calculus and the VSC. Secondly, it also differs between various refinements of Plotkin's calculus that can properly manage open terms, as discussed by Accattoli et al. [7].

Contribution 5: Co-Genericity. A difference between the head CbN case and weak CbV case is given by an interesting class of terms, those evaluating to an infinite sequence of abstractions, that is, such that $t \to_{\beta}^{*} \lambda x.t'$ with t' having the same property. Such terms are $\to_{\mathtt{h}}$-diverging (thus head CbN meaningless), but $\to_{\mathtt{w}}$-normalizing (CbV meaningful), and hereditarily so. We prove that these *$\to_{\mathtt{w}}$-super (normalizing) terms* are maximum elements of the CbV contextual pre-order, and the statement of this fact is a new notion of co-genericity:

> ***Co-Genericity:*** *let t be $\to_{\mathtt{w}}$-super and C be a context such that $C\langle u\rangle$ is $\to_{\mathtt{w}}$-normalizing for some u. Then, $C\langle t\rangle$ is $\to_{\mathtt{w}}$-normalizing.*

We then show a strengthened collapsibility result: all $\to_{\mathtt{w}}$-diverging terms *and* all $\to_{\mathtt{w}}$-super terms can be consistently collapsed.

Contribution 6: Alternative Proofs via Applicative Bisimilarity. Lastly, we show a different route to proving light genericity and co-genericity—in CbV, but the technique is general—by exploiting the link with contextual pre-orders. Namely, we give a second proof that weak CbV evaluation is openable in the VSC *without using light genericity*, and then we use the soundness of CbV applicative bisimilarity with respect to the (closed) contextual pre-order for giving very simple proofs of light genericity and co-genericity.

Related Work. There are many proofs of CbN genericity in the literature (but they do not all prove the same statement[3]): a topological one by Barendregt [14,

[3] Sometimes, one finds the following *genericity as application* statement: *let u be $\to_{\mathtt{h}}$-diverging and s be such that $su \to_{\beta}^{*} n$ with n β-normal. Then, $st \to_{\beta}^{*} n$ for all*

Prop. 14.3.24], via intersection types by Ghilezan [24], rewriting-based ones by Takahashi [44], Kuper [30], Kennaway et al. [28], and Endrullis and de Vrijer [21], and via Taylor expansion by Barbarossa and Manzonetto [11]. Salibra studies a generalization to an infinitary λ-calculus [43]. García-Pérez and Nogueira prove *partial* genericity for Plotkin's CbV λ-calculus [23] using a different notion of meaningless terms, not as well-behaved as CbV inscrutable terms.

The most famous application of genericity is the collapsibility of meaningless terms. Another application is Folkerts's invertibility of terms for $\lambda\eta$ [22].

Independently, Arrial, Guerrieri, and Kesner developed an alternative study of genericity in both CbN and CbV [10].

Proofs. Most proofs are omitted and can be found in the technical report [8].

2 Preliminaries

In this paper, we consider two languages, the λ-calculus and the value substitution calculus. Here we give abstract definitions that apply to both. We then refer to a generic language $\mathcal{L}$ of host reduction $\to_{\mathcal{L}} \subseteq \mathcal{L} \times \mathcal{L}$ together with an evaluation strategy discussed below. Terms of both languages are considered modulo α-renaming. Capture-avoiding substitution is noted $t\{x{\leftarrow}u\}$.

Evaluation Strategies. An evaluation strategy for us is a relation $\to_{\mathsf{s}} \subseteq \to_{\mathcal{L}}$ which is either deterministic or has the diamond property, which, according to Dal Lago and Martini [18], is defined as follows: a relation $\to_{\mathsf{r}}$ is *diamond* if $u_1 \,{}_{\mathsf{r}}{\leftarrow}\, t \to_{\mathsf{r}} u_2$ and $u_1 \neq u_2$ imply $u_1 \to_{\mathsf{r}} s \,{}_{\mathsf{r}}{\leftarrow}\, u_2$ for some s. If $\to_{\mathsf{r}}$ is diamond then it is confluent, all its reductions to normal form (if any) have the same length, and if there is one such reduction from t then there are no diverging reductions from t; essentially, the diamond property is a weakened notion of determinism.

We refer to a generic evaluation strategy with $\to_{\mathsf{s}}$ or simply with s, and we also simply call it a *strategy*, and usually we omit the underlying language. The *conversion relation* $=_{\mathsf{s}}$ associated to a strategy s is the smallest equivalence relation containing $\to_{\mathsf{s}}$. We say that t is:

- s-*normal*: if $t \not\to_{\mathsf{s}}$;
- s-*normalizing*: if there exists u such that $t \to_{\mathsf{s}}^{*} u$ and u is s-normal;
- s-*diverging*: if t is not s-normalizing.

We say that s is:

- *Consistent*: if there exist two closed terms t and u such that t is s-normalizing and u is s-diverging;
- *Normalizing*: if $t \to_{\mathcal{L}}^{*} u$ with u s-normal implies that t is s-normalizing;
- *Stabilizing*: if t s-normal and $t \to_{\mathcal{L}}^{*} u$ imply u s-normal;
- *Weak*: if there are no s-redexes under abstraction.

t. Genericity as application is weaker than heavy/light genericity, and cannot be directly used to infer the collapsibility of $\to_{\mathsf{h}}$-diverging terms.

Contexts. An essential tool in our study shall be *contexts*, which are terms where a sub-term has been replaced by a hole $\langle\cdot\rangle$. For instance, for the λ-calculus they are defined as follows: $C, C' ::= \langle\cdot\rangle \mid tC \mid Ct \mid \lambda x.C$. The basic operation on contexts is the *plugging* $C\langle t\rangle$ of a term t in C, which simply replaces $\langle\cdot\rangle$ with t in C, possibly capturing variables. For instance, $(\lambda x.\langle\cdot\rangle)\langle xy\rangle = \lambda x.xy$. Note that plugging cannot be expressed as capture-avoiding substitution since $(\lambda x.z)\{z{\leftarrow}xy\} = \lambda x'.xy \neq \lambda x.xy$.

Contextual Equivalences and Pre-Orders. The standard of reference for program equivalences is contextual equivalence. The following definition slightly generalizes the standard one as to catch also the open case studied in this paper.

Definition 1 (Open and Closed Contextual Pre-Order and Equivalence). *Given an evaluation strategy* $\mathtt{s}$, *we define the* open contextual pre-order $\precsim_{CO}^{\mathtt{s}}$ *and* open contextual equivalence $\simeq_{CO}^{\mathtt{s}}$ *as follows:*

- $t \precsim_{CO}^{\mathtt{s}} t'$ *if, for all contexts* C, $C\langle t\rangle$ *is* $\mathtt{s}$*-normalizing implies that* $C\langle t'\rangle$ *is* $\mathtt{s}$*-normalizing;*
- $t \simeq_{CO}^{\mathtt{s}} t'$ *is the equivalence relation induced by* $\precsim_{CO}^{\mathtt{s}}$, *that is,* $t \simeq_{CO}^{\mathtt{s}} t'$ *if* $t \precsim_{CO}^{\mathtt{s}} t'$ *and* $t' \precsim_{CO}^{\mathtt{s}} t$.

The closed variants, simply called contextual pre-order $\precsim_{C}^{\mathtt{s}}$ *and* contextual equivalence $\simeq_{C}^{\mathtt{s}}$, *are defined as above but restricting to contexts* C *such that* $C\langle t\rangle$ *and* $C\langle t'\rangle$ *are closed terms. We say that* $\mathtt{s}$ *is* openable *if* $\precsim_{CO}^{\mathtt{s}}$ *and* $\precsim_{C}^{\mathtt{s}}$ *coincide.*

It follows from the definitions that $\precsim_{CO}^{\mathtt{s}} \subseteq \precsim_{C}^{\mathtt{s}}$, and similarly for the equivalences, while the other direction is not obvious, and can indeed fail. For instance, if $\mathtt{p_w}$ is weak evaluation in Plotkin's CbV λ-calculus (to be defined in Sect. 5) and $\delta := \lambda z.zz$ then we have $\Omega_l := (\lambda x.\delta)(yy)\delta \precsim_{C}^{\mathtt{p_w}} \delta\delta =: \Omega$ but $\Omega_l \not\precsim_{CO}^{\mathtt{p_w}} \Omega$. That is, $\mathtt{p_w}$ is not openable. To our knowledge, the notion of openable strategy is new.

(In)Equational Theories. A relation is *compatible* if $t \mathrel{\mathcal{R}} u$ implies $C\langle t\rangle \mathrel{\mathcal{R}} C\langle u\rangle$ for any context C and any terms t and u. A term t is minimum for a pre-order $\leq$ if for all $u \in \mathcal{L}$, $t \leq u$. We denote abstract inequational theories with the symbol $\leq_{\mathcal{T}}$ to distinguish them from known program pre-orders, denoted with $\precsim_{\mathcal{P}}$.

Definition 2 (Inequational s-theory). *Let* $\mathtt{s}$ *be an evaluation strategy. An inequational* $\mathtt{s}$*-theory* $\leq_{\mathcal{T}}^{\mathtt{s}}$ *is a compatible pre-order on terms containing* $\mathtt{s}$*-conversion. An inequational* $\mathtt{s}$*-theory* $\leq_{\mathcal{T}}^{\mathtt{s}}$ *is called:*

- Consistent*: whenever it does not relate all terms;*
- $\mathtt{s}$-ground*: if* $\mathtt{s}$*-diverging terms are minimum terms for* $\leq_{\mathcal{T}}^{\mathtt{s}}$*;*
- $\mathtt{s}$-adequate*: if* $t \leq_{\mathcal{T}}^{\mathtt{s}} u$ *and* t *is* $\mathtt{s}$*-normalizing implies* u *is* $\mathtt{s}$*-normalizing.*

The notions of $\mathtt{s}$-ground and $\mathtt{s}$-adequate theories generalize to an abstract and inequational framework the λ-calculus notions of *sensible* and *semi-sensible theories* (whose non-abstract inequational versions are studied in particular in the recent book by Barendregt and Manzonetto [16]), up to a very minor difference: the definitions in the literature sometimes also ask for consistency which we treat independently. An equational theory is a symmetric inequational theory.

Remark 1. Any open contextual pre-order $\precsim_{CO}^{s}$ is s-adequate: if $t \precsim_{CO}^{s} u$ then, by considering the empty context, t s-normalizing implies u s-normalizing. Closed contextual pre-orders, instead, are not necessarily adequate: for weak evaluation p_w in Plotkin's calculus, $\Omega_l \precsim_{C}^{p_w} \Omega$, Ω_l is p_w-normal, and Ω is p_w-diverging.

Lastly, we show under which conditions the contextual pre-orders $\precsim_{CO}^{s}$ and $\precsim_{C}^{s}$ are consistent inequational s-theories.

Proposition 1. *Let $\mathcal{L}$ be a confluent language and* s *be a normalizing and stabilizing strategy. Then $\precsim_{CO}^{s}$ and $\precsim_{C}^{s}$ (resp. $\simeq_{CO}^{s}$ and $\simeq_{C}^{s}$) are inequational (resp. equational)* s*-theories. Moreover, if* s *is consistent then $\precsim_{CO}^{s}$, $\precsim_{C}^{s}$, $\simeq_{CO}^{s}$, and $\simeq_{C}^{s}$ are consistent.*

3 Light Genericity and Collapsibility

As working notion of genericity, we adopt the following abstract light version.

Definition 3 (Light genericity). *Let* s *be an evaluation strategy.* Light s-genericity *is the following property: if u is* s*-diverging and C is a context such that $C\langle u\rangle$ is* s*-normalizing, then $C\langle t\rangle$ is* s*-normalizing for all t. Concisely:* s*-diverging terms are minimums for $\precsim_{CO}^{s}$. Very concisely: $\precsim_{CO}^{s}$ is* s*-ground.*

We now show that light genericity is enough to obtain the main application of Barendregt's heavier notion, that is, that s-diverging terms can be consistently equated (when s is consistent, which is a very mild hypothesis verified by all strategies of interest), by showing that they are contextually equivalent. In both the closed and open variants, independently of whether the strategy is openable.

Proposition 2 (Collapsibility). *Let* s *be a consistent evaluation strategy satisfying light genericity. Then:*

1. Open*: $\simeq_{CO}^{s}$ equates all* s*-diverging terms and it is consistent;*
2. Closed*: $\simeq_{C}^{s}$ equates all* s*-diverging terms and it is consistent.*

Proof. 1. By light genericity, s-diverging terms are minimums for $\precsim_{CO}^{s}$. Since then any two s-diverging terms are $\precsim_{CO}^{s}$-smaller than each other, s-diverging terms are $\simeq_{CO}^{s}$-equivalent. Since s is consistent, $\simeq_{CO}^{s}$ is consistent by Prop. 1.
2. Since $\precsim_{CO}^{s} \subseteq \precsim_{C}^{s}$, we obtain that light genericity implies that s-diverging terms are minimums for $\precsim_{C}^{s}$, and so $\simeq_{C}^{s}$ equates all s-diverging terms. Since s is consistent, $\simeq_{C}^{s}$ is consistent by Prop. 1. □

Proposition 3 (Characterization of minimum terms for $\precsim_{CO}^{s}$). *Let* s *be a consistent evaluation strategy satisfying light genericity. Then the minimum terms for $\precsim_{CO}^{s}$ are exactly the* s*-diverging terms.*

Proof. By light genericity, s-diverging terms are minimums for $\precsim_{CO}^{s}$. Conversely, by consistency of s there exists a s-diverging term t. Let u be a minimum for $\precsim_{CO}^{s}$. Then $u \precsim_{CO}^{s} t$, hence u is s-diverging by s-adequacy of $\precsim_{CO}^{s}$ (given by Remark 1). □

LANGUAGE	BETA RULE
TERMS $\Lambda \ni t, u, s ::= x \mid \lambda x.t \mid tu$	$\dfrac{}{(\lambda x.t)u \mapsto_\beta t\{x{\leftarrow}u\}}$

HEAD REDUCTION

WEAK HEAD CONTEXTS $P ::= \langle\cdot\rangle \mid Pt$ HEAD CONTEXTS $H ::= \lambda x.H \mid P$	$\dfrac{t \mapsto_\beta t'}{H\langle t\rangle \to_{\mathtt{h}} H\langle t'\rangle}$
RIGID TERMS $r, r' ::= x \mid rt$	HEAD NORMAL FORMS $h, h' ::= \lambda x.h \mid r$

Fig. 1. Call-by-Name calculus.

The characterization of minimum terms does not hold in the closed case, because the closed contextual pre-order is not necessarily adequate (Remark 1). For weak evaluation $\mathtt{p_w}$ in Plotkin's calculus, indeed, Ω_l is a minimum term for $\precsim_C^{\mathtt{p_w}}$ and it is $\mathtt{p_w}$-normal. The characterization lifts when $\mathtt{s}$ is openable.

4 The Head Call-by-Name Case

Here we revisit two results from the theory of the λ-calculus, and use them to prove that head evaluation is openable. The first result is genericity for unsolvable terms—that is, head-diverging terms—for which we give a proof of light genericity. The second result is the maximality of the open head contextual pre-order.

The host language $\mathcal{L}$ here is the λ-calculus and the evaluation strategy $\mathtt{s}$ is the head strategy $\mathtt{h}$. Both are defined in Fig. 1.

Solvability and Head Reduction. In the literature, the original notion of meaningful terms are the solvable ones, characterized by Wadsworth as those terminating for head reduction [46]; meaningless terms are their complement.

Definition 4 ((Un)Solvable terms). *A term t is* solvable *if there is a head context H such that $H\langle t\rangle \to_\beta^* \mathtt{I} = \lambda x.x$, and* unsolvable *otherwise.*

Theorem 1 (Operational characterization of solvability, [46]). *t is solvable (resp. unsolvable) if and only if t is $\mathtt{h}$-normalizing (resp. $\mathtt{h}$-diverging).*

Apart from the proof of Thm. 4.1 below, we shall always use the operational characterization and never refer to solvability itself.

Head Contextual Pre-Orders are Inequational. It is well-known that the λ-calculus is confluent, that head normal forms are stable by reduction (that is, $\mathtt{h}$ is stabilizing), and that the following normalization theorem holds (for a recent simple proof of this classic result see Accattoli et al. [3]). These facts and Prop. 1 give that the contextual pre-orders are inequational $\mathtt{h}$-theories.

Theorem 2 (Head normalization). *If $t \to_\beta^* t'$ and t' is $\mathtt{h}$-normal then t is $\mathtt{h}$-normalizing.*

Proposition 4. *The head pre-orders $\precsim_{CO}^{\mathtt{h}}$ and $\precsim_C^{\mathtt{h}}$ are inequational $\mathtt{h}$-theories.*

Proofs of Genericity. In his book [14], Barendregt gives two proofs that h-diverging terms can be consistently equated, both using heavy genericity (defined in the introduction). A first one [14, Lemma 16.1.8-thm 16.1.9] uses it to show that the minimal equational theory equating them, noted $\mathcal{H}$, is consistent. This proof is where the heavy part of genericity is used. A second proof [14, Lemma 16.2.3] exploits the consistency of $\simeq^{\mathtt{h}}_{\mathcal{CO}}$ (noted $\mathcal{H}^*$ in [14]), which is trivial, and uses genericity to show that $\mathcal{H} \subseteq \simeq^{\mathtt{h}}_{\mathcal{CO}}$, i.e. that $\simeq^{\mathtt{h}}_{\mathcal{CO}}$ equates all h-diverging terms.

The second proof in [14] uses heavy genericity, but the heavy aspect is in fact not needed for the proof to go through. The abstract result of the previous section, indeed, follows essentially the same reasoning and uses only light genericity.

We now prove light genericity for head reduction, via a direct proof, using the rewriting properties of head reduction.

Head Light Genericity via Takahashi's Technique. Our proof of light genericity adapts Takahashi's simple technique for heavy genericity [44]. We stress that two standard and crucial properties of head reduction are at work in Takahashi's proof, despite the fact that she does not point them out, namely the head normalization theorem (Theorem 2) and the following property.

Proposition 5 (Head substitutivity). *If $t \to_{\mathtt{h}}^* u$ then $t\{x{\leftarrow}s\} \to_{\mathtt{h}}^* u\{x{\leftarrow}s\}$, for all t, u, s.*

Firstly, we prove genericity for h-normal forms, via a simple induction on the structure of normal forms, using an auxiliary lemma [8, Lemma 4].

Proposition 6 (Normal genericity). *Let u be* h*-diverging and s be any term.*

1. *If r is a rigid term and $r\{x{\leftarrow}u\}$ is* h*-normalizing then $r\{x{\leftarrow}s\}$ is a rigid term.*
2. *If h is* h*-normal and $h\{x{\leftarrow}u\}$ is* h*-normalizing then $h\{x{\leftarrow}s\}$ is* h*-normal.*

We can now prove (light) genericity, which is done in two steps. The first one simply lifts h-normal genericity to non-h-normal terms, obtaining a substitution-based version of genericity. The second one turns the substitution-based statement into a context-based statement, and its proof is what we shall refer to as *Takahashi's trick*. For the sake of clarity, note that the two statements are not immediately equivalent, since substitution is a *capture-avoiding* operation while context plugging *may capture* free variables.

Theorem 3 (Light genericity). *Let u be* h*-diverging and s be any term.*

1. Light genericity as substitution: *if t is a term and $t\{x{\leftarrow}u\}$ is* h*-normalizing then $t\{x{\leftarrow}s\}$ is* h*-normalizing.*
2. Light genericity as context: *if C is a context and $C\langle u\rangle$ is* h*-normalizing then $C\langle s\rangle$ is* h*-normalizing.*

Proof. 1. It follows from Prop. 5 (precisely, via a lemma in [8, Lemma 4]), that if $t\{x{\leftarrow}u\}$ is h-normalizing then so is t. Then $t \to_{\mathtt{h}}^* h$ for some h-normal h. Again, by stability of head reduction under substitutions, we have both $t\{x{\leftarrow}u\} \to_{\mathtt{h}}^* h\{x{\leftarrow}u\}$ and $t\{x{\leftarrow}s\} \to_{\mathtt{h}}^* h\{x{\leftarrow}s\}$. Note that $t\{x{\leftarrow}u\}$ h-normalizing implies $h\{x{\leftarrow}u\}$ h-normalizing. By normal genericity (Prop. 6), $h\{x{\leftarrow}s\}$ is h-normal. Therefore, $t\{x{\leftarrow}s\}$ is h-normalizing.

2. Let $\mathtt{fv}(u) \cup \mathtt{fv}(s) = \{x_1, \ldots, x_k\}$, and y be a variable fresh with respect to $\mathtt{fv}(u) \cup \mathtt{fv}(s) \cup \mathtt{fv}(C)$ and not captured by C. Note that $\bar{u} := \lambda x_1. \ldots \lambda x_k.u$ is a closed term. Consider $t := C\langle y x_1 \ldots x_k \rangle$, and note that:

$$t\{y{\leftarrow}\bar{u}\} = C\langle \bar{u} x_1 \ldots x_k \rangle = C\langle (\lambda x_1. \ldots \lambda x_k.u) x_1 \ldots x_k \rangle \to_\beta^k C\langle u \rangle.$$

The fact that u is h-diverging implies that $\bar{u}$ is also h-diverging. If $C\langle u \rangle$ is h-normalizing then so is $t\{y{\leftarrow}\bar{u}\}$ by the h-normalization theorem (Theorem 2). By *genericity as substitution*, $t\{y{\leftarrow}s'\}$ is h-normalizing for every s'. In particular, take $s' := \bar{s} = \lambda x_1. \ldots \lambda x_k.s$, then $t\{y{\leftarrow}\bar{s}\}$ h-normalizes to some h and note that $t\{y{\leftarrow}\bar{s}\} \to_\beta^* C\langle s \rangle$. Since β is confluent and h is stabilizing, there exists a h-normal form h' such that $h \to_\beta^* h'$ and $C\langle s \rangle \to_\beta^* h'$. By the h-normalization theorem (Theorem 2), $C\langle s \rangle$ is h-normalizing. □

Maximality of $\precsim_{CO}^{\mathtt{h}}$. Barendregt shows that $\simeq_{CO}^{\mathtt{h}}$ is a *maximal consistent* theory, that is, that equating more terms would yield an inconsistent theory [14, Thm 16.2.6]. Later on, Barendregt and Manzonetto refine the result for $\precsim_{CO}^{\mathtt{h}}$ [16], by using the same technique, which relies on Böhm theorem. We present here a new proof of maximality based only on light genericity and not needing Böhm theorem, which is a heavier property, thus obtaining an arguably simpler proof. It is inspired by the proof of maximality for CbV by Egidi et al. [19].

Theorem 4. *1. Let $\mathcal{T}$ be an inequational* h*-theory that is* h*-ground but not* h*-adequate. Then $\mathcal{T}$ is inconsistent.*
2. Maximality of $\precsim_{CO}^{\mathtt{h}}$: *$\precsim_{CO}^{\mathtt{h}}$ is a maximal consistent inequational* h*-theory.*

Proof. 1. Since $\mathcal{T}$ is not h-adequate, there are t h-normalizing and u h-diverging such that $t \leq_{\mathcal{T}} u$. Since t is h-normalizing, by solvability there is a head context H sending it to the identity I. By the definition of inequational theory, we have $\mathtt{I} =_{\mathcal{T}} H\langle t \rangle \leq_{\mathcal{T}} H\langle u \rangle$. Now, let s be a term. Then $s =_{\mathcal{T}} \mathtt{I}s$ because $=_\beta \subseteq \mathcal{T}$ by definition of inequational theory. By the context closure of theories and $\mathtt{I} \leq_{\mathcal{T}} H\langle u \rangle$, we obtain $\mathtt{I}s \leq_{\mathcal{T}} H\langle u \rangle s$. Since u is h-diverging, thus unsolvable, $H\langle u \rangle$ is h-diverging. Since $\mathcal{T}$ is h-ground and both $H\langle u \rangle$ and $H\langle u \rangle s$ are h-diverging, $H\langle u \rangle s =_{\mathcal{T}} H\langle u \rangle$. Summing up, $s =_{\mathcal{T}} \mathtt{I}s \leq_{\mathcal{T}} H\langle u \rangle s =_{\mathcal{T}} H\langle u \rangle$ and, by the fact that $\mathcal{T}$ is h-ground, $H\langle u \rangle \leq_{\mathcal{T}} s$. Hence, $s =_{\mathcal{T}} H\langle u \rangle$ for every term s, that is, $\mathcal{T}$ is inconsistent.

2. Any theory $\mathcal{T}$ extending $\precsim_{CO}^{\mathtt{h}}$ is such that $t \leq_{\mathcal{T}} u$ with $t \not\precsim_{CO}^{\mathtt{h}} u$, i.e. such that $C\langle t \rangle$ is h-normalizing and $C\langle u \rangle$ is h-diverging for some C. By compatibility of $\mathcal{T}$, $C\langle t \rangle \leq_{\mathcal{T}} C\langle u \rangle$. Hence $\mathcal{T}$ is not h-adequate. Since $\precsim_{CO}^{\mathtt{h}}$ is h-ground by head light genericity (Theorem 3), every theory $\mathcal{T}$ extending $\precsim_{CO}^{\mathtt{h}}$ is also h-ground. Then $\mathcal{T}$ is h-ground and not h-adequate. By Point 1, $\mathcal{T}$ is inconsistent. □

Maximality and Head is Openable. From maximality of $\precsim_{CO}^{\mathtt{h}}$ it elegantly follows that $\precsim_{CO}^{\mathtt{h}}$ and $\precsim_{C}^{\mathtt{h}}$ coincide. To our knowledge, there is no such result in the

Language	Root rule
Terms $\Lambda \ni t, u, s \ ::= \ v \mid tu$ Values $v, v' \ ::= \ x \mid \lambda x.t$	$(\lambda x.t)v \ \mapsto_{\beta_v} \ t\{x {\leftarrow} v\}$

Weak Evaluation $\mathtt{p_w}$	Strong Evaluation
Weak Ctxs $E ::= \langle\cdot\rangle \mid Et \mid tE$	Strong Ctxs $C ::= \langle\cdot\rangle \mid Ct \mid tC \mid \lambda x.C$
$\dfrac{t \mapsto_{\beta_v} t'}{E\langle t\rangle \to_{\mathtt{p_w}} E\langle t'\rangle}$	$\dfrac{t \mapsto_{\beta_v} t'}{C\langle t\rangle \to_{\beta_v} C\langle t'\rangle}$

Fig. 2. Plotkin's CbV and Weak Evaluation $\mathtt{p_w}$.

literature but it is folklore for CbN. Note that, despite the apparently trivial proof that we provide below, the equivalence of $\precsim^{\mathtt{h}}_{\mathcal{CO}}$ and $\precsim^{\mathtt{h}}_{\mathcal{C}}$ is *not* a trivial fact, as the crucial inclusion $\precsim^{\mathtt{h}}_{\mathcal{C}} \subseteq \precsim^{\mathtt{h}}_{\mathcal{CO}}$ cannot be proved directly from the definitions of the pre-orders—in our proof, the non-trivial aspect is encapsulated in the use of maximality. Paolini proves that closed theories can be uniquely extended to open terms [37], but this does not imply that the extension of the closed contextual pre-order coincides with the open contextual pre-order.

Proposition 7 (Head evaluation is openable). *Open and closed head contextual pre-orders coincide:* $\precsim^{\mathtt{h}}_{\mathcal{CO}} = \precsim^{\mathtt{h}}_{\mathcal{C}}$.

Proof. Firstly, $\precsim^{\mathtt{h}}_{\mathcal{CO}} \subseteq \precsim^{\mathtt{h}}_{\mathcal{C}}$ follows from the definitions. Secondly, by maximality of $\precsim^{\mathtt{h}}_{\mathcal{CO}}$ (Theorem 4) and since $\precsim^{\mathtt{h}}_{\mathcal{C}}$ is consistent (because $\mathtt{I} \not\precsim^{\mathtt{h}}_{\mathcal{C}} \Omega$), we have that the two pre-orders must coincide, *i.e.* $\precsim^{\mathtt{h}}_{\mathcal{CO}} = \precsim^{\mathtt{h}}_{\mathcal{C}}$. □

5 Weak Call-by-Value and the VSC

We now turn our attention to the CbV case, for which the literature has already extensively discussed two issues that arise when adapting the CbN case to Plotkin's CbV λ-calculus, recalled after the definition of the calculus.

Plotkin's CbV λ-Calculus. Plotkin's CbV λ-calculus is defined in Fig. 2, following the modern presentation by Dal Lago and Martini [18] rather than Plotkin's original one [40]. We also define its weak evaluation strategy $\to_{\mathtt{p_w}}$.

Issue 1: CbV Unsolvable Terms Are Not Collapsible. As pointed out by Accattoli and Guerrieri [6], the CbV variant of unsolvable terms is not a good notion of meaningless terms, as their identification induces an inconsistent equational theory. The solution amounts to switching to a different notion of meaningless terms, the *inscrutable ones* (that coincide with the *non-potentially valuable terms* of Paolini and Ronchi della Rocca [38,36,41]), which are collapsible [6].

Definition 5 (Testing contexts). *Testing contexts are defined by the following grammar* $T ::= \langle\cdot\rangle \mid (\lambda x.T)t \mid Tt$.

Definition 6 ((In)Scrutable terms). *A term t is* scrutable *if there is a testing context T and a value v such that $T\langle t\rangle \to^*_{\beta_v} v$, and* inscrutable *otherwise.*

LANGUAGE				ROOT RULES
TERMS	$\Lambda \ni t, u, s$	$::=$	$v \mid tu \mid t[x{\leftarrow}u]$	$S\langle \lambda x.t\rangle u \mapsto_{\mathsf{m}} S\langle t[x{\leftarrow}u]\rangle$
VALUES	v, v'	$::=$	$x \mid \lambda x.t$	$t[x{\leftarrow}S\langle v\rangle] \mapsto_{\mathsf{e}} S\langle t\{x{\leftarrow}v\}\rangle$
SUB. CTXS	S, S'	$::=$	$\langle\cdot\rangle \mid S[x{\leftarrow}u]$	

WEAK CBV REDUCTION

WEAK CTXS $E ::= \langle\cdot\rangle \mid Et \mid tE \mid t[x{\leftarrow}E] \mid E[x{\leftarrow}u]$

$$\frac{t \mapsto_{\mathsf{a}} t'}{E\langle t\rangle \to_{\mathsf{wa}} E\langle t'\rangle}\ \mathsf{a}{\in}\{\mathsf{m},\mathsf{e}\}$$

NOTATION $\to_{\mathsf{w}} := \to_{\mathsf{wm}} \cup \to_{\mathsf{we}}$

STRONG CBV REDUCTION

STRONG CTXS $C ::= \langle\cdot\rangle \mid Ct \mid tC \mid t[x{\leftarrow}C] \mid C[x{\leftarrow}u] \mid \lambda x.C$

$$\frac{t \mapsto_{\mathsf{a}} t'}{C\langle t\rangle \to_{\mathsf{a}} C\langle t'\rangle}\ \mathsf{a}{\in}\{\mathsf{m},\mathsf{e}\}$$

NOTATION $\to_{\mathsf{vsc}} := \to_{\mathsf{m}} \cup \to_{\mathsf{e}}$

Fig. 3. Weak Value Substitution Calculus.

Issue 2: CbV Inscrutable Terms Have No Operational Characterization in Plotkin's CbV. The term $\Omega_l := (\lambda x.\delta)(yy)\delta$ is inscrutable but $\to_{\beta_v}$-normal. Therefore, in Plotkin's CbV there cannot be any operational characterization of inscrutable terms via a notion of divergence, as instead happens in CbN (Thm. 1). This fact is a real drawback, and boils down to the well-known inability of Plotkin's calculus to deal with open terms, which is also the reason why—as we have pointed out in Sect. 2—the closed and open contextual notions induced by weak evaluation in Plotkin's calculus do not coincide.

The solution amounts to switching to a refined CbV λ-calculus, extending Plotkin's as to better deal with open terms while retaining the same notion of contextual equivalence, as we now explain.

The VSC. Accattoli and Paolini's value substitution calculus (VSC) [9], defined in Figure 3, is exactly one such framework.

Intuitively, the VSC is a CbV λ-calculus extended with let-expressions, as is common for CbV λ-calculi such as Moggi's one [33,34]. We do however replace a let-expression $\mathsf{let}\ x = u\ \mathsf{in}\ t$ with a more compact *explicit substitution* (ES for short) notation $t[x{\leftarrow}u]$, which binds x in t and that has precedence over abstraction and application (that is, $\lambda x.t[y{\leftarrow}u]$ stands for $\lambda x.(t[y{\leftarrow}u])$ and $ts[y{\leftarrow}u]$ for $t(s[y{\leftarrow}u])$). Moreover, our let/ES does not fix an order of evaluation between t and u, in contrast to many papers in the literature (*e.g.* Sabry and Wadler [42] or Levy et al. [32]) where u is evaluated first.

The reduction rules of VSC are slightly unusual as they use *contexts* both to allow one to reduce redexes located in sub-terms, which is standard, *and* to define the redexes themselves, which is less standard—these kind of rules is called *at a distance*. The rationale is that the rewriting rules are designed to mimic cut-elimination on proof nets, via Girard's CbV translation $(A \Rightarrow B)^v = !(A^v \multimap B^v)$ of intuitionistic logic into linear logic [25], see Accattoli [2].

Examples of steps: $(\lambda x.y)[y{\leftarrow}t]u \mapsto_{\mathsf{m}} y[x{\leftarrow}u][y{\leftarrow}t]$ and $(\lambda z.xx)[x{\leftarrow}y[y{\leftarrow}t]] \mapsto_{\mathsf{e}} (\lambda z.yy)[y{\leftarrow}t]$. One with on-the-fly α-renaming is $(\lambda x.y)[y{\leftarrow}t]y \mapsto_{\mathsf{m}} z[x{\leftarrow}y][z{\leftarrow}t]$.

A key point is that β-redexes are decomposed via ESs, indeed $\mapsto_{\beta_v}$ is simulated as $(\lambda x.t)v \mapsto_{\mathsf{m}} t[x{\leftarrow}v] \mapsto_{\mathsf{e}} t\{x{\leftarrow}v\}$. Note that the *by-value* restriction is on ES-redexes, *not* on β-redexes, because only values can be substituted. The VSC is a conservative refinement for both closed and open terms: its weak evaluation on closed terms terminates if and only if Plotkin's $\rightarrow_{\mathtt{p_w}}$ does, hence the closed contextual pre-orders coincide (Prop. 8.3 below). On open terms, the VSC can simulate every $\rightarrow_{\mathtt{p_w}}$ step, but not vice-versa (which is why we adopt the VSC).

The Characterization of Inscrutable Terms. In the VSC, (in)scrutable terms admit an operational characterization, due to Accattoli and Paolini [9].

Theorem 5 (Operational characterization of (in)scrutability, [9]). *t is scrutable (resp. inscrutable) if and only if t is* w*-normalizing (resp.* w*-diverging).*

Apart from the proof of Thm. 8 below and Prop. 15 in Section 10, we shall always use the operational characterization and never refer to scrutability itself.

Weak Contextual Pre-Orders Are Inequational. The VSC is confluent and its weak strategy w is diamond [9]. Moreover, w is stabilizing and the normalization theorem below holds. These facts and Prop. 1 give that the contextual pre-orders are inequational w-theories. Moreover, the closed pre-order coincides with the one on Plotkin's calculus[4].

Proposition 8. *1.* Weak normalization, [6]*: if $t \rightarrow^*_{\mathtt{vsc}} t'$ and t' is* w*-normal then t is* w*-normalizing.*

2. Inequational theories*: $\precsim^{\mathtt{w}}_{\mathcal{CO}}$ and $\precsim^{\mathtt{w}}_{\mathcal{C}}$ are inequational* w*-theories.*

3. VSC and Plotkin's contextual pre-orders coincide, [6]*: on λ-terms, $\precsim^{\mathtt{w}}_{\mathcal{C}} = \precsim^{\mathtt{p_w}}_{\mathcal{C}}$.*

6 Light Genericity for Weak Call-by-Value

Here, we prove a new result: light genericity for weak evaluation in the VSC.

Takahashi's Technique Does Not Really Scale Up. Proving CbV light genericity via Takahashi's technique is not as elegant as for CbN. We did develop such a proof, but it is considerably more involved than for CbN. There are various reasons. Firstly, the substitutivity property of Prop. 5 does not hold in CbV. Substitutivity for values does hold, but one really needs general substitutivity. Secondly, Takahashi's *trick* lifting genericity as substitutions to genericity as contexts also breaks, because it is based on adding abstractions, which do not change unsolvability but do affect inscrutability. Thirdly, head reduction reduces only on the head, while weak reduction reduces in all sub-terms out of abstractions, which raises additional difficulties. Therefore, we follow a different approach.

[4] The closed CbV contextual pre-order in Carraro and Guerrieri's shuffling calculus [17], studied by Kerinec et al. in [29], also coincides with $\precsim^{\mathtt{w}}_{\mathcal{C}} = \precsim^{\mathtt{p_w}}_{\mathcal{C}}$. Moreover, the open pre-order of the shuffling calculus coincides with the one of the VSC. These facts follow easily from results relating the three calculi in [26,4,6].

LINEAR TYPES $L, L' ::= M \to N$ MULTI TYPES $M, N ::= [L_1, \ldots, L_n]\ n \geq 0$

$$\frac{}{x:[L] \vdash x:L}\ \mathsf{ax} \qquad \frac{\Gamma, x:M \vdash t:N}{\Gamma \vdash \lambda x.t:M \to N}\ \lambda \qquad \frac{(\Gamma_i \vdash v:L_i)_{i\in I} \quad I \text{ finite}}{\biguplus_{i\in I} \Gamma_i \vdash v:\biguplus_{i\in I}[L_i]}\ \mathsf{many}$$

$$\frac{\Gamma \vdash t:[M \to N] \quad \Delta \vdash u:M}{\Gamma \uplus \Delta \vdash tu:N}\ @ \qquad \frac{\Gamma, x:M \vdash t:N \quad \Delta \vdash u:M}{\Gamma \uplus \Delta \vdash t[x{\leftarrow}u]:N}\ \mathsf{es}$$

Fig. 4. Call-by-Value Multi Type System for VSC.

Light Genericity via Multi Types. We provide a proof of light genericity relying on Accattoli and Guerrieri's characterization of w-diverging terms [6] via Ehrhard's CbV multi types [20] (multi types are also known as *non-idempotent intersection types*). The idea behind the proof is very simple: we show that multi types induce a pre-order $\precsim_{type}$ contained in the open contextual pre-order, that is, $\precsim_{type} \subseteq \precsim_{CO}^{\mathtt{w}}$, *and* that w-diverging terms are minimum elements for $\precsim_{type}$, which implies that they are minimums for $\precsim_{CO}^{\mathtt{w}}$. The proof itself is very simple as well. What is less simple is the characterization of w-diverging terms via multi types, which however we use as a black box from the literature. The same technique can be used also in CbN, since h-diverging terms can also be characterized via multi types.

Our argument via multi types is similar to Ghilezan's one based on intersection types for CbN [24], even if the details are quite different: she proves a different statement, namely heavy genericity in its *as-application* variant (see the footnote at page 5), and she uses intersection types (which are idempotent, or non-linear). We use multi types because the result from the literature that we exploit is based on them, but the proof technique could also be based on intersection types (once the result from the literature is adapted, which is possible).

CbV Multi Types. We introduce the bare minimum about CbV multi types, since here they are used only as a tool, not as an object of study. For more, see [5,6].

The definition of the multi type system for the VSC is in Figure 4. Multi types M are defined by mutual induction with linear types L. Multi types are finite multi-sets $[L_1, \ldots, L_n]$, which intuitively denote the intersection $L_1 \cap \ldots \cap L_n$, where the intersection $\cap$ is a commutative, associative and non-idempotent ($A \cap A \neq A$) operator, the neutral element of which is $[\,]$, the empty multi set. Note that there is no ground type, its role is played by the empty multi type $[\,]$.

Typing judgments have shape $\Gamma \vdash t:T$ where T is a linear or a multi type and Γ is a typing context, that is, an assignment of multi types to a finite set of variables ($\Gamma = x_1:M_1, \ldots, x_n:M_n$). A typing derivation $\pi \triangleright \Gamma \vdash t:M$ is a tree built from the rules in Figure 4 which ends with the typing judgment $\Gamma \vdash t:M$.

Typing Rules. Linear types only type values, via the rules ax and λ. To give a multi type to value v, one has to use the many rule, turning an indexed family of linear types for v into a multi type. Note that any value can be typed with the empty multi type $[\,]$. The symbol $\uplus$ is the disjoint union operator on multi sets (corresponding to the intersection operator when intersections are multi-sets).

Characterization of Termination. The key property of CbV multi types is that typability characterizes termination with respect to weak evaluation $\to_{\mathtt{w}}$; therefore w-diverging terms are simply the untypable ones. The characterization is proved via subject reduction and expansion.

Theorem 6 (Characterization of termination, [6]).

1. Subject reduction and expansion*: let* $t \to_{\mathtt{vsc}} u$*. Then* $\Gamma \vdash t{:}M$ *iff* $\Gamma \vdash u{:}M$*.*
2. t *is* $\to_{\mathtt{w}}$*-normalizing if and only if there exists* Γ *and* M *such that* $\Gamma \vdash t{:}M$*.*

Type Pre-order. The type pre-order is defined as follows.

Definition 7 (Type pre-order). *The type pre-order* $t \precsim_{type} t'$ *holds if* $\Gamma \vdash t{:}M$ *implies* $\Gamma \vdash t'{:}M$ *for all* Γ *and* M*.*

Point 2 of Thm. 6 ensures that $\precsim_{type}$ is both w-ground—which is the key point of the proof technique—and w-adequate. We also show that $\precsim_{type}$ is an inequational w-theory. Point 1 of Thm. 6 implies that $\precsim_{type}$ contains w-conversion. Compatibility holds because $\precsim_{type}$ is defined via a *compositional* type system.

Proposition 9. *The type pre-order* $\precsim_{type}$ *is a* w*-ground,* w*-adequate, and consistent inequational* w*-theory.*

Adequacy and compatibility of $\precsim_{type}$ imply that $\precsim_{type} \subseteq \precsim^{\mathtt{w}}_{CO}$, hence minimum elements of $\precsim_{type}$ are minimum for $\precsim^{\mathtt{w}}_{CO}$.

Theorem 7. *Light genericity for* w*:* $\precsim^{\mathtt{w}}_{CO}$ *is* w*-ground.*

7 CbV Maximality

Here, we use light genericity to prove maximality of $\precsim^{\mathtt{w}}_{CO}$ and the fact that w is openable, adapting the proofs for the head case.

Maximality of $\precsim^{\mathtt{w}}_{CO}$. The following result adapts to our setting a result of Accattoli and Guerrieri [6, Thm 6.5], itself adapting a result by Egidi et al. [19, Prop 35].

Theorem 8. *1. Any* w*-ground inequational theory* $\mathcal{T}$ *that is not* w*-adequate is inconsistent.*
2. Maximality of $\precsim^{\mathtt{w}}_{CO}$*:* $\precsim^{\mathtt{w}}_{CO}$ *is a maximal consistent inequational theory.*

Proof. 1. Since $\mathcal{T}$ is not w-adequate, there are t w-normalizing and u w-diverging such that $t \leq_{\mathcal{T}} u$. Since t is w-normalizing, t is scrutable, that is, there is a testing context T sending it to a value v. By the definition of inequational w-theory, we have $v =_{\mathcal{T}} T\langle t\rangle \leq_{\mathcal{T}} T\langle u\rangle$. Now, let s be a term and $y \notin \mathtt{fv}(s)$. Then $s =_{\mathcal{T}} (\lambda y.s)v$ because $=_{\mathtt{vsc}} \subseteq =_{\mathcal{T}}$ by definition of inequational theory. By the compatibility of theories and $v \leq_{\mathcal{T}} T\langle u\rangle$, we obtain $(\lambda y.s)v \leq_{\mathcal{T}} (\lambda y.s)T\langle u\rangle$. Since u is w-diverging, thus inscrutable, $T\langle u\rangle$ is also w-diverging. Since $\mathcal{T}$ is w-ground and both $T\langle u\rangle$ and $(\lambda y.s)T\langle u\rangle$ are w-diverging, $(\lambda y.s)T\langle u\rangle =_{\mathcal{T}} T\langle u\rangle$. Summing up, $s =_{\mathcal{T}} (\lambda y.s)v \leq_{\mathcal{T}} (\lambda y.s)T\langle u\rangle \leq_{\mathcal{T}} T\langle u\rangle$ and, since $\mathcal{T}$ is w-ground, $T\langle u\rangle \leq_{\mathcal{T}} s$. Hence, $s =_{\mathcal{T}} T\langle u\rangle$ for every term s, that is, $\mathcal{T}$ is inconsistent.

2. From Point 1 and CbV light genericity (Thm. 7.3), as in the head case. □

The proof of Thm. 8.1 is *similar* to the one of the CbN case, but it is *not* the *same argument*: the CbN one relies on solvability, reduction to the identity, and head context closure; the CbV one relies on scrutability, reduction to a value, a different context closure, and on the fact that diverging arguments cannot be erased in CbV. Therefore, our proofs of maximality cannot be done abstractly.

The fact that weak evaluation is openable then follows as in the head case.

Proposition 10 (Weak evaluation is openable in the VSC). *Open and closed weak contextual pre-orders coincide:* $\precsim^{\mathtt{w}}_{\mathcal{CO}} = \precsim^{\mathtt{w}}_{\mathcal{C}}$.

8 Co-Genericity

Here, we study a new notion dual to light genericity, which we dub *co-genericity.*

s-*Super Terms.* In the λ-calculus (both in CbN and CbV) there are terms reducing to an infinite sequence of abstractions using strong evaluation. For instance, let $\delta_\lambda := \lambda x.\lambda y.xx$, then $\Omega_\lambda := \delta_\lambda\delta_\lambda$ is one such term. Indeed its weak evaluation gives $\Omega_\lambda \mapsto_{\beta_v} \lambda y.\Omega_\lambda$. Now, the new copy of Ω_λ shall itself (strongly) reduce to $\lambda y.\Omega_\lambda$, and so on, producing $\lambda y.\lambda y.\lambda y.\ldots$. Such a behavior, when seen with respect to weak evaluation, is a form of hereditary, or *super* normalization.

Note that the example can be generalized by using $\delta_{k\lambda} := \lambda x.\lambda y_1.\ldots.\lambda y_k.xx$ instead of δ_λ, obtaining a family of terms $\Omega_{k\lambda} := \delta_{k\lambda}\delta_{k\lambda}$ all producing infinitely many head abstractions and with no (finite) reduct in common. As for meaningless terms, it is natural to wonder whether these super meaningful terms can all be consistently collapsed. In the literature, super terms appear in weak CbN as maximum ($\top$) elements in Lévy-Longo trees [31]—but we are not aware of a proof that these $\top$-enriched Lévy-Longo trees induce a consistent equational theory—and in the hierarchy of unsolvable terms [35,1] as unsolvable terms of order ∞. In CbV, we believe that super terms have not been studied.

Here we connect the collapsibility of super terms to a sort of dual variant of light genericity. We start by setting up the concept of super normalization *abstractly.* It is specific to weak strategies and makes sense also for weak CbN.

Definition 8 (s-super terms). *Let* s *be a weak strategy. A term t is* s*-super (normalizing) if, co-inductively,* $t \to_{\mathtt{s}}^{*} \lambda x.t'$ *and* t' *is* s*-super.*

Co-genericity is the property stating that s-super terms are maximum elements for $\precsim^{\mathtt{s}}_{\mathcal{CO}}$, that shall be captured by the following notion of *being* s*-roof.* As expected, a term t is maximum for a pre-order $\leq$ if for all $u \in \mathcal{L}$, $u \leq t$.

Definition 9. *Let* s *be a weak strategy. An inequational* s*-theory* $\leq^{\mathtt{s}}_{\mathcal{T}}$ *is called:*

1. s-roof*: if* s*-super terms are maximum terms for* $\leq^{\mathtt{s}}_{\mathcal{T}}$*;*
2. Super s-adequate*: if* $t \leq^{\mathtt{s}}_{\mathcal{T}} u$ *and t is* s*-super entails u is* s*-super.*

Definition 10 (Co-genericity). *Let* s *be a weak strategy.* Co-s-genericity *is the following property: if u is* s*-super and C is a context such that $C\langle t\rangle$ is* s*-normalizing for some t, then $C\langle u\rangle$ is* s*-normalizing. Concisely:* s*-super terms are maximum for $\precsim^{\mathtt{s}}_{CO}$. Very concisely: $\precsim^{\mathtt{s}}_{CO}$ is* s*-roof.*

Note that there cannot be a *heavy* co-genericity property mentioning strong normal forms because s-super terms are diverging for strong s-evaluation, by definition. Co-genericity is thus *enabled* by the switch from heavy to light genericity.

As for light genericity, co-genericity is enough to prove that s-super terms can be consistently equated (as soon as s is consistent).

Proposition 11 (Co-collapsibility). *Let* s *be a consistent weak strategy satisfying co-genericity. Then $\simeq^{\mathtt{s}}_{CO}$ equates all* s*-super terms and it is consistent.*

A weak strategy s is *super consistent* if there exists a s-super term.

Proposition 12 (Characterization of maximum terms for $\precsim^{\mathtt{s}}_{CO}$). *Let* s *be a super consistent weak strategy satisfying co-genericity. If $\precsim^{\mathtt{s}}_{CO}$ is super* s*-adequate then the maximum terms for $\precsim^{\mathtt{s}}_{CO}$ are exactly the* s*-super terms.*

Proof. By co-genericity, s-super terms are maximal for $\precsim^{\mathtt{s}}_{CO}$. For the other direction, let t be a s-super term, which exists by super consistency of s, and let u be maximal for $\precsim^{\mathtt{s}}_{CO}$. Then $t \precsim^{\mathtt{s}}_{CO} u$. By super s-adequacy, u is s-super. □

The two following sections present independent proofs of co-genericity for weak evaluation in the VSC. We do not use multi types for good reasons: w-super terms are *not* maximum for $\precsim_{type}$, see the technical report [8, Prop. 18].

9 CbV Co-Genericity via Takahashi's Technique

In this section, we prove co-genericity for weak evaluation in the VSC adapting Takahashi's technique for genericity.

Co-Genericity via Normal Forms. The proof of co-genericity for CbV is based on a key property of w-super terms with respect to w-normal forms, akin to the normal genericity lemma of the CbN case. Then co-genericity follows via Takahashi's trick, which is not problematic here, since w-super terms are stable by adding head abstractions. Another difficulty arises in CbV, however, which is discussed in the technical report [8, p.27] before the proof of the following lemma.

Lemma 1 (Key property of w-super terms). *Let s be a* w*-super term. If n is a* w*-normal form then $n\{x{\leftarrow}s\}$ is* w*-normalizing.*

As CbV evaluation only validate value-substitutivity (substitutivity restricted to values: if $t \to_{\mathtt{w}} u$ then for all v $t\{x{\leftarrow}v\} \to_{\mathtt{w}} u\{x{\leftarrow}v\}$), the statement of co-genericity as substitution is split into two points.

Lemma 2 (Co-genericity). *Let u be any term, s be a* w*-super term, v be any value, and v' be a* w*-super value.*

1. Co-genericity as v-substitution*: if $t\{x{\leftarrow}v\}$ is* w*-normalizing then so is $t\{x{\leftarrow}v'\}$.*
2. Co-genericity as substitution*: if $t\{x{\leftarrow}u\}$ is* w*-normalizing then so is $t\{x{\leftarrow}s\}$.*
3. Co-genericity as context*: if $C\langle u\rangle$ is* w*-normalizing then so is $C\langle s\rangle$.*

Super w*-Adequacy for* $\precsim^{\mathtt{w}}_{CO}$. Co-genericity states that w-super terms are maximal for $\precsim^{\mathtt{w}}_{CO}$. For the full characterization (Prop. 12), we need *super adequacy* and *super consistency.* Super consistency is easily verified as Ω_λ exists.

Proposition 13 (Super w-Adequacy).

1. Super adequacy*:* $\precsim^{\mathtt{w}}_{CO}$ *is super* w*-adequate.*
2. Characterization of maximum terms for $\precsim^{\mathtt{w}}_{CO}$*: maximum terms for* $\precsim^{\mathtt{w}}_{CO}$ *are exactly* w*-super terms.*

10 CbV (Co-)Genericity via Applicative Similarity

In this section, we present alternative proofs of genericity and co-genericity for weak evaluation in the VSC. We use a well-known tool developed to study Plotkin's CbV contextual equivalence $\simeq^{\mathtt{Pw}}_{C}$, namely the CbV variant [27,39] of Abramsky's *applicative (bi)similarity* [1].

The following definition differs slightly from the literature on two points. Firstly, we use a well known equivalent definition that does *not* ask that the results of evaluation are similar (which is a fact needed for the definition of applicative simulations, but not for applicative similarity). Secondly, we replace Plotkin's CbV by the VSC, which are equivalent for closed terms.

Definition 11 (Applicative similarity [1]). *Applicative similarity* $t \precsim^{\mathtt{w}}_{AS} u$ *is the relation on closed terms defined by: if* $t\, v_1 \dots v_n$ *is* w*-normalizing then* $u\, v_1 \dots v_n$ *is* w*-normalizing, for all* $n \in \mathbb{N}$ *and* $v_1, \dots, v_n$ *closed values. Applicative similarity is extended to open terms via closing substitutions:* $t \precsim^{\mathtt{w}}_{AS} u$ *if* $t\sigma \precsim^{\mathtt{w}}_{AS} u\sigma$ *for all substitutions of values* σ *closing* t *and* u*.*

From the following lemma, it follows easily that w-diverging and w-super terms are minimum and maximum for $\precsim^{\mathtt{w}}_{AS}$.

Lemma 3. *If* t *is* w*-diverging (resp.* w*-super) then so are* $t\{x{\leftarrow}v\}$ *and* tv*.*

Proposition 14. *1.* Minimums: w*-diverging terms are minimum for* $\precsim^{\mathtt{w}}_{AS}$*.*
2. Maximums: w*-super terms are maximum for* $\precsim^{\mathtt{w}}_{AS}$*.*

Proof. 1. Let t be a w-diverging term and u any term. Then by Lemma 3, for any closing substitution σ of t and u and for any n and any $v_1, \dots, v_n$ we still have that $t\sigma\, v_1, \dots, v_n$ is w-diverging. Hence $t \precsim^{\mathtt{w}}_{AS} u$ for any term u, that is, t is a minimum term for $\precsim^{\mathtt{w}}_{AS}$.
2. Let t be a w-super term and u any term. For any closing substitution σ of t and u and for any n and values $v_1, \dots, v_n$, either $u\sigma\, v_1, \dots, v_n$ is w-diverging or $u\sigma\, v_1, \dots, v_n$ is w-normalizing. In both cases, by Lemma 3, we still have that $t\sigma\, v_1, \dots, v_n$ is w-super, hence w-normalizing. Thus, $u \precsim^{\mathtt{w}}_{AS} t$. □

Proving (co-)genericity amounts to show that the results of the previous proposition transfer to $\precsim^{\mathtt{w}}_{CO}$. This can be done by showing $\precsim^{\mathtt{w}}_{AS} \subseteq \precsim^{\mathtt{w}}_{CO}$ via:

1. The soundness of applicative similarity $\precsim^{\mathtt{w}}_{AS}$ for Plotkin's pre-order $\precsim^{\mathtt{Pw}}_{C}$, that is, that $\precsim^{\mathtt{w}}_{AS}\subseteq\precsim^{\mathtt{Pw}}_{C}$ (completeness holds as well, but it is not useful here);
2. The equivalence $\precsim^{\mathtt{Pw}}_{C}=\precsim^{\mathtt{w}}_{C}$, given by Prop. 8.3;
3. The openability of w-evaluation, that is, $\precsim^{\mathtt{w}}_{C}=\precsim^{\mathtt{w}}_{CO}$.

Soundness of $\precsim^{\mathtt{w}}_{AS}$ is a non-trivial result in the literature, established by Howe's method [27,39], which we here use as a black box. About openability, we proved it in Sect. 7 but that proof uses light genericity (and maximality), which is our goal here, so we have to re-prove openability without using light genericity.

w is Openable without Light Genericity. We know that $\precsim^{\mathtt{w}}_{CO}\subseteq\precsim^{\mathtt{w}}_{C}$, thus we only have to show the other inclusion, which follows from w-adequacy of $\precsim^{\mathtt{w}}_{C}$.

Proposition 15. *The inequational theory $\precsim^{\mathtt{w}}_{C}$ is* w*-adequate, hence* w *is openable.*

Proof. The proof is in [8, p.32], here we only give the idea for w-adequacy. Let $t \precsim^{\mathtt{w}}_{C} u$ with t w-normalizing. Then, we use the operational characterization of scrutability (Thm. 5) to build a closing context C such that $C\langle t\rangle$ is w-normalizing and such that if u were w-diverging, so would be $C\langle u\rangle$. □

(Co-)genericity via Applicative Similarity. The three points above are established, and so we obtain new proofs of light genericity and co-genericity.

Proposition 16 (CbV light (co-)genericity). $\precsim^{\mathtt{w}}_{CO}$ *is* w*-ground and* w*-roof.*

11 Conclusions

We develop in this paper a theory of *light* genericity, which is as powerful as heavy genericity for proving the collapsibility of meaningless terms, it is connected to contextual pre-orders, and dualizable as *co-genericity.*

We also provide light proofs of the *maximality* of open contextual pre-orders, which in turn provide an elegant proof of the fact that the closed and open contextual pre-orders coincide. Lastly, we show that CbV applicative similarity can be used for alternative simple proofs of light (co-)genericity. These simple proofs via applicative similarity are easily adaptable to the (weak) CbN case.

Summing up, our work paints Barendregt's genericity with a fresh, modern hue, connecting it to program equivalences and maximality, following an abstract approach and providing neat proofs.

Acknowledgements. To Giulio Manzonetto and Gabriele Vanoni for feedback on a first draft, and to Victor Arrial for helpful discussions about genericity.

References

1. Abramsky, S., Ong, C.L.: Full abstraction in the lazy lambda calculus. Inf. Comput. **105**(2), 159–267 (1993). https://doi.org/10.1006/inco.1993.1044
2. Accattoli, B.: Proof nets and the call-by-value λ-calculus. Theor. Comput. Sci. **606**, 2–24 (2015). https://doi.org/10.1016/j.tcs.2015.08.006
3. Accattoli, B., Faggian, C., Guerrieri, G.: Factorization and normalization, essentially. In: Lin, A.W. (ed.) Programming Languages and Systems - 17th Asian Symposium, APLAS 2019, Nusa Dua, Bali, Indonesia, December 1-4, 2019, Proceedings. Lecture Notes in Computer Science, vol. 11893, pp. 159–180. Springer (2019). https://doi.org/10.1007/978-3-030-34175-6_9
4. Accattoli, B., Guerrieri, G.: Open call-by-value. In: Igarashi, A. (ed.) Programming Languages and Systems - 14th Asian Symposium, APLAS 2016, Hanoi, Vietnam, November 21-23, 2016, Proceedings. Lecture Notes in Computer Science, vol. 10017, pp. 206–226 (2016). https://doi.org/10.1007/978-3-319-47958-3_12, https://doi.org/10.1007/978-3-319-47958-3_12
5. Accattoli, B., Guerrieri, G.: Types of fireballs. In: Ryu, S. (ed.) Programming Languages and Systems - 16th Asian Symposium, APLAS 2018, Wellington, New Zealand, December 2-6, 2018, Proceedings. Lecture Notes in Computer Science, vol. 11275, pp. 45–66. Springer (2018). https://doi.org/10.1007/978-3-030-02768-1_3
6. Accattoli, B., Guerrieri, G.: The theory of call-by-value solvability. Proc. ACM Program. Lang. **6**(ICFP), 855–885 (2022). https://doi.org/10.1145/3547652
7. Accattoli, B., Guerrieri, G., Leberle, M.: Strong call-by-value and multi types. In: Ábrahám, E., Dubslaff, C., Tarifa, S.L.T. (eds.) Theoretical Aspects of Computing - ICTAC 2023 - 20th International Colloquium, Lima, Peru, December 4-8, 2023, Proceedings. Lecture Notes in Computer Science, vol. 14446, pp. 196–215. Springer (2023). https://doi.org/10.1007/978-3-031-47963-2_13
8. Accattoli, B., Lancelot, A.: Light Genericity (Jan 2024), https://hal.science/hal-04406343, technical report
9. Accattoli, B., Paolini, L.: Call-by-value solvability, revisited. In: Schrijvers, T., Thiemann, P. (eds.) Functional and Logic Programming - 11th International Symposium, FLOPS 2012, Kobe, Japan, May 23-25, 2012. Proceedings. Lecture Notes in Computer Science, vol. 7294, pp. 4–16. Springer (2012). https://doi.org/10.1007/978-3-642-29822-6_4
10. Arrial, V., Guerrieri, G., Kesner, D.: Genericity through stratification (2024), https://arxiv.org/abs/2401.12212
11. Barbarossa, D., Manzonetto, G.: Taylor subsumes Scott, Berry, Kahn and Plotkin. Proc. ACM Program. Lang. **4**(POPL), 1:1–1:23 (2020). https://doi.org/10.1145/3371069
12. Barendregt, H.P.: Some extensional term models for combinatory logics and λ-calculi. Ph.D. thesis, Univ. Utrecht (1971)
13. Barendregt, H.P.: Solvability in lambda-calculi. In: Guillaume, M. (ed.) Colloque international de logique : Clermont-Ferrand, 18-25 juillet 1975. pp. 209–219. Éditions du C.N.R.S., Paris (1977)
14. Barendregt, H.P.: The Lambda Calculus – Its Syntax and Semantics, Studies in logic and the foundations of mathematics, vol. 103. North-Holland (1984)
15. Barendregt, H.: Representing 'undefined' in lambda calculus. J. Funct. Program. **2**(3), 367–374 (1992). https://doi.org/10.1017/S0956796800000447
16. Barendregt, H., Manzonetto, G.: A Lambda Calculus Satellite. College Publications (2022), https://www.collegepublications.co.uk/logic/mlf/?00035

17. Carraro, A., Guerrieri, G.: A semantical and operational account of call-by-value solvability. In: Muscholl, A. (ed.) Foundations of Software Science and Computation Structures - 17th International Conference, FOSSACS 2014, Grenoble, France, April 5-13, 2014, Proceedings. Lecture Notes in Computer Science, vol. 8412, pp. 103–118. Springer (2014). https://doi.org/10.1007/978-3-642-54830-7_7, https://doi.org/10.1007/978-3-642-54830-7_7
18. Dal Lago, U., Martini, S.: The weak lambda calculus as a reasonable machine. Theor. Comput. Sci. **398**(1-3), 32–50 (2008). https://doi.org/10.1016/j.tcs.2008.01.044
19. Egidi, L., Honsell, F., Ronchi Della Rocca, S.: Operational, denotational and logical descriptions: a case study. Fundam. Inform. **16**(1), 149–169 (1992)
20. Ehrhard, T.: Collapsing non-idempotent intersection types. In: Cégielski, P., Durand, A. (eds.) Computer Science Logic (CSL'12) - 26th International Workshop/21st Annual Conference of the EACSL, CSL 2012, September 3-6, 2012, Fontainebleau, France. LIPIcs, vol. 16, pp. 259–273. Schloss Dagstuhl - Leibniz-Zentrum für Informatik (2012). https://doi.org/10.4230/LIPIcs.CSL.2012.259
21. Endrullis, J., de Vrijer, R.C.: Reduction under substitution. In: Voronkov, A. (ed.) Rewriting Techniques and Applications, 19th International Conference, RTA 2008, Hagenberg, Austria, July 15-17, 2008, Proceedings. Lecture Notes in Computer Science, vol. 5117, pp. 425–440. Springer (2008). https://doi.org/10.1007/978-3-540-70590-1_29
22. Folkerts, E.: Invertibility in lambda-eta. In: Thirteenth Annual IEEE Symposium on Logic in Computer Science, Indianapolis, Indiana, USA, June 21-24, 1998. pp. 418–429. IEEE Computer Society (1998). https://doi.org/10.1109/LICS.1998.705676
23. García-Pérez, Á., Nogueira, P.: No solvable lambda-value term left behind. Log. Methods Comput. Sci. **12**(2) (2016). https://doi.org/10.2168/LMCS-12(2:12)2016
24. Ghilezan, S.: Full intersection types and topologies in lambda calculus. J. Comput. Syst. Sci. **62**(1), 1–14 (2001). https://doi.org/10.1006/jcss.2000.1703
25. Girard, J.Y.: Linear Logic. Theoretical Computer Science **50**, 1–102 (1987). https://doi.org/10.1016/0304-3975(87)90045-4
26. Guerrieri, G., Paolini, L., Ronchi Della Rocca, S.: Standardization and conservativity of a refined call-by-value lambda-calculus. Logical Methods in Computer Science **13**(4) (2017). https://doi.org/10.23638/LMCS-13(4:29)2017
27. Howe, D.J.: Proving congruence of bisimulation in functional programming languages. Inf. Comput. **124**(2), 103–112 (1996). https://doi.org/10.1006/inco.1996.0008
28. Kennaway, R., van Oostrom, V., de Vries, F.: Meaningless terms in rewriting. J. Funct. Log. Program. **1999**(1) (1999), http://danae.uni-muenster.de/lehre/kuchen/JFLP/articles/1999/A99-01/A99-01.html
29. Kerinec, A., Manzonetto, G., Pagani, M.: Revisiting call-by-value böhm trees in light of their taylor expansion. Log. Methods Comput. Sci. **16**(3) (2020), https://lmcs.episciences.org/6638
30. Kuper, J.: Proving the genericity lemma by leftmost reduction is simple. In: Hsiang, J. (ed.) Rewriting Techniques and Applications, 6th International Conference, RTA-95, Kaiserslautern, Germany, April 5-7, 1995, Proceedings. Lecture Notes in Computer Science, vol. 914, pp. 271–278. Springer (1995). https://doi.org/10.1007/3-540-59200-8_63
31. Lassen, S.B.: Bisimulation in untyped lambda calculus: Böhm trees and bisimulation up to context **20**, 346–374 (1999). https://doi.org/10.1016/S1571-0661(04)80083-5, https://doi.org/10.1016/S1571-0661(04)80083-5

32. Levy, P.B., Power, J., Thielecke, H.: Modelling environments in call-by-value programming languages. Inf. Comput. **185**(2), 182–210 (2003). https://doi.org/10.1016/S0890-5401(03)00088-9
33. Moggi, E.: Computational λ-Calculus and Monads. LFCS report ECS-LFCS-88-66, University of Edinburgh (1988), http://www.lfcs.inf.ed.ac.uk/reports/88/ECS-LFCS-88-66/ECS-LFCS-88-66.pdf
34. Moggi, E.: Computational λ-Calculus and Monads. In: Proceedings of the Fourth Annual Symposium on Logic in Computer Science (LICS '89), Pacific Grove, California, USA, June 5-8, 1989. pp. 14–23. IEEE Computer Society (1989). https://doi.org/10.1109/LICS.1989.39155
35. Ong, C.L.: Lazy lambda calculus: Theories, models and local structure characterization (extended abstract). In: Kuich, W. (ed.) Automata, Languages and Programming, 19th International Colloquium, ICALP92, Vienna, Austria, July 13-17, 1992, Proceedings. Lecture Notes in Computer Science, vol. 623, pp. 487–498. Springer (1992). https://doi.org/10.1007/3-540-55719-9_98
36. Paolini, L.: Call-by-value separability and computability. In: Theoretical Computer Science, 7th Italian Conference, ICTCS 2001, Torino, Italy, October 4-6, 2001, Proceedings. pp. 74–89 (2001). https://doi.org/10.1007/3-540-45446-2_5
37. Paolini, L.: Parametric λ-theories. Theoretical Computer Science **398**(1), 51–62 (2008). https://doi.org/https://doi.org/10.1016/j.tcs.2008.01.021, calculi, Types and Applications: Essays in honour of M. Coppo, M. Dezani-Ciancaglini and S. Ronchi Della Rocca
38. Paolini, L., Ronchi Della Rocca, S.: Call-by-value solvability. RAIRO Theor. Informatics Appl. **33**(6), 507–534 (1999). https://doi.org/10.1051/ita:1999130
39. Pitts, A.M.: Howe's method for higher-order languages. In: Sangiorgi, D., Rutten, J.J.M.M. (eds.) Advanced Topics in Bisimulation and Coinduction, Cambridge tracts in theoretical computer science, vol. 52, pp. 197–232. Cambridge University Press (2012). https://doi.org/10.1017/CBO9780511792588.006
40. Plotkin, G.D.: Call-by-name, call-by-value and the λ-calculus. Theoretical Computer Science **1**(2), 125–159 (1975). https://doi.org/https://doi.org/10.1016/0304-3975(75)90017-1
41. Ronchi Della Rocca, S., Paolini, L.: The Parametric λ-Calculus – A Metamodel for Computation. Texts in Theoretical Computer Science. An EATCS Series, Springer (2004). https://doi.org/10.1007/978-3-662-10394-4
42. Sabry, A., Wadler, P.: A Reflection on Call-by-Value. ACM Trans. Program. Lang. Syst. **19**(6), 916–941 (1997). https://doi.org/10.1145/267959.269968
43. Salibra, A.: On the algebraic models of lambda calculus. Theor. Comput. Sci. **249**(1), 197–240 (2000). https://doi.org/10.1016/S0304-3975(00)00059-1
44. Takahashi, M.: A simple proof of the genericity lemma. In: Jones, N.D., Hagiya, M., Sato, M. (eds.) Logic, Language and Computation, Festschrift in Honor of Satoru Takasu. Lecture Notes in Computer Science, vol. 792, pp. 117–118. Springer (1994). https://doi.org/10.1007/BFb0032397
45. Wadsworth, C.P.: Semantics and pragmatics of the lambda-calculus. PhD Thesis, University of Oxford (1971)
46. Wadsworth, C.P.: The Relation Between Computational and Denotational Properties for Scott's D_∞-Models of the Lambda-Calculus. SIAM J. Comput. **5**(3), 488–521 (1976). https://doi.org/10.1137/0205036

Logical Predicates in Higher-Order Mathematical Operational Semantics

Sergey Goncharov[1,⋆], Alessio Santamaria[2], Lutz Schröder[1,⋆⋆], Stelios Tsampas[1(✉),⋆⋆⋆] and Henning Urbat[1,†]

[1] Friedrich-Alexander-Universität Erlangen-Nürnberg, Erlangen, Germany
{sergey.goncharov,lutz.schroder,stelios.tsampas@fau.de, henning.urbat}@fau.de

[2] University of Sussex, Brighton, UK
a.santamaria@sussex.ac.uk

Abstract. We present a systematic approach to logical predicates based on universal coalgebra and higher-order abstract GSOS, thus making a first step towards a unifying theory of logical relations. We start with the observation that logical predicates are special cases of *coalgebraic invariants* on mixed-variance functors. We then introduce the notion of a *locally maximal logical refinement* of a given predicate, with a view to enabling inductive reasoning, and identify sufficient conditions on the overall setup in which locally maximal logical refinements canonically exist. Finally, we develop induction-up-to techniques that simplify inductive proofs via logical predicates on systems encoded as (certain classes of) higher-order GSOS laws by identifying and abstracting away from their boiler-plate part.

1 Introduction

Logical relations are arguably the most widely used method for reasoning on higher-order languages. Historically, early examples of logical relations [44,46,47,51,55,56,58,59] were based on denotational semantics, before the method evolved into logical relations based on operational semantics [7,17,34,50,52,53]. Today, operationally-based logical relations are ubiquitous and serve purposes ranging from strong normalization proofs [6] and safety properties [21,22] to reasoning about contextual equivalence [5,60] and formally verified compilation [8,33,45,48], in a variety of settings such as effectful [37], probabilistic [4,10,63], and differential programming [15,40,41].

Unfortunately, despite the extensive literature, there is a distinct lack of a general formal theory of (operational) logical relations. As a reasoning method, logical relations are applied in a largely empirical manner, more so because their core principles are well understood on an intuitive level. For example, there is typically no formal notion of a logical predicate or relation; instead, if a predicate or relation is defined by induction on

⋆ Supported by Deutsche Forschungsgemeinschaft (DFG, German Research Foundation) – project number 501369690

⋆⋆ Supported by Deutsche Forschungsgemeinschaft (DFG, German Research Foundation) – project numbers 419850228

⋆⋆⋆ Supported by Deutsche Forschungsgemeinschaft (DFG, German Research Foundation) – project numbers 419850228 and 527481841

† Supported by Deutsche Forschungsgemeinschaft (DFG, German Research Foundation) – project number 470467389

N. Kobayashi and J. Worrell (Eds.): FoSSaCS 2024, LNCS 14575, pp. 47–69, 2024.
https://doi.org/10.1007/978-3-031-57231-9_3

types and maps "related inputs to related outputs", it then meets the informal criteria to be called "logical". However, the empirical character of logical relations is problematic for two main reasons: (i) complex machinery associated to logical relations needs to be re-established anew on a per-case basis, and (ii) it is hard to abstract and simplify said machinery, even though certain parts of proofs via logical relations seem generic.

Recently, *Higher-order Mathematical Operational Semantics* [24], or *higher-order abstract GSOS*, has emerged as a unifying approach to the operational semantics of higher-order languages. In this framework, languages are represented as *higher-order GSOS laws*, a form of distributive law of a syntax functor Σ over a mixed-variance behaviour bifunctor B. In further work [62], an abstract form of *Howe's method* [16,31,32] for higher-order abstract GSOS has been identified, in which an otherwise complex and application-specific operational technique is, at the same time, lifted to an appropriate level of generality and reduced to a simple *lax bialgebra* condition.

In the present paper, we work towards establishing a theory of logical relations based on coalgebra and higher-order abstract GSOS, starting from *logical predicates*, understood as unary logical relations. In more detail, we present the following contributions:

(i) A systematization of the method of logical predicates (Section 3), achieved by
  (a) identifying logical predicates as certain coalgebraic invariants (Definition 12), parametric in a predicate lifting of the underlying mixed-variance bifunctor,
  (b) introducing the *locally maximal logical refinement* $\Box P$ of a predicate P (Definition 14), which enables inductive proofs of $\Box P$, and
  (c) identifying an abstract setting in which locally maximal logical refinements of predicates exist and are unique (Section 3.3).

(ii) The development of efficient reasoning techniques on logical predicates, which we call *induction up-to* (Theorems 34 and 36), for higher-order GSOS laws satisfying a *relative flatness* condition (Definition 30).

We illustrate (ii) by providing proofs of strong normalization for typed combinatory logic and type safety for the simply typed λ-calculus which, thanks to the use of our up-to techniques, are significantly shorter and simpler than standard arguments found in the literature. Finally, we exploit the genericity of our framework to study strong normalization on the level of higher-order GSOS laws (Theorem 42). We note that the implementation of typed languages as higher-order GSOS laws as such is also novel.

Full proofs and additional details can be found in the arXiv version [25] of our paper.

Related work While denotational logical relations have been studied in categorical generality, e.g. [27,28,29,38], general abstract foundations of operational logical relations are far less developed. In recent work [13,14], Dagnino and Gavazzo introduce a categorical notion of operational logical relations that is largely orthogonal to ours, in particular regarding the parametrization of the framework: In *op. cit.*, the authors work with a fixed *fine-grain call-by-value* language [42], parametrized by a signature of generic effects, while the notion of logical relation is kept variable and in fact is parametrized over a fibration; contrastingly, we keep to the traditional notion of logical relation but parametrize over the syntax and semantics of the language. Moreover, we work with a small-step operational semantics, whereas the semantics used in *op. cit.* is an axiomatically defined categorical evaluation semantics.

2 Preliminaries

2.1 Category Theory

Familiarity with basic category theory [43] (e.g. functors, natural transformations, (co)limits, monads) is assumed. We review some concepts and notation.

Notation. Given objects X_1, X_2 in a category C, we write $X_1 \times X_2$ for the product and $\langle f_1, f_2 \rangle \colon X \to X_1 \times X_2$ for the pairing of $f_i \colon X \to X_i$, $i = 1, 2$. We let $X_1 + X_2$ denote the coproduct, $\mathsf{inl} \colon X_1 \to X_1 + X_2$ and $\mathsf{inr} \colon X_2 \to X_1 + X_2$ the injections, $[g_1, g_2] \colon X_1 + X_2 \to X$ the copairing of $g_i \colon X_i \to X$, $i = 1, 2$, and $\nabla = [\mathsf{id}_X, \mathsf{id}_X] \colon X + X \to X$ the codiagonal. The *slice category* C/X, where $X \in C$, has as objects all pairs (Y, p_Y) of an object $Y \in C$ and a morphism $p_Y \colon Y \to X$, and a morphism from (Y, p_Y) to (Z, p_Z) is a morphism $f \colon Y \to Z$ of C such that $p_Y = p_Z \cdot f$. The *coslice category* X/C is defined dually.

Extensive categories. A category C is *(finitely) extensive* [12] if it has finite coproducts and for every finite family of objects X_i ($i \in I$) the functor $E \colon \prod_{i \in I} C/X_i \to C/\coprod_{i \in I} X_i$ sending $(p_i \colon Y_i \to X_i)_{i \in I}$ to $\coprod_{i \in I} p_i \colon \coprod_i Y_i \to \coprod_i X_i$ is an equivalence of categories. A *countably extensive* category satisfies the analogous property for countable coproducts. In extensive categories, coproduct injections $\mathsf{inl}, \mathsf{inr}$ are monic, and coproducts of monomorphisms are monic; generally, coproducts behave like disjoint unions of sets.

Example 1. Examples of countably extensive categories include the category **Set** of sets and functions; the category $\mathbf{Set}^C$ of presheaves on a small category C and natural transformations; and the categories of posets and monotone maps, nominal sets and equivariant maps, and metric spaces and non-expansive maps, respectively.

Algebras. Given an endofunctor F on a category C, an F*-algebra* is a pair (A, a) consisting of an object A and a morphism $a \colon FA \to A$ (the *structure*). A *morphism* from (A, a) to an F-algebra (B, b) is a morphism $h \colon A \to B$ of C such that $h \cdot a = b \cdot Fh$. Algebras for F and their morphisms form a category $\mathbf{Alg}(F)$, and an *initial* F-algebra is simply an initial object in that category. We denote the initial F-algebra by μF if it exists, and its structure by $\iota \colon F(\mu F) \to \mu F$. Initial algebras admit the *structural induction principle*: the algebra μF has no proper subalgebras, that is, every F-algebra monomorphism $m \colon (A, a) \rightarrowtail (\mu F, \iota)$ is an isomorphism.

More generally, a *free* F*-algebra* on an object X of C is an F-algebra $(F^\star X, \iota_X)$ together with a morphism $\eta_X \colon X \to F^\star X$ of C such that for every algebra (A, a) and every $h \colon X \to A$ in C, there exists a unique F-algebra morphism $h^\sharp \colon (F^\star X, \iota_X) \to (A, a)$ such that $h = h^\sharp \cdot \eta_X$. If free algebras exist on every object, their formation induces a monad $F^\star \colon C \to C$, the *free monad* generated by F. Every F-algebra (A, a) yields an Eilenberg-Moore algebra $\widehat{a} \colon F^\star A \to A$ as the free extension of $\mathsf{id}_A \colon A \to A$.

The most familiar example of functor algebras are algebras for a signature. Given a set S of *sorts*, an S*-sorted algebraic signature* consists of a set Σ of operation symbols together with a map $\mathsf{ar} \colon \Sigma \to S^\star \times S$ associating to every $\mathsf{f} \in \Sigma$ its *arity*. We write $\mathsf{f} \colon s_1 \times \cdots \times s_n \to s$ if $\mathsf{ar}(\mathsf{f}) = (s_1, \ldots, s_n, s)$, and $\mathsf{f} \colon s$ if $n = 0$ (in which case f is called a *constant*). Every signature Σ induces a polynomial functor on the category $\mathbf{Set}^S$ of S-sorted sets, denoted by the same letter Σ, given by $(\Sigma X)_s = \coprod_{\mathsf{f} \colon s_1 \cdots s_n \to s} \prod_{i=1}^n X_{s_i}$ for $X \in \mathbf{Set}^S$ and $s \in S$. An algebra for the functor Σ is precisely an algebra for the signature Σ, viz. an S-sorted set $A = (A_s)_{s \in S}$ in $\mathbf{Set}^S$ equipped with an operation $\mathsf{f}^A \colon \prod_{i=1}^n A_{s_i} \to A_s$ for every $\mathsf{f} \colon s_1 \cdots s_n \to s$ in Σ. Morphisms of Σ-algebras are S-sorted maps respecting the algebraic structure. Given an S-sorted set X of variables, the free algebra $\Sigma^\star X$ is the Σ-algebra of Σ-terms with variables from X; more precisely, $(\Sigma^\star X)_s$ is inductively defined by $X_s \subseteq (\Sigma^\star X)_s$ and $\mathsf{f}(t_1, \ldots, t_n) \in (\Sigma^\star X)_s$ for all $\mathsf{f} \colon s_1 \cdots s_n \to s$ and $t_i \in (\Sigma^\star X)_{s_i}$. In particular, the free algebra on the empty set is the initial algebra $\mu\Sigma$; it is formed by all *closed terms* of the signature. For every Σ-algebra (A, a), the induced Eilenberg-Moore algebra $\widehat{a} \colon \Sigma^\star A \to A$ is given by the map that evaluates terms over A in the algebra A.

Coalgebras. Dual to the notion of algebra, a *coalgebra* for an endofunctor F on C is a pair (C, c) of an object C (the *state space*) and a morphism $c \colon C \to FC$ (the *structure*).

2.2 Higher-Order Abstract GSOS

We summarize the framework of higher-order abstract GSOS [24], which extends the original, first-order counterpart introduced by Turi and Plotkin [61]. In higher-order abstract GSOS, the operational semantics of a higher-order language is presented in the form of a *higher-order GSOS law*, a categorical structure parametric in

(1) a category C with finite products and coproducts;

(2) an object $V \in C$ of *variables*;

(3) an endofunctor $\Sigma \colon C \to C$, where $\Sigma = V + \Sigma'$ for some endofunctor Σ', such that free Σ-algebras exist on every object (hence Σ generates a free monad $\Sigma^\star$);

(4) a mixed-variance bifunctor $B \colon C^{\mathsf{op}} \times C \to C$.

The functors Σ and B represent the *syntax* and the *behaviour* of a higher-order language. The motivation behind B having two arguments is that transitions have labels, which behave contravariantly, and poststates, which behave covariantly; in term models the objects of labels and states will coincide. The presence of an object V of variables is a technical requirement for the modelling of languages with variable binding [19,20], such as the λ-calculus. An object of V/C, the coslice category of *V-pointed objects*, is thought of as a set X of programs with an embedding $p_X \colon V \to X$ of the variables. In point-free calculi, e.g. **xTCL** as introduced below, we put $V = 0$ (the initial object).

Definition 2. A *(V-pointed) higher-order GSOS law* of Σ over B is a family of morphisms (1) that is dinatural in $(X, p_X) \in V/C$ and natural in $Y \in C$:

$$\varrho_{(X,p_X),Y} \colon \Sigma(X \times B(X, Y)) \to B(X, \Sigma^\star(X + Y)) \tag{1}$$

Notation 3. (i) In (1), we have implicitly applied the forgetful functor $V/C \to C$ at (X, p_X). In addition, we write $\varrho_{X,Y}$ for $\varrho_{(X,p_X),Y}$ if the point p_X is clear from the context.
(ii) For $(A, a) \in \mathbf{Alg}(\Sigma)$, we view A as V-pointed by $p_A = (V \xrightarrow{\mathsf{inl}} V + \Sigma' A = \Sigma A \xrightarrow{a} A)$.

Informally, $\varrho_{X,Y}$ assigns to an operation of the language with formal arguments from X having specified next-step behaviours in $B(X, Y)$ (i.e. with labels in X and formal poststates in Y) a next-step behaviour in $B(X, \Sigma^\star(X + Y))$, i.e. with the same labels, and with poststates being program terms mentioning variables from both X and Y. Every

$$\frac{}{\mathsf{e} \xrightarrow{\checkmark}} \qquad \frac{}{S_{\tau_1,\tau_2,\tau_3} \xrightarrow{t} S'_{\tau_1,\tau_2,\tau_3}(t)} \qquad \frac{}{S'_{\tau_1,\tau_2,\tau_3}(p) \xrightarrow{t} S''_{\tau_1,\tau_2,\tau_3}(p,t)}$$

$$\frac{}{S''_{\tau_1,\tau_2,\tau_3}(p,q) \xrightarrow{t} (p\,t)\,(q\,t)} \qquad \frac{}{K_{\tau_1,\tau_2} \xrightarrow{t} K'_{\tau_1,\tau_2}(t)} \qquad \frac{}{K'_{\tau_1,\tau_2}(p) \xrightarrow{t} p}$$

$$\frac{}{I_\tau \xrightarrow{t} t} \qquad \frac{p \to p'}{p\,q \to p'\,q} \qquad \frac{p \xrightarrow{q} p'}{p\,q \to p'}$$

Fig. 1. (Call-by-name) operational semantics of **xTCL**.

higher-order GSOS law (1) induces a canonical *operational model* $\gamma\colon \mu\Sigma \to B(\mu\Sigma, \mu\Sigma)$, viz. a $B(\mu\Sigma, -)$-coalgebra on the initial algebra $\mu\Sigma$, defined by *primitive recursion* [36, Prop. 2.4.7] as the unique morphism γ making the following diagram commute:

$$\begin{array}{ccccc}
\Sigma(\mu\Sigma) & & \xrightarrow{\qquad\qquad\iota\qquad\qquad} & & \mu\Sigma \\
\Big\downarrow{\scriptstyle \Sigma\langle \mathsf{id}, \gamma\rangle} & & & & \Big\downarrow{\scriptstyle \gamma} \\
\Sigma(\mu\Sigma \times B(\mu\Sigma, \mu\Sigma)) & \xrightarrow{\varrho_{\mu\Sigma,\mu\Sigma}} & B(\mu\Sigma, \Sigma^\star(\mu\Sigma + \mu\Sigma)) & \xrightarrow{B(\mu\Sigma, \hat{\iota}\cdot\Sigma^\star\nabla)} & B(\mu\Sigma, \mu\Sigma)
\end{array}$$

Here, we regard the initial algebra $(\mu\Sigma, \iota)$ as V-pointed as explained in Notation 3.

Simply Typed SKI Calculus. We illustrate the ideas behind higher-order abstract GSOS with an extended version of the simply typed SKI calculus [30], a typed combinatory logic which we call **xTCL**. It is expressively equivalent to the simply typed λ-calculus but does not use variables; hence it avoids the complexities associated to variable binding and substitution in the λ-calculus, which we treat in Section 4.2. The set Ty of *types* is inductively defined as

$$\mathsf{Ty} ::= \mathsf{unit} \mid \mathsf{Ty} \Rightarrow \mathsf{Ty}. \tag{2}$$

The constructor $\Rightarrow$ is right-associative, i.e. $\tau_1 \Rightarrow \tau_2 \Rightarrow \tau_3$ is parsed as $\tau_1 \Rightarrow (\tau_2 \Rightarrow \tau_3)$. The terms of **xTCL** are formed over the Ty-sorted signature Σ whose operation symbols are listed below, with $\tau, \tau_1, \tau_2, \tau_3$ ranging over all types in Ty:

$$\begin{array}{ll}
\mathsf{e}\colon \mathsf{unit} & \mathsf{app}_{\tau_1,\tau_2}\colon (\tau_1 \Rightarrow \tau_2) \times \tau_1 \to \tau_2 \\
S_{\tau_1,\tau_2,\tau_3}\colon (\tau_1 \Rightarrow \tau_2 \Rightarrow \tau_3) \Rightarrow (\tau_1 \Rightarrow \tau_2) \Rightarrow \tau_1 \Rightarrow \tau_3 & K_{\tau_1,\tau_2}\colon \tau_1 \Rightarrow \tau_2 \Rightarrow \tau_1 \\
S'_{\tau_1,\tau_2,\tau_3}\colon (\tau_1 \Rightarrow \tau_2 \Rightarrow \tau_3) \to ((\tau_1 \Rightarrow \tau_2) \Rightarrow \tau_1 \Rightarrow \tau_3) & K'_{\tau_1,\tau_2}\colon \tau_1 \to (\tau_2 \Rightarrow \tau_1) \\
S''_{\tau_1,\tau_2,\tau_3}\colon (\tau_1 \Rightarrow \tau_2 \Rightarrow \tau_3) \times (\tau_1 \Rightarrow \tau_2) \to (\tau_1 \Rightarrow \tau_3) & I_\tau\colon \tau \Rightarrow \tau
\end{array}$$

We let $\mathsf{Tr} = \mu\Sigma$ denote the Ty-sorted set of closed Σ-terms. Informally, app represents function application (we write $s\,t$ for $\mathsf{app}(s,t)$), and the constants I_τ, K_{τ_1,τ_2}, S_{τ_1,τ_2,τ_3} represent the λ-terms $\lambda t.\, t$, $\lambda t.\, \lambda s.\, t$ and $\lambda t.\, \lambda s.\, \lambda u.\, (s\,u)\,(t\,u)$, respectively. The operational semantics of **xTCL** involves three kinds of transitions: $\xrightarrow{\checkmark}$, $\xrightarrow{t}$ and $\longrightarrow$. It is presented in Figure 1; here, p, p', q, t range over terms in Tr of appropriate type. Intuitively, $s \xrightarrow{\checkmark}$ identifies s as an explicitly irreducible term; $s \xrightarrow{t} r$ states that s acts as a function mapping t to r; and $s \to t$ indicates that s reduces to t. Our use of labelled transitions in higher-order operational semantics is inspired by work on bisimilarity in the λ-calculus [1,26]. The use of K', S' and S'' does not impact the behaviour of programs, except for possibly adding more unlabelled transitions. For example, the standard rule $S\,t\,s\,e \to (te)(se)$ for the S-combinator is rendered as the chain of transitions $S\,t\,s\,e \to S'(t)\,s\,e \to S''(t,s)\,e \to (te)(se)$. The transition system for **xTCL** is deterministic: for every term s, either $s \xrightarrow{\checkmark}$, or there exists a unique t such that $s \to t$, or for each appropriately typed t there exists a unique s_t such that $s \xrightarrow{t} s_t$. Therefore, given

$$B_\tau(X,Y) = Y_\tau + D_\tau(X,Y), \tag{3}$$

$$D_{\mathsf{unit}}(X,Y) = 1 = \{*\} \qquad \text{and} \qquad D_{\tau_1 \multimap \tau_2}(X,Y) = Y_{\tau_2}^{X_{\tau_1}}, \tag{4}$$

the operational rules in Figure 1 determine a $\mathbf{Set}^{\mathsf{Ty}}$-morphism $\gamma\colon \mathsf{Tr} \to B(\mathsf{Tr},\mathsf{Tr})$:

$$\begin{aligned} \gamma_{\mathsf{unit}}(s) &= \mathsf{inr}(*) && \text{if } s \xrightarrow{\checkmark} \text{ where } s\colon \mathsf{unit},\\ \gamma_\tau(s) &= \mathsf{inl}(t) && \text{if } s \longrightarrow t \text{ where } s,t\colon \tau,\\ \gamma_{\tau_1 \multimap \tau_2}(s) &= \mathsf{inr}(\lambda t.\, s_t) && \text{if } s \xrightarrow{t} s_t \text{ for } s\colon \tau_1 \multimap \tau_2 \text{ and } t\colon \tau_1. \end{aligned} \tag{5}$$

Proposition 4. *The object assignments* (3) *and* (4) *extend to mixed-variance bifunctors*

$$B, D\colon (\mathbf{Set}^{\mathsf{Ty}})^{\mathsf{op}} \times \mathbf{Set}^{\mathsf{Ty}} \to \mathbf{Set}^{\mathsf{Ty}}. \tag{6}$$

The semantics of **xTCL** in Figure 1 corresponds to a (0-pointed) higher-order GSOS law of the syntax functor Σ over the behaviour bifunctor B, i.e. to a family of maps (1) dinatural in $X \in \mathbf{Set}^{\mathsf{Ty}}$ and natural in $Y \in \mathbf{Set}^{\mathsf{Ty}}$. The maps $\varrho_{X,Y}$ are cotuples defined by distinguishing cases on the constructors $\mathsf{e}, S, S', S'', K, K', I, \mathsf{app}$ of **xTCL**, and each component of ϱ is determined by the rules that apply to the corresponding constructor. We provide a few illustrative cases; see [25, p. 25], for a complete definition.

$$\varrho_{X,Y}\colon \Sigma(X \times B(X,Y)) \to B(X, \Sigma^\star(X+Y)) \tag{7}$$

$$\varrho_{X,Y}\,(S''_{\tau_1,\tau_2,\tau_3}((p,f),(q,g))) = \lambda t.\,(p\,t)\,(q\,t) \tag{8}$$

$$\varrho_{X,Y}\,((p,f)\,(q,g)) = f(q) \qquad \text{if } f\colon Y_{\tau_2}^{X_{\tau_1}} \tag{9}$$

$$\varrho_{X,Y}\,((p,f)\,(q,g)) = f\,q \qquad \text{if } f\colon Y_{\tau_1 \multimap \tau_2} \tag{10}$$

The operational model $\gamma\colon \mathsf{Tr} \to B(\mathsf{Tr},\mathsf{Tr})$ of ϱ coincides with the coalgebra (5).

Remark 5. The rules for application in Figure 1 implement the call-by-name evaluation strategy. Other strategies can be captured by varying the rules and consequently the corresponding higher-order GSOS law. For the call-by-value strategy, one replaces the last rule with (11) and (12) below and modifies clause (9) in the definition of ϱ accordingly. One can also model the traditional view of combinatory logic as a rewrite system [30] where any redex can be reduced, no matter how deeply. This amounts to specifying a maximally nondeterministic strategy by adding the rule (13) below to Figure 1. Notably, this makes the operational model nondeterministic, and hence the corresponding higher-order GSOS law relies on the behaviour functor $\mathcal{P}B$ instead of the original B given by (3), where $\mathcal{P}$ is the powerset functor.

$$\frac{p \xrightarrow{t} p' \quad q \to q'}{p\,q \to p\,q'}\ (11) \qquad \frac{p \xrightarrow{q} p' \quad q \xrightarrow{t} q'}{p\,q \to p'}\ (12) \qquad \frac{q \to q'}{p\,q \to p\,q'}\ (13)$$

3 Coalgebraic Logical Predicates

3.1 Predicate Lifting

Predicates and relations on coalgebras are often most conveniently modelled through *predicate* and *relation liftings* [39] of the underlying type functors. In the following we introduce a framework of predicate liftings for mixed-variance bifunctors, adapting existing notions of relation lifting [62], which enables reasoning about "higher-order" coalgebras, such as operational models of higher-order GSOS laws. The following global assumptions ensure that predicates and relations behave in an expected manner:

Assumptions 6. From now on, we fix $\mathbb{C}$ to be a complete, well-powered and extensive category in which, additionally, strong epimorphisms are stable under pullbacks.

The categories of Example 1 satisfy these assumptions. Since $\mathbb{C}$ is complete and well-powered, every morphism f admits a (strong epi, mono)-factorization $f = m \cdot e$ [11, Prop. 4.4.3]; we call m the *image* of f. The category $\mathbf{Pred}(\mathbb{C})$ of *predicates* over $\mathbb{C}$ has as objects all monics (predicates) $P \rightarrowtail X$ from $\mathbb{C}$, and as morphisms $(p\colon P \rightarrowtail X) \to (q\colon Q \rightarrowtail Y)$ all pairs $(f\colon X \to Y, f|_P\colon P \to Q)$ such that $q \cdot f|_P = f \cdot p$ (so $f|_P$ is uniquely determined by f). (Co)products in $\mathbf{Pred}(\mathbb{C})$ are lifted from $\mathbb{C}$. The *fiber* $\mathbf{Pred}_X(\mathbb{C})$ is the subcategory of all monics $P \rightarrowtail X$ for fixed X and morphisms $(\mathsf{id}_X, -)$. It is is preordered by $p \le q$ if p factors through q; identifying p, q if $p \le q$ and $q \le p$, we regard $\mathbf{Pred}_X(\mathbb{C})$ as a poset. Since $\mathbb{C}$ is complete and well-powered, $\mathbf{Pred}_X(\mathbb{C})$ is a complete lattice; we write $\bigwedge$ for meets (i.e. pullbacks) and $\bigvee$ for joins. We will also write $f^\star[P]$ for the *inverse image* of a predicate $p\colon P \rightarrowtail X$ under $f\colon Y \to X$, i.e. the pullback of p along f. The *direct image* $f_\star[Q]$ of $q\colon Q \rightarrowtail Y$ under $f\colon Y \to X$ is the image of the composite $f \cdot p\colon Q \to X$. This yields an adjunction between $\mathbf{Pred}_X(\mathbb{C})$ and $\mathbf{Pred}_Y(\mathbb{C})$, i.e. $Q \le f^\star[P]$ iff $f_\star[Q] \le P$.

A *predicate lifting* of an endofunctor $\Sigma\colon \mathbb{C} \to \mathbb{C}$ is an endofunctor $\overline{\Sigma}\colon \mathbf{Pred}(\mathbb{C}) \to \mathbf{Pred}(\mathbb{C})$ making the left-hand diagram below commute; similarly, a *predicate lifting* of a mixed-variance bifunctor $B\colon \mathbb{C}^{\mathsf{op}} \times \mathbb{C} \to \mathbb{C}$ is a bifunctor $\overline{B}\colon \mathbf{Pred}(\mathbb{C})^{\mathsf{op}} \times \mathbf{Pred}(\mathbb{C}) \to \mathbf{Pred}(\mathbb{C})$ making the right-hand diagram below commute. Here $|{-}|$ is the forgetful functor sending $p\colon P \rightarrowtail X$ to X.

$$
\begin{array}{ccc}
\mathbf{Pred}(\mathbb{C}) & \xrightarrow{\overline{\Sigma}} & \mathbf{Pred}(\mathbb{C}) \\
{\scriptstyle |-|}\downarrow & & \downarrow{\scriptstyle |-|} \\
\mathbb{C} & \xrightarrow{\Sigma} & \mathbb{C}
\end{array}
\qquad
\begin{array}{ccc}
\mathbf{Pred}(\mathbb{C})^{\mathsf{op}} \times \mathbf{Pred}(\mathbb{C}) & \xrightarrow{\overline{B}} & \mathbf{Pred}(\mathbb{C}) \\
{\scriptstyle |-|^{\mathsf{op}} \times |-|}\downarrow & & \downarrow{\scriptstyle |-|} \\
\mathbb{C}^{\mathsf{op}} \times \mathbb{C} & \xrightarrow{B} & \mathbb{C}
\end{array}
\tag{14}
$$

We denote by $\overline{\Sigma}$ both the action on predicates and on the corresponding objects in $\mathbb{C}$, i.e. $\overline{\Sigma}(p\colon P \rightarrowtail X)\colon \overline{\Sigma}P \rightarrowtail \Sigma X$.

Every endofunctor Σ on $\mathbb{C}$ admits a canonical predicate lifting $\overline{\Sigma}$ mapping $p\colon P \rightarrowtail X$ to the image $\overline{\Sigma}p\colon \overline{\Sigma}P \rightarrowtail \Sigma X$ of $\Sigma p\colon \Sigma P \to \Sigma X$ [36]. Note that $\overline{\Sigma}p = \Sigma p$ if Σ preserves monos. In the remainder we will only consider canonical liftings of endofunctors.

Proposition 7. *If Σ preserves strong epis, then $\overline{\Sigma}^\star = \overline{\Sigma^\star}$.*

The canonical predicate liftings for mixed-variance bifunctors are slightly more complex. Similarly to the case of relation liftings of such functors developed in recent work [62], their construction involves suitable pullbacks.

Proposition 8. *Every bifunctor $B\colon C^{op} \times C \to C$ admits a canonical predicate lifting $\overline{B}\colon \mathbf{Pred}(C)^{op} \times \mathbf{Pred}(C) \to \mathbf{Pred}(C)$ sending $(p\colon P \rightarrowtail X,\ q\colon Q \rightarrowtail Y)$ to the predicate $m_{P,Q}\colon \overline{B}(P,Q) \rightarrowtail B(X,Y)$, the image of the morphism $r_{P,Q}$ given by the pullback below:*

$$
\begin{array}{ccccc}
 & & T_{P,Q} & \xrightarrow{s_{P,Q}} & B(P,Q) \\
 & \overset{e_{P,Q}}{\swarrow} & \big\downarrow r_{P,Q} \ \lrcorner & & \big\downarrow B(\mathrm{id},q) \\
\overline{B}(P,Q) & \underset{m_{P,Q}}{\searrow} & B(X,Y) & \xrightarrow{B(p,\mathrm{id})} & B(P,Y)
\end{array}
\tag{15}
$$

If B preserves monos in the covariant argument, then $B(\mathrm{id},q)$ is monic and, since monos are pullback-stable, $\overline{B}(P,Q)$ is simply the predicate $r_{P,Q}\colon T_{P,Q} \rightarrowtail B(X,Y)$.

Example 9. The bifunctors B and D of (3) and (4) have canonical predicate liftings

$$\overline{B}_\tau(P,Q) = Q_\tau + \overline{D}_\tau(P,Q) \quad \text{where} \tag{16}$$

$$\overline{D}_{\mathsf{unit}}(P,Q) = 1, \quad \overline{D}_{\tau_1 \to \tau_2}(P,Q) = \{f\colon X_{\tau_1} \to Y_{\tau_2} \mid \forall x \in P_{\tau_1}.\, f(x) \in Q_{\tau_2}\} \subseteq Y_{\tau_2}^{X_{\tau_1}}. \tag{17}$$

Predicate liftings allow us to generalize *coalgebraic invariants* [36, §6.2], viz. predicates on the state space of a coalgebra that are closed under the coalgebra structure in a suitable sense, from endofunctors to mixed-variance bifunctors:

Notation 10. For the remainder of the paper, we fix a mixed-variance bifunctor $B\colon C^{op} \times C \to C$ and a predicate lifting $\overline{B}\colon \mathbf{Pred}(C)^{op} \times \mathbf{Pred}(C) \to \mathbf{Pred}(C)$.

Definition 11 (Coalgebraic invariant)**.** Let $c\colon Y \to B(X,Y)$ be a $B(X,-)$-coalgebra. Given predicates $S \rightarrowtail X$, $P \rightarrowtail Y$, we say that P is an *S-relative ($\overline{B}$-)invariant (for c)* if $P \le c^\star[\overline{B}(S,P)]$, equivalently, $c_\star[P] \le \overline{B}(S,P)$. (Mention of $\overline{B}$ is usually omitted.)

Coalgebraic invariants will feature centrally in our notion of logical predicate.

3.2 Logical Predicates via Lifted Bifunctors

As a reasoning device, the method of logical predicates (which are unary logical relations) typically applies to the following scenario: One has an operational semantics on an inductively defined set $\mu\Sigma$ of Σ-terms and a target predicate $P \rightarrowtail \mu\Sigma$ to be proved, in the sense that one wants to show $P = \mu\Sigma$. Logical predicates come into play when a direct proof of $P = \mu\Sigma$ by structural induction is not possible. The classical example of such a predicate is *strong normalization* [23,59]. The idea is to strengthen P, obtaining a predicate featuring a certain "logical" structure that does allow for a proof by induction. We now develop this scenario in our abstract bifunctorial setting.

Definition 12 (Coalgebraic logical predicate)**.** Suppose that $c\colon X \to B(X,X)$ is a $B(X,-)$ coalgebra with state space X. A predicate $P \rightarrowtail X$ is *logical (for c)* if it is a P-relative $\overline{B}$-invariant (as per Def. 11), i.e. $P \le c^\star[\overline{B}(P,P)]$, equivalently, $c_\star[P] \le \overline{B}(P,P)$.

In applications, c is the operational model $\gamma\colon \mu\Sigma \to B(\mu\Sigma,\mu\Sigma)$ of a higher-order language, or some coalgebra derived from it. The self-referential nature of logical predicates (as relative to themselves) is meant to cater for the property that "inputs in P are mapped to outputs in P". The following example from **xTCL** illustrates this:

Example 13. For B given by (3) and its canonical lifting $\overline{B}$, a predicate $P \rightarrowtail \mathsf{Tr}$ is logical for the operational model $\gamma\colon \mathsf{Tr} \to B(\mathsf{Tr},\mathsf{Tr})$ from (5) if $\gamma_\star[P] \leq \overline{B}(P,P)$, that is,

$$(\gamma_{\mathsf{unit}})_\star[P_{\mathsf{unit}}] \leq P_{\mathsf{unit}} + 1,$$
$$\forall \tau_1,\tau_2.\,(\gamma_{\tau_1\to\tau_2})_\star[P_{\tau_1\to\tau_2}] \leq P_{\tau_1\to\tau_2} + \{f\colon \mathsf{Tr}_{\tau_1} \to \mathsf{Tr}_{\tau_2} \mid \forall s \in P_{\tau_1}.\, f(s) \in P_{\tau_2}\},$$

using the description of $\overline{B}$ from Example 9. More explicitly, this means that

- if $s \in P_\tau$ and $s \to t$ then $t \in P_\tau$;
- if $s \in P_{\tau_1\to\tau_2}$ and $s \xrightarrow{t} u$, then $t \in P_{\tau_1}$ implies $u \in P_{\tau_2}$.

As we can see in the second clause, function terms that satisfy P produce outputs that satisfy P on all inputs that satisfy P. This is the key property of any logical predicate.

Defining a suitable logical predicate (or relation) is the centerpiece of various sophisticated arguments in higher-order settings. One standard application of logical predicates are proofs of strong normalization, which we now illustrate in the case of **xTCL**. For the operational model $\gamma\colon \mathsf{Tr} \to B(\mathsf{Tr},\mathsf{Tr})$ and terms r, s, t of compatible type, put

- $s \Rightarrow t$ if $s = s_0 \to s_1 \to \cdots \to s_n = t$ for some $n \geq 0$ and terms $s_0, \ldots, s_n$;
- $s \overset{t}{\Rightarrow} r$ if $s \Rightarrow s'$ and $s' \xrightarrow{t} r$ for some (unique) s';
- $\Downarrow(s)$ if $s \Rightarrow s'$ and $\gamma(s') \in D(\mathsf{Tr},\mathsf{Tr})$ for some (unique) s'.

Coalgebraically, this associates a *weak operational model* $\widetilde{\gamma}\colon \mathsf{Tr} \to \mathcal{P}B(\mathsf{Tr},\mathsf{Tr})$ to γ, where $\widetilde{\gamma}(t) = \{t' \mid t \Rightarrow t'\} \cup \{\gamma(t') \mid t \Rightarrow t', \gamma(t') \in D(\mathsf{Tr},\mathsf{Tr})\}$.

Strong normalization of **xTCL** asserts that $\Downarrow = \mathsf{Tr}$: every term eventually reduces to a function or explicitly terminates. We now devise three different logical predicates on Tr, each of which provides a proof of that property. The idea is to refine the target predicate $\Downarrow \rightarrowtail \mathsf{Tr}$ to a logical predicate, for which showing that it is totally true will be facilitated by its invariance w.r.t. a corresponding coalgebra structure. Our first example will be based on the following notion of refinement:

Definition 14 (Locally maximal logical refinement)**.** Let $c\colon X \to B(X,X)$ be a coalgebra and let $P \rightarrowtail X$ be a predicate. A predicate $\Box P \rightarrowtail X$ is a *locally maximal logical refinement of* P if (i) $\Box P \leq P$, (ii) $\Box P$ is logical (i.e. a $\Box P$-relative $\overline{B}$-invariant), and (iii) for every predicate $Q \leq P$ that is a $\Box P$-relative $\overline{B}$-invariant, one has $Q \leq \Box P$.

Example 15. We define the predicate $\Box\Downarrow \rightarrowtail \mathsf{Tr}$, i.e. a family of subsets $\Box\Downarrow_\tau \subseteq \mathsf{Tr}_\tau$ ($\tau \in \mathsf{Ty}$), by induction on the structure of the type τ: we put $\Box\Downarrow_{\mathsf{unit}} = \Downarrow_{\mathsf{unit}}$, and we take $\Box\Downarrow_{\tau_1\to\tau_2}$ to be the greatest subset of $\mathsf{Tr}_{\tau_1\to\tau_2}$ satisfying

$$\Box\Downarrow_{\tau_1\to\tau_2}(t) \implies \Downarrow_{\tau_1\to\tau_2}(t) \wedge \begin{cases} \Box\Downarrow_{\tau_1\to\tau_2}(t') & \text{if } t \to t' \\ \Box\Downarrow_{\tau_1}(s) \implies \Box\Downarrow_{\tau_2}(t') & \text{if } t \xrightarrow{s} t' \end{cases}$$

From this definition it is not difficult to verify by induction on the type that

$$\Box\Downarrow \text{ is a locally maximal logical refinement of } \Downarrow. \tag{18}$$

Our goal is to show that $\Box\Downarrow$ is a subalgebra of $\mu\Sigma$, equivalently $\overline{\Sigma}(\Box\Downarrow) \leq \iota^\star[\Box\Downarrow]$, which then implies $\Box\Downarrow = \mathsf{Tr}$ and hence $\Downarrow = \mathsf{Tr}$ by structural induction. Taking the partition $\Sigma = \Xi + \Delta$ where Ξ is the part of the signature for application and Δ is the part of the signature for the remaining term constructors, we separately prove $\overline{\Xi}(\Box\Downarrow) \leq \iota^\star[\Box\Downarrow]$ and $\overline{\Delta}(\Box\Downarrow) \leq \iota^\star[\Box\Downarrow]$. It suffices to come up with $\Box\Downarrow$-relative invariants $A, C \subseteq \Downarrow$ such that $\overline{\Xi}(\Box\Downarrow) \leq \iota^\star[A]$ and $\overline{\Delta}(\Box\Downarrow) \leq \iota^\star[C]$. Then by (18) we can conclude $A, C \subseteq \Box\Downarrow$, so

$$\overline{\Xi}(\Box\Downarrow) \leq \iota^\star[A] \leq \iota^\star[\Box\Downarrow] \qquad \text{and} \qquad \overline{\Delta}(\Box\Downarrow) \leq \iota^\star[C] \leq \iota^\star[\Box\Downarrow].$$

Let us record for further reference what it means for $Q \rightarrowtail \mathsf{Tr}$ to be a $\Box\Downarrow$-relative invariant contained in $\Downarrow$. Given $t \in Q_\tau$, the following must hold:

$$\text{(1) } \Downarrow_\tau t, \text{ (2) if } t \to t' \text{ then } Q_\tau(t'), \text{ (3) if } t\colon \tau_1 \twoheadrightarrow \tau_2 \text{ and } t \xrightarrow{s} t' \text{ and } \Box\Downarrow_{\tau_1} s \text{ then } Q_{\tau_2}(t').$$

We first put $A = \Box\Downarrow \vee (\iota \cdot \mathsf{inl})_\star[\overline{\Xi}\Box\Downarrow]$, and prove (1)–(3) for $Q = A$. So let $t \in A_\tau$; we distinguish cases on the disjunction defining A. If $\Box\Downarrow_\tau t$, then (1)–(3) follow easily by definition. Otherwise, we have $t = p\,q$ such that $\Box\Downarrow_{\tau_1 \twoheadrightarrow \tau_2} p$ and $\Box\Downarrow_{\tau_1} q$.

(1) By definition, $\Box\Downarrow_{\tau_1 \twoheadrightarrow \tau_2} p$ and $\Box\Downarrow_{\tau_1} q$ entail that $p \overset{q}{\Rightarrow} p'$ for a (unique) term p', and that $\Box\Downarrow_{\tau_2} p'$, hence $\Downarrow_{\tau_2} p'$. Since $p\,q \Rightarrow p'$, it follows that $\Downarrow_{\tau_2} p\,q$.

(2) We distinguish cases over the semantic rules for application:

(a) $p\,q \to p'\,q$ where $p \to p'$. Then $\Box\Downarrow_{\tau_1 \twoheadrightarrow \tau_2} p'$, hence $A_{\tau_2}(p'\,q)$.

(b) $p\,q \to p'$ where $p \xrightarrow{q} p'$. Since $\Box\Downarrow_{\tau_1 \twoheadrightarrow \tau_2} p$ and $\Box\Downarrow_{\tau_1} q$, we have $\Box\Downarrow_{\tau_2} p'$, so $A_{\tau_2}(p')$.

(3) t does not have labelled transitions, hence this case is void.

Next, we show that $C = \Box\Downarrow \vee (\iota \cdot \mathsf{inr})_\star[\overline{\Delta}(\Box\Downarrow)]$ is a $\Box\Downarrow$-relative invariant. We consider two representative cases; the remaining cases are handled similarly.

- Case $I_\tau\colon \tau \twoheadrightarrow \tau$. Since I terminates immediately, property (1) holds by definition of $\Downarrow$ and (2) holds vacuously. For (3), if $I \xrightarrow{s} t'$ and $\Box\Downarrow_\tau s$, then $t' = s \in \Box\Downarrow_\tau \subseteq C_\tau$.
- Case $S''_{\tau_1,\tau_2,\tau_3}(t, s)\colon \tau_1 \twoheadrightarrow \tau_3$ with $\Box\Downarrow_{\tau_1 \twoheadrightarrow \tau_2 \twoheadrightarrow \tau_3} t$ and $\Box\Downarrow_{\tau_1 \twoheadrightarrow \tau_2} s$. Again, (1) holds because $S''(t, s)$ terminates immediately, and (2) holds vacuously. For (3), suppose that $\Box\Downarrow_{\tau_1} r$; we have to show $(t\,r)\,(s\,r) \in C_{\tau_3}$. This follows from the inequality $\overline{\Xi}(\Box\Downarrow) \leq \iota_\star[\Box\Downarrow]$ shown above, because $\Box\Downarrow_{\tau_2 \twoheadrightarrow \tau_3}(t\,r)$, $\Box\Downarrow_{\tau_2}(s\,r)$ by definition of $\Box\Downarrow$.

Note that the definition of $\Box\Downarrow$ uses both induction (over the structure of types) and coinduction (by taking at every type the greatest predicate satisfying some property).

Example 16. We give an alternative logical predicate defined purely inductively. It resembles Plotkin's original concept of logical relation [55]. We define $\Downarrow\!\!\Downarrow \rightarrowtail \mathsf{Tr}$ by

$$\begin{aligned} &\Downarrow\!\!\Downarrow_{\mathsf{unit}}(t) \iff \Downarrow_{\mathsf{unit}}(t), \\ &\Downarrow\!\!\Downarrow_{\tau_1 \twoheadrightarrow \tau_2}(t) \iff \Downarrow_{\tau_1 \twoheadrightarrow \tau_2} t \wedge (\forall s\colon \tau_1.\, t \overset{s}{\Rightarrow} t' \wedge \Downarrow\!\!\Downarrow_{\tau_1}(s) \implies \Downarrow\!\!\Downarrow_{\tau_2}(t')). \end{aligned} \tag{19}$$

It is evidently logical for the restriction $\widetilde{\overline{\gamma}}\colon \mathsf{Tr} \to \mathcal{P}D(\mathsf{Tr}, \mathsf{Tr})$ of the weak operational model to labelled transitions, given by $\widetilde{\overline{\gamma}}(t) := \{\gamma(t')\}$ if $t \Rightarrow t'$ and $\gamma(t') \in D(\mathsf{Tr}, \mathsf{Tr})$, and $\widetilde{\overline{\gamma}}(t) := \emptyset$ otherwise. A proof of strong normalization using $\Downarrow\!\!\Downarrow$ is given in [25, App. A].

Example 17. A more popular (cf. [57,58]) and subtly different variant of $\Downarrow\!\!\Downarrow$ for proving strong normalization goes back to Tait [59]. We define $\mathsf{SN} \rightarrowtail \mathsf{Tr}$ by

$$\begin{aligned} \mathsf{SN}_{\mathsf{unit}}(t) &\iff \Downarrow_{\mathsf{unit}}(t) \\ \mathsf{SN}_{\tau_1 \twoheadrightarrow \tau_2}(t) &\iff \Downarrow_{\tau_1 \twoheadrightarrow \tau_2}(t) \wedge (\forall s\colon \tau_1 .\, \mathsf{SN}_{\tau_1}(s) \implies \mathsf{SN}_{\tau_2}(t\,s)) \end{aligned} \tag{20}$$

Unlike $\Downarrow\!\!\Downarrow$, it is not immediate that SN is logical for $\widetilde{\overline{\gamma}}$ (see [25, App. A]). For a proof of strong normalization based on SN in the context of the λ-calculus, see [57, Sec. 2].

While all three logical predicates $\square\Downarrow$, $\Downarrow\!\!\Downarrow$, SN are eligible for proving strong normalization, with proofs of similar length and complexity, the predicate $\square\Downarrow$ arguably has the most generic flavour, as it depends neither on a system-specific notion of weak transition (which appears in the definition of $\Downarrow\!\!\Downarrow$) nor on the syntax of the language (such as the application operator appearing in the definition of SN). Thus, our abstract categorical approach to logical predicates will be based on a generalization of $\square\Downarrow$.

3.3 Constructing Logical Predicates

Our abstract coalgebraic notion of logical predicate (Definition 12) is parametric in the bifunctor B and its lifting $\overline{B}$ and decoupled from any specific syntax. Next, we develop a systematic construction that promotes a predicate P to a logical predicate, specifically to a locally maximal refinement of P, generalizing $\square\Downarrow$ in Example 15. The construction proceeds in two stages. First, we fix the contravariant argument of the lifted bifunctor $\overline{B}$ and construct a greatest coalgebraic invariant w.r.t. the resulting endofunctor [36, §6.3]:

Definition 18 (Relative henceforth). Let $c\colon Y \to B(X,Y)$ and let $S \rightarrowtail X$ be a predicate. The *(S-)relative henceforth modality* sends $P \rightarrowtail Y$ to $\square^{\overline{B},c}(S,P) \rightarrowtail Y$, which is the supremum in $\mathbf{Pred}_Y(\mathbb{C})$ of all S-relative invariants contained in P:

$$\square^{\overline{B},c}(S,P) = \bigvee\{Q \leq P \mid Q \text{ is an } S\text{-relative } \overline{B}\text{-invariant for } c\}. \tag{21}$$

We will omit the superscripts $\overline{B}, c$ when they are irrelevant or clear from the context.

Proposition 19. *The predicate $\square(S,P)$ is the greatest S-relative $\overline{B}$-invariant contained in P. Moreover, the map $(S,P) \mapsto \square(S,P)$ is antitone in S and monotone in P.*

Proof. The first statement follows from the Knaster-Tarski theorem since $\square(S,P)$ is the greatest fixed point $\square(S,P) = \nu G.\, P \wedge c^\star[\overline{B}(S,G)]$ in the complete lattice $\mathbf{Pred}_Y(\mathbb{C})$. The second statement holds due to the mixed variance of the predicate lifting $\overline{B}$. $\square$

The relative henceforth modality only yields relative invariants. To obtain a logical predicate, i.e. an invariant relative to itself, we move to the second stage of our construction, which is based on ultrametric semantics, see e.g. [9]. Let us briefly recall some terminology. A metric space $(X, d\colon X \times X \to \mathbb{R})$ is *1-bounded* if $d(x,y) \leq 1$ for all x, y, an *ultrametric space* if $d(x,y) \leq \max\{d(x,z), d(z,y)\}$ for all x,y,z, and *complete* if every Cauchy sequence converges. A map $f\colon (X,d) \to (X',d')$ between metric spaces is *nonexpansive* if $d'(f(x), f(y)) \leq d(x,y)$ for all x, y, and *contractive* if there exists

$c \in [0, 1)$, called a *contraction factor*, such that $d'(f(x), f(y)) \le c \cdot d(x, y)$ for all x, y. A family of maps $(f_i : X \to X')_{i \in I}$ is *uniformly contractive* if there exists $c \in [0, 1)$ such that each f_i is contractive with factor c. By Banach's fixed point theorem, every contractive endomap $f : X \to X$ on a non-empty complete metric space has a unique fixed point.

Definition 20. The category C is *predicate-contractive* if

(1) every $\mathbf{Pred}_X(C)$ carries the structure of a complete 1-bounded ultrametric space;
(2) for every $f : X \to Y$ in C, the map $f^\star[-] : \mathbf{Pred}_Y(C) \to \mathbf{Pred}_X(C)$ is non-expansive;
(3) for any two co-well-ordered families $(P^i \rightarrowtail X)_{i \in I}$ and $(Q^i \rightarrowtail X)_{i \in I}$ of predicates,

$$d(\textstyle\bigwedge_{i \in I} P^i, \bigwedge_{i \in I} Q^i) \le \sup_{i \in I} d(P^i, Q^i).$$

Here $(P^i \rightarrowtail X)_{i \in I}$ is *co-well-ordered* if each nonempty subfamily has a greatest element.

Example 21. The category $C = \mathbf{Set}^{\mathsf{Ty}}$ is predicate-contractive when equipped with the ultrametric on $\mathbf{Pred}_X(C)$ given by $d(P, Q) = 2^{-n}$ for $P, Q \rightarrowtail X$, where $n = \inf\{\sharp\tau \mid P_\tau \neq Q_\tau\}$ and $\sharp\tau$ is the size of τ, defined by $\sharp\mathsf{unit} = 1$ and $\sharp(\tau_1 \twoheadrightarrow \tau_2) = \sharp\tau_1 + \sharp\tau_2$. By convention, $\inf \emptyset = \infty$ and $2^{-\infty} = 0$. To see predicate-contractivity, first note that a function $\mathcal{F} : \mathbf{Pred}_Y(C) \to \mathbf{Pred}_X(C)$ is non-expansive iff

$$\inf\{\sharp\tau \mid (\mathcal{F}P)_\tau \neq (\mathcal{F}Q)_\tau\} \ge \inf\{\sharp\tau \mid P_\tau \neq Q_\tau\} \qquad \text{for all } P, Q \rightarrowtail Y,$$

and contractive (necessarily with factor at most 1/2) iff that inequality holds strictly.

This immediately implies clause (2) of Definition 20: inverse images in $\mathbf{Set}^{\mathsf{Ty}}$ are computed pointwise, and $f_\tau^\star[P_\tau] \neq f_\tau^\star[Q_\tau]$ implies $P_\tau \neq Q_\tau$ for $f : X \to Y$ and $P, Q \rightarrowtail Y$. Similarly, since intersections are computed pointwise, clause (3) amounts to

$$\inf\Big\{\sharp\tau \mid \bigcap_{i \in I} P^i_\tau \neq \bigcap_{i \in I} Q^i_\tau\Big\} \ge \inf\{\sharp\tau \mid \exists i \in I : P^i_\tau \neq Q^i_\tau\},$$

which is clearly true, for if $\bigcap_{i \in I} P^i_\tau \neq \bigcap_{i \in I} Q^i_\tau$ then $P^i_\tau \neq Q^i_\tau$ for some $i \in I$.

Definition 22 (Contractive lifting). Suppose that C is predicate-contractive. The predicate lifting $\overline{B} : \mathbf{Pred}(C)^{\mathsf{op}} \times \mathbf{Pred}(C) \to \mathbf{Pred}(C)$ is *contractive* if for every $S \rightarrowtail X$ the map $\overline{B}(S, -)$ is non-expansive, and the family $(\overline{B}(-, P))_{P \rightarrowtail X}$ is uniformly contractive.

Proposition 23. *Let $\overline{B}$ be contractive and $c : X \to B(X, X)$. For every $S \rightarrowtail X$, the map $\square^{\overline{B},c}(S, -)$ is non-expansive, and the family $(\square^{\overline{B},c}(-, P))_{P \rightarrowtail X}$ is uniformly contractive.*

Contractive liftings allow us to augment every predicate P to a logical predicate:

Definition 24 (Henceforth). Let $\overline{B}$ be contractive and $c : X \to B(X, X)$. For each predicate $P \rightarrowtail X$ we define $\square^{\overline{B},c} P \rightarrowtail X$ (where we usually omit the superscripts) to be the unique fixed point of the contractive endomap

$$S \mapsto \square^{\overline{B},c}(S, P) \quad \text{on} \quad \mathbf{Pred}_X(C). \tag{22}$$

Theorem 25. *The predicate $\square P$ is the unique locally maximal logical refinement of P.*

Proof. By (22), $\square P$ is the unique predicate satisfying $\square P = \square(\square P, P)$. By (21), this equality says that $\square P$ is the greatest $\square P$-relative invariant contained in P, as needed. □

Example 26. Let B be the behaviour bifunctor on $\mathbf{Set}^{\mathsf{Ty}}$ given by (3). Its canonical lifting $\overline{B}$ (Example 9) is contractive because $\overline{B}_{\tau_1 \to \tau_2}(P, Q)$ depends only on P_{τ_1}, Q_{τ_2}, $Q_{\tau_1 \to \tau_2}$; in other words, $\overline{B}$ decreases the size of types in the contravariant argument and does not increase it in the covariant argument. Given a coalgebra $c\colon X \to B(X, X)$ and $P \rightarrowtail X$, the fixed point $\Box^{\overline{B},c} P$ is given by the Ty-indexed family of greatest fixed points

$$\Box P_{\mathsf{unit}} = \nu G.\, P_{\mathsf{unit}} \wedge c_{\mathsf{unit}}{}^{\star}[G + 1],$$
$$\Box P_{\tau_1 \to \tau_2} = \nu G.\, P_{\tau_1 \to \tau_2} \wedge c_{\tau_1 \to \tau_2}{}^{\star}[G + \{f\colon \mathsf{Tr}_{\tau_1} \to \mathsf{Tr}_{\tau_2} \mid \forall s \in \Box P_{\tau_1}.\, f(s) \in \Box P_{\tau_2}\}].$$

This follows from Theorem 25 since the above predicate is clearly a locally maximal refinement of P. By instantiating c to the operational model $\gamma\colon \mu\Sigma \to B(\mu\Sigma, \mu\Sigma)$ of **xTCL** and taking $P = \Downarrow$, we recover the definition of $\Box\Downarrow$ in Example 15.

Example 27. The logical predicate $\Downarrow\!\!\Downarrow \rightarrowtail \mathsf{Tr}$ of Example 16 is precisely $\Box\Downarrow$ for $\mathcal{P}D$ w.r.t. its canonical lifting and the coalgebra $\widetilde{\widetilde{\gamma}}\colon \mathsf{Tr} \to \mathcal{P}D(\mathsf{Tr}, \mathsf{Tr})$. More generally, for a coalgebra $c\colon X \to \mathcal{P}D(X, X)$, the predicate $\Box P$ is inductively defined as follows:

$$\Box P_{\mathsf{unit}} = P_{\mathsf{unit}},$$
$$\Box P_{\tau_1 \to \tau_2} = P_{\tau_1 \to \tau_2} \wedge c_{\tau_1 \to \tau_2}{}^{\star}[\{F \subseteq X_{\tau_2}^{X_{\tau_1}} \mid \forall f \in F.\, s \in \Box P_{\tau_1} \implies f(s) \in \Box P_{\tau_2}\}].$$

Remark 28. The construction of logical predicates for typed languages is enabled by the "type-decreasing" nature of the associated behaviour bifunctors. In untyped settings, e.g. for $B(X, Y) = Y + Y^X$ on **Set** modelling untyped combinatory logic [24], the canonical lifting $\overline{B}$ is not contractive, hence the fixed point $\Box P$ in general fails to exist.

Remark 29. The forgetful functor $|{-}|\colon \mathbf{Pred}(C) \to C$ forms a complete lattice fibration [35], equivalently a topological functor [2], and all notions and results of the present subsection extend to that level of generality. We leave the details for future work, as our reasoning techniques found in the upcoming sections are tailored to logical predicates.

We are now in a position to state precisely what a proof via logical predicates is in our framework. Given the operational model $\gamma\colon \mu\Sigma \to B(\mu\Sigma, \mu\Sigma)$ of a higher-order language, a predicate lifting $\overline{B}$, and a target predicate $P \rightarrowtail \mu\Sigma$, a *proof of P via logical predicates* is a proof that $\Box P$ forms a subalgebra of the initial algebra $\mu\Sigma$, which means

$$\overline{\Sigma}(\Box P) \leq \iota^{\star}[\Box P], \quad \text{equivalently} \quad \iota_{\star}[\overline{\Sigma}(\Box P)] \leq \Box P. \tag{23}$$

Then $\Box P = \mu\Sigma$ by structural induction, whence $P = \mu\Sigma$ because $\Box P \leq P$.

Up to this point, we have streamlined and formalized coalgebraic logical predicates as a certain abstract construction on predicates (Definition 24) and presented proofs by coalgebraic logical predicates as standard structural induction on said construction. This presentation is indeed that of an abstract method: the various parts of the problem setting, namely the syntax, the behaviour and its predicate lifting, as well as the operational semantics, are all parameters. In the next section, we exploit the parametric and generic nature of this method in two main ways. First, we present up-to techniques that simplify the proof goal (23) as much as possible. Second, we look to instantiate our method to problems on *classes of higher-order languages*, as opposed to reasoning about operational models of individual languages such as **xTCL** or the λ-calculus.

4 Logical Predicates and Higher-Order Abstract GSOS

As indicated before, substantial parts of the proof of strong normalization in Example 15 look generic. Specifically, the properties (2) and (3) established for $Q = A$ and $Q = C$ are independent of the choice of predicate $P = \Downarrow$ in $\Box P$. Moreover, these steps are either obvious or follow immediately from the operational rules of **xTCL**: the predicates A and C being invariants can be attributed to the fact that except for terms of the form $S''(-,-)$, all terms evolve either to a variable or to some flat term such as $p'\,q$. The core of the proof, which is tailored to the choice of P, lies in proving property (1).

As it turns out, for a class of higher-order GSOS laws that we call *relatively flat higher-order GSOS laws*, conditions (2) and (3) are automatic. This insight leads us to a powerful up-to technique that simplifies proofs via logical predicates.

4.1 Relatively Flat Higher-Order GSOS Laws

The following definition abstracts the restricted nature of the rules of **xTCL** to the level of higher-order GSOS laws. For simplicity, we confine ourselves to 0-pointed laws, however all the results of this subsection easily extend to the V-pointed case.

Definition 30. Let $\Sigma\colon C \to C$ be a syntax functor of the form $\Sigma = \coprod_{j\in J}\Sigma_j$, where $(J,\prec)$ is a non-empty well-founded strict partial order, and put $\Sigma_{\prec k} = \coprod_{j\prec k}\Sigma_j$. A *relatively flat (0-pointed) higher-order GSOS law* of Σ over B is a J-indexed family of morphisms

$$\varrho^j_{X,Y}\colon \Sigma_j(X\times B(X,Y)) \to B(X, \Sigma^\star_{\prec j}(X+Y) + \Sigma_j\Sigma^\star_{\prec j}(X+Y)) \tag{24}$$

dinatural in $X\in C$ and natural in $Y \in C$.

We put $e_{j,X} = [\mathsf{in}^\sharp_{\prec j}, \iota\cdot \mathsf{in}_j\cdot\Sigma_j(\mathsf{in}^\sharp_{\prec j})]\colon \Sigma^\star_{\prec j}X + \Sigma_j\Sigma^\star_{\prec j}X \to \Sigma^\star X$ where $\mathsf{in}_{\prec j}\colon \Sigma_{\prec j}\to\Sigma$ and $\mathsf{in}_j\colon\Sigma_j\to\Sigma$ are the coproduct injections, with free extensions $\mathsf{in}^\sharp_{\prec j}\colon \Sigma^\star_{\prec j}\to\Sigma^\star$ and $\mathsf{in}^\sharp_j\colon\Sigma^\star_j\to\Sigma^\star$. Every relatively flat higher-order GSOS law (24) determines an ordinary higher-order GSOS law of Σ over B with components

$$\varrho_{X,Y} = \coprod_{j\in J}\Sigma_j(X\times B(X,Y)) \xrightarrow{\coprod_{j\in J}\varrho^j_{X,Y}} \coprod_{j\in J} B(X,\Sigma^\star_{\prec j}(X+Y)+\Sigma_j\Sigma^\star_{\prec j}(X+Y)) \xrightarrow{[B(X,e_{j,X+Y})]_{j\in J}} B(X,\Sigma^\star(X+Y)).$$

When we interpret a higher-order GSOS law as a set of operational rules, relative flatness means that the operations of the language can be ranked in a way that every term $\mathsf{f}(-,\cdots,-)$ with f of rank j evolves into a term that uses only operations of strictly lower rank, except possibly its head symbol which may have the same rank j.

Example 31. **xTCL** is relatively flat: put $J = \{0 \prec 1\}$, let Σ_0 contain application, and let Σ_1 contain all other operation symbols. This is immediate from the rules in Figure 1.

Definition 32. Suppose that each Σ_j preserves strong epimorphisms. A *predicate lifting* of (24) is a relatively flat 0-pointed higher-order GSOS law $(\overline{\varrho}^j)_{j\in J}$ of $\overline{\Sigma} = \coprod_j\overline{\Sigma}_j$ over $\overline{B}$ where for every $P \rightarrowtail X$ and $Q\rightarrowtail Y$ the $\mathbf{Pred}(C)$-morphism $\overline{\varrho}^j_{P,Q}$ is carried by $\varrho^j_{X,Y}$.

Remark 33. (1) The condition on Σ_j ensures $\overline{\Sigma_j}^{\star} = \overline{\Sigma}_j^{\star}$ (Proposition 7), so that the first component of $\overline{\varrho}^j_{P,Q}$ has type $\Sigma_j(X \times B(X,Y)) \to B(X, \Sigma^\star_{<j}(X+Y) + \Sigma_j\Sigma^\star_{<j}(X+Y))$.
(2) Liftings are unique if they exist: since $\overline{\varrho}^j_{P,Q}$ is a **Pred**(C)-morphism, it is determined by its first component $\varrho^j_{X,Y}$. Moreover, the (di)naturality of $\overline{\varrho}^j$ follows from that of ϱ^j.
(3) For the canonical lifting $\overline{B}$, a lifting $(\overline{\varrho}^j)_{j\in J}$ of $(\varrho^j)_{j\in J}$ always exists [25, App. D].

The following theorem establishes a sound up-to technique for logical predicates. It states that for operational models of relatively flat laws, the proof goal (23) can be established by checking a substantially relaxed property.

Theorem 34 (Induction up to $\Box$). *Let $\gamma\colon \mu\Sigma \to B(\mu\Sigma, \mu\Sigma)$ be the operational model of a relatively flat 0-pointed higher-order GSOS law that admits a predicate lifting. Then for every predicate $P \rightarrowtail \mu\Sigma$ and every locally maximal logical refinement $\Box^{\gamma,\overline{B}}P$,*

$$\overline{\Sigma}(\Box^{\gamma,\overline{B}}P) \le \iota^\star[P] \qquad \textit{implies} \quad \overline{\Sigma}(\Box^{\gamma,\overline{B}}P) \le \iota^\star[\Box^{\gamma,\overline{B}}P] \quad (\textit{hence } P = \mu\Sigma).$$

We stress that the theorem applies to any refinement $\Box^{\gamma,\overline{B}}P$ and does not assume a specific construction (e.g. that of Section 3.3). The up-to technique facilitates proofs via logical predicates quite dramatically. For illustration, we revisit strong normalization:

Example 35. We give an alternative proof of strong normalization of **xTCL** (cf. Example 15) via induction up to $\Box$. Hence it suffices to prove

$$\overline{\Sigma}(\Box\Downarrow) \le \iota^\star[\Downarrow],$$

which states that a term is terminating if all of its subterms are in the logical predicate $\Box\Downarrow$. This is clear for terms that are not applications, since they immediately terminate (cf. Figure 1). Now consider an application $p\,q$ such that $\Box_{\tau_1\to\tau_2}\Downarrow(p)$ and $\Box_{\tau_1}\Downarrow(q)$. Since $\Box\Downarrow$ is a logical predicate contained in $\Downarrow$, this entails that $p \overset{q}{\Rightarrow} p'$ for a (unique) term p', and that $\Box\Downarrow_{\tau_2} p'$, hence $\Downarrow_{\tau_2} p'$. Since $p\,q \Rightarrow p'$, it follows that $\Downarrow_{\tau_2} p\,q$.

Analogous reasoning shows that **xTCL** is strongly normalizing under the call-by-value and the maximally nondeterministic evaluation strategy (Remark 5). In the latter case, strong normalization means that every term must eventually terminate, independently of the order of evaluation.

The reader should compare the above compact argument to the laborious original proof given in Example 15. Our up-to technique can be seen to precisely isolate the non-trivial core of the proof, while providing its generic parts for free. For a further application – type safety of the simply typed λ-calculus – see Section 4.2.

4.2 λ-Laws

We proceed to explain how our theory of logical predicates applies to languages with variables and binders. We highlight the core ideas and technical challenges in the case of the λ-calculus, and briefly sketch their categorical generalization; a full exposition can be found in [25, App. E]. Let **STLC** be the simply typed call-by-name λ-calculus with the set Ty of types given by (2) and operational rules

$$\frac{t \to t'}{t\,s \to t'\,s} \qquad \frac{}{(\lambda x\colon \tau_1.\,t)\,s \to t[s/x]} \tag{25}$$

where s, t, t' range over λ-terms of appropriate type, and $[-/-]$ denotes capture-avoiding substitution. To model **STLC** in higher-order abstract GSOS, we follow ideas by Fiore [18]. Our base category C is the presheaf category $(\mathbf{Set}^{\mathbb{F}/\mathsf{Ty}})^{\mathsf{Ty}}$ where $\mathbb{F}$ denotes the category of finite cardinals and functions, and the set Ty is regarded as a discrete category. An object $\Gamma\colon n \to \mathsf{Ty}$ of $\mathbb{F}/\mathsf{Ty}$ is a *typed context*, associating to each variable $x \in n$ a type; we put $|\Gamma| := n$. A presheaf $X \in (\mathbf{Set}^{\mathbb{F}/\mathsf{Ty}})^{\mathsf{Ty}}$ associates to each context Γ and each type τ a set $X_\tau(\Gamma)$ whose elements we think of as terms of type τ in context Γ.

The syntax of **STLC** is captured by the functor $\Sigma\colon (\mathbf{Set}^{\mathbb{F}/\mathsf{Ty}})^{\mathsf{Ty}} \to (\mathbf{Set}^{\mathbb{F}/\mathsf{Ty}})^{\mathsf{Ty}}$ where

$$\begin{aligned} \Sigma_{\mathsf{unit}} X &= V_{\mathsf{unit}} + K_1 + \coprod\nolimits_{\tau\in\mathsf{Ty}} X_{\tau\to\mathsf{unit}} \times X_\tau, \\ \Sigma_{\tau_1\to\tau_2} X &= V_{\tau_1\to\tau_2} + \delta^{\tau_1}_{\tau_2} X + \coprod\nolimits_{\tau\in\mathsf{Ty}} X_{\tau\to\tau_1\to\tau_2} \times X_\tau. \end{aligned} \tag{26}$$

Here $K_1 \in \mathbf{Set}^{\mathbb{F}/\mathsf{Ty}}$ is the constant presheaf on 1, V is given by $V_\tau(\Gamma) = \{x \in |\Gamma| \mid \Gamma(x) = \tau\}$, and δ by $(\delta^{\tau_1}_{\tau_2} X)(\Gamma) = X_{\tau_2}(\Gamma + \check{\tau}_1)$ with $(-) + \check{\tau}_1$ denoting context extension by a variable of type τ_1. Informally, K_1, V and δ represent the constant $\mathsf{e}\colon \mathsf{unit}$, variables, and λ-abstraction, respectively. The initial algebra for Σ is the presheaf Λ of λ-terms, i.e. $\Lambda_\tau(\Gamma)$ is the set of λ-terms (modulo α-equivalence) of type τ in context Γ [18].

The behaviour bifunctor $B^\lambda\colon ((\mathbf{Set}^{\mathbb{F}/\mathsf{Ty}})^{\mathsf{Ty}})^{\mathsf{op}} \times (\mathbf{Set}^{\mathbb{F}/\mathsf{Ty}})^{\mathsf{Ty}} \to (\mathbf{Set}^{\mathbb{F}/\mathsf{Ty}})^{\mathsf{Ty}}$ for **STLC** has two separate components: it is given by a product

$$B^\lambda(X, Y) = \langle\!\langle X, Y\rangle\!\rangle \times B(X, Y) \tag{27}$$

where

$$\begin{aligned} \langle\!\langle X, Y\rangle\!\rangle_\tau(\Gamma) &= \mathbf{Set}^{\mathbb{F}/\mathsf{Ty}}\Big(\prod\nolimits_{x\in|\Gamma|} X_{\Gamma(x)}, Y_\tau\Big), \\ B(X, Y) &= (K_1 + Y + D(X, Y)), \\ D_{\mathsf{unit}}(X, Y) = K_1 \qquad &\text{and} \qquad D_{\tau_1\to\tau_2}(X, Y) = Y_{\tau_2}^{X_{\tau_1}}, \end{aligned}$$

and $Y_{\tau_2}^{X_{\tau_1}}$ is an exponential object in $\mathbf{Set}^{\mathbb{F}/\mathsf{Ty}}$. The bifunctor $\langle\!\langle -, -\rangle\!\rangle$ models an abstract substitution structure; for instance, every λ-term $t \in \Lambda_\tau(\Gamma)$ induces a natural transformation $\prod_{x\in|\Gamma|} \Lambda_{\Gamma(x)} \to \Lambda_\tau$ in $\langle\!\langle \Lambda, \Lambda\rangle\!\rangle_\tau(\Gamma)$ mapping a tuple $(t_1, \dots, t_{|\Gamma|})$ to the term obtained by simultaneous substitution of the terms t_i for the variables of t. The summands of the bifunctor B abstract from the possible operational behaviour of λ-terms: a term may explicitly terminate, reduce, get stuck (e.g. if it is a variable), or act as a function.

The operational rules (25) of **STLC** can be encoded into a V-pointed higher-order GSOS law of Σ over B^λ, similar to the untyped λ-calculus treated in earlier work [24]. The operational model $\langle\phi, \gamma\rangle\colon \Lambda \to \langle\!\langle \Lambda, \Lambda\rangle\!\rangle \times B(\Lambda, \Lambda)$ is the coalgebra whose components ϕ, γ describe the substitution structure and the operational behaviour of λ-terms.

At this point, a key technical issue can be observed: the canonical predicate lifting $\overline{\langle\!\langle -, -\rangle\!\rangle}$ is not contractive. Indeed, given $P \rightarrowtail X$, $Q \rightarrowtail Y$, the predicate $\overline{\langle\!\langle P, Q\rangle\!\rangle}_\tau$ consists of all natural transformations $\prod_{x\in|\Gamma|} X_{\Gamma(x)} \to Y_\tau$ that restrict to $\prod_{x\in|\Gamma|} P_{\Gamma(x)} \to Q_\tau$, and this expression depends on $P_{\Gamma(x)}$ where the type $\Gamma(x)$ may be of higher complexity than τ. In particular, we conclude that $\overline{B^\lambda}$ is not contractive. In contrast, the canonical lifting $\overline{B}$ is contractive and hence $\Box^{\gamma,\overline{B}}P$ exists for every $P \rightarrowtail \Lambda$ (Definition 24). However, it is well-known that logical predicates do not do the trick for inductive proofs in the λ-calculus, see e.g. [57, p. 9] and [49, p. 150]; rather, one needs to prove the *open extension* of the logical predicate, which is the larger predicate

$$\boxdot^{\gamma,\overline{B}}P = \phi^\star[\overline{\langle\!\langle \Box^{\gamma,\overline{B}}P, \Box^{\gamma,\overline{B}}P \rangle\!\rangle}].$$

The standard proof method is then to show $\boxdot^{\gamma,\overline{B}}P = \Lambda$ directly by structural induction. However, this can be greatly simplified by the following up-to-principle, which works with the original predicate $\Box^{\gamma,\overline{B}}P$ and forms a counterpart of Theorem 34 for the λ-calculus:

Theorem 36 (Induction up to $\boxdot$). *Let $P \rightarrowtail \Lambda$ be a predicate. Then*

$$\overline{\Sigma}(\Box^{\gamma,\overline{B}}) \leq \iota^\star[P] \quad \textit{implies} \quad \overline{\Sigma}(\boxdot^{\gamma,\overline{B}}P) \leq \iota^\star[\boxdot^{\gamma,\overline{B}}P] \quad (\textit{hence } P = \Lambda).$$

Remark 37. Concretely, the theorem states that to prove $P = \Lambda$, it suffices to prove that (1) variables satisfy P, (2) the unit expression $\mathsf{e}\colon \mathsf{unit}$ satisfies P, (3) for all application terms $p\,q$ such that $\Box_{\tau_1\to\tau_2}P(\Gamma)(p)$ and $\Box_{\tau_1}P(\Gamma)(q)$, we have $P_{\tau_2}(\Gamma)(p\,q)$, and (4) for all λ-abstractions $\lambda x\colon \tau_1.\,t$ such that $t \in \Box_{\tau_2}P(\Gamma, x)$, we have $P_{\tau_1\to\tau_2}(\Gamma)(\lambda x\colon \tau_1.\,t)$.

Example 38. We prove type safety for **STLC** via induction up to $\boxdot$. Thus consider the predicate Safe $\rightarrowtail \Lambda$ that is constantly true on open terms and given by

$$t \in \mathrm{Safe}_\tau(\varnothing) \iff (\forall e.\, t \Rightarrow e \implies (e \text{ is not an application}) \vee \exists r.\, e \to r),$$

on closed terms. We only need to check the conditions (1)–(4) of Remark 37. Conditions (1), (2), (4) are clear since variables are open terms and the term $\mathsf{e}\colon \mathsf{unit}$ and λ-abstractions do not reduce. The only interesting clause is (3) for the empty context. Thus let $p\,q$ be a closed application term with $p \in \Box\mathrm{Safe}_{\tau_1\to\tau_2}(\varnothing)$ and $q \in \Box\mathrm{Safe}_{\tau_1}(\varnothing)$; we need to show $p\,q \in \mathrm{Safe}_{\tau_2}(\varnothing)$. We proceed by case distinction on $p\,q \Rightarrow e$:

(a) $p \Rightarrow p'$ and $e = p'\,q$. Then $p' \in \Box\mathrm{Safe}_{\tau_1\to\tau_2}(\varnothing)$ by invariance, in particular p' is safe, so p' is either not an application or reduces. In the former case, p' is necessarily a λ-abstraction since it is closed and not of type unit. Thus, in both cases, e reduces.

(b) $p \Rightarrow \lambda x.p'$ and $p'[q/x] \Rightarrow e$. Since $\Box$Safe is a logical predicate, from $p \in \Box\mathrm{Safe}_{\tau_1\to\tau_2}(\varnothing)$ and $q \in \Box_{\tau_1}\mathrm{Safe}(\varnothing)$ we can deduce $p'[q/x] \in \Box_{\tau_2}\mathrm{Safe}(\varnothing)$, whence $e \in \Box_{\tau_2}\mathrm{Safe}(\varnothing)$. In particular, e is safe, which implies that e is either not an application or reduces.

As an exercise, we invite the reader to prove strong normalization of **STLC** via induction up to $\boxdot$. The reader should compare these short and simple proofs with more traditional ones, see e.g. [57].

All the above results and observations for **STLC** can be generalized and developed at the level of general higher-order abstract GSOS laws. To this end, we first abstract the behaviour functor (27) to a functor of the form $B(X, Y) = (X \twoheadrightarrow Y) \times B'(X, Y)$, where

$(-) \multimap (-)$ is the internal hom-functor of a suitable closed monoidal structure on the base category C. In the case of **STLC**, this structure is given by Fiore's *substitution tensor* [18]. Second, we observe that the higher-order GSOS law of **STLC** is an instance of a special kind of law that we coin *relatively flat λ-laws*. The induction-up-to-$\boxdot$ technique of Theorem 36 then can be shown to hold for operational models of relatively flat λ-laws. More details can be found in [25, App. E].

5 Strong Normalization for Deterministic Systems, Abstractly

The high level of generality in which the theory of logical predicates is developed above enables reasoning uniformly about whole families of languages and behaviours. In this section, we narrow our focus to deterministic systems and establish a general strong normalization criterion, which can be checked in concrete instances by mere inspection of the operational rules corresponding to higher-order abstract GSOS laws.

Throughout this section, we fix a 0-pointed higher-order GSOS law ϱ of a signature endofunctor $\Sigma\colon C \to C$ over a behaviour bifunctor $B\colon C^{\mathsf{op}} \times C \to C$, where

$$B(X,Y) = Y + D(X,Y) \quad \text{for some} \quad D\colon C^{\mathsf{op}} \times C \to C.$$

For instance, the type functor (3) for **xTCL** is of that form. The operational model $\gamma\colon \mu\Sigma \to \mu\Sigma + D(\mu\Sigma,\mu\Sigma)$ has an n-step extension $\gamma^{(n)}\colon \mu\Sigma \to \mu\Sigma + D(\mu\Sigma,\mu\Sigma)$, for each $n \in \mathbb{N}$, where $\gamma^{(0)}$ is the left coproduct injection and $\gamma^{(n+1)}$ is the composite

$$\mu\Sigma \xrightarrow{\gamma} \mu\Sigma + D(\mu\Sigma,\mu\Sigma) \xrightarrow{\gamma^{(n)}+\mathsf{id}} \mu\Sigma + D(\mu\Sigma,\mu\Sigma) + D(\mu\Sigma,\mu\Sigma) \xrightarrow{\mathsf{id}+\nabla} \mu\Sigma + D(\mu\Sigma,\mu\Sigma).$$

We regard $D(\mu\Sigma,\mu\Sigma)$ as a predicate on $B(\mu\Sigma,\mu\Sigma)$ via the right coproduct injection, which is monic by extensivity of C, and define the following predicates on $\mu\Sigma$:

$$\Downarrow_n = (\gamma^{(n)})^\star[D(\mu\Sigma,\mu\Sigma)] \qquad \text{and} \qquad \Downarrow = \bigvee_n \Downarrow_n .$$

In **xTCL**, these are the predicates of strong normalization or strong normalization after at most n steps, resp. Accordingly, we define strong normalization abstractly as follows:

Definition 39. The higher-order GSOS law ϱ is *strongly normalizing* if $\Downarrow = \mu\Sigma$.

We next identify two natural conditions on the law ϱ that together ensure strong normalization. The first roughly asserts that for a term $t = \mathsf{f}(x_1,\dots,x_n)$ whose variables x_i are non-progressing, the term t is either non-progressing or it progresses to a variable.

Definition 40. The higher-order GSOS law ϱ is *simple* if its components $\varrho_{X,Y}$ restrict to morphisms $\varrho^0_{X,Y}$ as in the diagram below, where η is the unit of the free monad $\Sigma^\star$:

$$\begin{array}{ccc}
\Sigma(X \times D(X,Y)) & \xrightarrow{\varrho^0_{X,Y}} & X + Y + D(X,\Sigma^\star(X+Y)) \\
{\scriptstyle \Sigma(\mathsf{id}\times\mathsf{inr})}\downarrow & & \downarrow{\scriptstyle \eta_{X+Y}+\mathsf{id}} \\
\Sigma(X \times (Y + D(X,Y))) & \xrightarrow{\varrho_{X,Y}} & \Sigma^\star(X+Y) + D(X,\Sigma^\star(X+Y))
\end{array}$$

The second condition asserts that the rules represented by the higher-order GSOS law remain sound when strong transitions are replaced by weak ones. In the following, the *graph* of a morphism $f\colon A \to B$ is the image $\mathsf{gra}(f) \rightarrowtail A \times B$ of $\langle \mathsf{id}, f\rangle\colon A \to A \times B$.

Definition 41. The higher-order GSOS law ϱ *respects weak transitions* if for every $n \in \mathbb{N}$, the graph of the composite below is contained in $\bigvee_k \mathsf{gra}(\gamma^{(k)} \cdot \iota)$.

$$\Sigma(\mu\Sigma) \xrightarrow{\Sigma\langle \mathsf{id}, \gamma^{(n)}\rangle} \Sigma(\mu\Sigma \times B(\mu\Sigma, \mu\Sigma)) \xrightarrow{\varrho_{\mu\Sigma,\mu\Sigma}} B(\mu\Sigma, \Sigma^\star(\mu\Sigma + \mu\Sigma)) \xrightarrow{B(\mathsf{id}, \hat{\iota}\cdot\Sigma^\star\nabla)} B(\mu\Sigma, \mu\Sigma)$$

Note that the higher-order GSOS law for **xTCL** is simple and respects weak transitions. Thus, strong normalization of **xTCL** is an instance of the following strong normalization theorem for higher-order abstract GSOS. Concerning its conditions, an *ω-directed union* is a colimit of an ω-chain $X_0 \rightarrowtail X_1 \rightarrowtail X_2 \rightarrowtail \cdots$ of monics. We say that monos in $\mathbb{C}$ are *ω-smooth* if any such colimit has monic injections, and moreover for every compatible cocone of monos, the mediating morphism is monic. This property holds in every locally finitely presentable category [3, Prop. 1.62], e.g. sets, posets, or presheaves.

Theorem 42 (Strong normalization). *Suppose that the following conditions hold:*

(1) *On top of Assumptions 6, $\mathbb{C}$ is countably extensive, and monos are ω-smooth.*

(2) *Σ preserves ω-directed unions, and D preserves monos in the second component.*

(3) *ϱ is relatively flat, simple, and respects weak transitions.*

(4) *$\Downarrow$ has a locally maximal logical refinement w.r.t. γ and the canonical lifting $\overline{B}$.*

Then the higher-order GSOS law ϱ is strongly normalizing.

Recall that condition (4) holds if $\overline{B}$ is contractive (Theorem 25). The proof uses the induction-up-to-$\Box$ technique and a careful categorical abstraction of Example 35.

6 Conclusion and Future Work

Our work presents the initial steps towards a unifying, efficient theory of logical relations for higher-order languages based on higher-order abstract GSOS. This theory can be broadened in various directions. One obvious direction would be to extend our theory from predicates to relations. Binary logical relations are often utilized as sound (and sometimes complete) relations w.r.t. *contextual equivalence*. Additional generalizations are suggested by the large amount of existing work on logical relations. One important direction is to generalize the type system to cover, e.g., recursive types, parametric polymorphism, or dependent types. Supporting recursive types will presumably require an adaptation of the method of step-indexing [17] to our abstract setting. Another point of interest is to apply and extend our framework to effectful (e.g. probabilistic) settings [40,54], including e.g. an effectful version of the criterion of Section 5.

As indicated in Remark 29, large parts of our development in Section 3 can be reformulated in fibrational terms. This has the potential merit of enabling abstract reasoning about higher-order programs in metric and differential settings as done in previous work on fine-grain call-by-value [13,14]. In future work, we aim to develop such a generalization, and to explore the connection between our weak transition semantics and the general evaluation semantics used in *op. cit.*

References

1. Abramsky, S.: The lazy λ-calculus. In: Research topics in Functional Programming, pp. 65–117. Addison Wesley (1990)
2. Adámek, J., Herrlich, H., Strecker, G.E.: Abstract and Concrete Categories. Wiley (1990), republished in: Reprints in Theory and Applications of Categories 17 (2006), pp. 1-507, http://www.tac.mta.ca/tac/reprints/articles/17/tr17abs.html
3. Adámek, J., Rosický, J.: Locally Presentable and Accessible Categories. London Mathematical Society Lecture Note Series, Cambridge University Press (1994). https://doi.org/10.1017/CBO9780511600579
4. Aguirre, A., Birkedal, L.: Step-indexed logical relations for countable nondeterminism and probabilistic choice. In: 50th ACM SIGPLAN Symposium on Principles of Programming Languages (POPL 2023). Proc. ACM Program. Lang., vol. 7. ACM (2023). https://doi.org/10.1145/3571195
5. Ahmed, A.: Step-indexed syntactic logical relations for recursive and quantified types. In: 15th European Symposium on Programming (ESOP 2006). LNCS, vol. 3924, pp. 69–83. Springer (2006). https://doi.org/10.1007/11693024_6
6. Altenkirch, T., Kaposi, A.: Normalisation by evaluation for dependent types. In: 1st International Conference on Formal Structures for Computation and Deduction (FSCD 2016). LIPIcs, vol. 52, pp. 6:1–6:16. Schloss Dagstuhl – Leibniz-Zentrum fuer Informatik (2016). https://doi.org/10.4230/LIPIcs.FSCD.2016.6
7. Appel, A.W., McAllester, D.A.: An indexed model of recursive types for foundational proof-carrying code. ACM Trans. Program. Lang. Syst. **23**(5), 657–683 (2001). https://doi.org/10.1145/504709.504712
8. Benton, N., Hur, C.K.: Biorthogonality, step-indexing and compiler correctness. In: 14th ACM SIGPLAN International Conference on Functional Programming (ICFP 2009). p. 97–108. ACM (2009). https://doi.org/10.1145/1596550.1596567
9. Birkedal, L., Støvring, K., Thamsborg, J.: The category-theoretic solution of recursive metric-space equations. Theoretical Computer Science **411**(47), 4102–4122 (2010). https://doi.org/10.1016/j.tcs.2010.07.010
10. Bizjak, A., Birkedal, L.: Step-indexed logical relations for probability. In: Pitts, A. (ed.) 18th International Conference on Foundations of Software Science and Computation Structures (FoSSaCS 2015). LNCS, vol. 9034, pp. 279–294. Springer (2015). https://doi.org/10.1007/978-3-662-46678-0_18
11. Borceux, F.: Handbook of Categorical Algebra: Volume 1: Basic Category Theory, Encyclopedia of Mathematics and Its Applications, vol. 1. Cambridge University Press (1994). https://doi.org/10.1017/CBO9780511525858
12. Carboni, A., Lack, S., Walters, R.F.C.: Introduction to extensive and distributive categories. Journal of Pure and Applied Algebra **84**(2), 145–158 (Feb 1993). https://doi.org/10.1016/0022-4049(93)90035-R
13. Dagnino, F., Gavazzo, F.: A fibrational tale of operational logical relations. In: 7th International Conference on Formal Structures for Computation and Deduction (FSCD 2022). LIPIcs, vol. 228, pp. 3:1–3:21. Schloss Dagstuhl – Leibniz-Zentrum für Informatik (2022). https://doi.org/10.4230/LIPIcs.FSCD.2022.3
14. Dagnino, F., Gavazzo, F.: A Fibrational Tale of Operational Logical Relations: Pure, Effectful and Differential. CoRR (2023). https://doi.org/10.48550/arXiv.2303.03271
15. Dal Lago, U., Gavazzo, F.: Differential logical relations, part ii increments and derivatives. Theor. Comput. Sci. **895**(C), 34–47 (2021). https://doi.org/10.1016/j.tcs.2021.09.027

16. Dal Lago, U., Gavazzo, F., Levy, P.B.: Effectful applicative bisimilarity: Monads, relators, and Howe's method. In: 32nd Annual ACM/IEEE Symposium on Logic in Computer Science (LICS 2017). pp. 1–12. IEEE Computer Society (2017). https://doi.org/10.1109/LICS.2017.8005117
17. Dreyer, D., Ahmed, A., Birkedal, L.: Logical step-indexed logical relations. In: 24th Annual IEEE Symposium on Logic In Computer Science (LICS 2009). pp. 71–80. IEEE Computer Society (2009). https://doi.org/10.1109/LICS.2009.34
18. Fiore, M.: Semantic analysis of normalisation by evaluation for typed lambda calculus. Math. Struct. Comput. Sci. **32**(8), 1028–1065 (2022). https://doi.org/10.1017/S0960129522000263
19. Fiore, M.P., Plotkin, G.D., Turi, D.: Abstract syntax and variable binding. In: 14th Annual IEEE Symposium on Logic in Computer Science (LICS 1999). pp. 193–202. IEEE Computer Society (1999). https://doi.org/10.1109/LICS.1999.782615
20. Fiore, M.P., Turi, D.: Semantics of name and value passing. In: 16th Annual IEEE Symposium on Logic in Computer Science (LICS 2001). pp. 93–104. IEEE Computer Society (2001). https://doi.org/10.1109/LICS.2001.932486
21. Georges, A.L., Guéneau, A., Van Strydonck, T., Timany, A., Trieu, A., Devriese, D., Birkedal, L.: Cerise: Program verification on a capability machine in the presence of untrusted code. J. ACM (2023). https://doi.org/10.1145/3623510
22. Giarrusso, P.G., Stefanesco, L., Timany, A., Birkedal, L., Krebbers, R.: Scala step-by-step: Soundness for dot with step-indexed logical relations in iris. In: 25th ACM SIGPLAN International Conference on Functional Programming (ICFP 2020). Proc. ACM Program. Lang., vol. 4. ACM (2020). https://doi.org/10.1145/3408996
23. Girard, J.Y., Taylor, P., Lafont, Y.: Proofs and types, vol. 7. Cambridge University Press (1989)
24. Goncharov, S., Milius, S., Schröder, L., Tsampas, S., Urbat, H.: Towards a higher-order mathematical operational semantics. In: 50th ACM SIGPLAN Symposium on Principles of Programming Languages (POPL 2023). Proc. ACM Program. Lang., vol. 7. ACM (2023). https://doi.org/10.1145/3571215
25. Goncharov, S., Santamaria, A., Schröder, L., Tsampas, S., Urbat, H.: Logical predicates in higher-order mathematical operational semantics (2024), https://arxiv.org/abs/2401.05872
26. Gordon, A.D.: Bisimilarity as a theory of functional programming. Theor. Comput. Sci. **228**(1-2), 5–47 (1999). https://doi.org/10.1016/S0304-3975(98)00353-3
27. Goubault-Larrecq, J., Lasota, S., Nowak, D.: Logical relations for monadic types. Math. Struct. Comput. Sci. **18**(6), 1169–1217 (2008). https://doi.org/10.1017/S0960129508007172
28. Hermida, C., Reddy, U.S., Robinson, E.P.: Logical relations and parametricity - A Reynolds programme for category theory and programming languages. Electron. Notes Theor. Comput. Sci. **303**, 149–180 (2014). https://doi.org/10.1016/j.entcs.2014.02.008
29. Hermida, C.A.: Fibrations, logical predicates and indeterminates. Ph.D. thesis, University of Edinburgh (1993), https://era.ed.ac.uk/handle/1842/14057
30. Hindley, J.R., Seldin, J.P.: Lambda-Calculus and Combinators: An Introduction. Cambridge University Press, 2 edn. (2008). https://doi.org/10.1017/CBO9780511809835
31. Howe, D.J.: Equality in lazy computation systems. In: 4th Annual Symposium on Logic in Computer Science (LICS 1989). pp. 198–203. IEEE Computer Society (1989). https://doi.org/10.1109/LICS.1989.39174
32. Howe, D.J.: Proving congruence of bisimulation in functional programming languages. Inf. Comput. **124**(2), 103–112 (1996). https://doi.org/10.1006/inco.1996.0008

33. Hur, C.K., Dreyer, D.: A Kripke Logical Relation between ML and Assembly. In: 38th Annual ACM SIGPLAN-SIGACT Symposium on Principles of Programming Languages (POPL 2023). p. 133–146. ACM (2011). https://doi.org/10.1145/1926385.1926402
34. Hur, C.K., Dreyer, D., Neis, G., Vafeiadis, V.: The Marriage of Bisimulations and Kripke Logical Relations. In: 39th Annual ACM SIGPLAN-SIGACT Symposium on Principles of Programming Languages (POPL 2012). SIGPLAN Not., vol. 47, p. 59–72. ACM (2012). https://doi.org/10.1145/2103621.2103666
35. Jacobs, B.: Categorical Logic and Type Theory. No. 141 in Studies in Logic and the Foundations of Mathematics, North Holland (1999)
36. Jacobs, B.: Introduction to Coalgebra: Towards Mathematics of States and Observation, Cambridge Tracts in Theoretical Computer Science, vol. 59. Cambridge University Press (2016). https://doi.org/10.1017/CBO9781316823187
37. Johann, P., Simpson, A., Voigtländer, J.: A generic operational metatheory for algebraic effects. In: 25th Annual IEEE Symposium on Logic in Computer Science (LICS 2010). pp. 209–218. IEEE Computer Society (2010). https://doi.org/10.1109/LICS.2010.29
38. Katsumata, S.: A generalisation of pre-logical predicates and its applications. Ph.D. thesis, University of Edinburgh (2005), http://hdl.handle.net/1842/850
39. Kurz, A., Velebil, J.: Relation lifting, a survey. Journal of Logical and Algebraic Methods in Programming **85**(4), 475–499 (2016). https://doi.org/10.1016/j.jlamp.2015.08.002
40. Lago, U.D., Gavazzo, F.: Effectful program distancing. In: 49th Annual ACM SIGPLAN Symposium on Principles of Programming Languages (POPL 2022). Proc. ACM Program. Lang., vol. 6, pp. 1–30 (2022). https://doi.org/10.1145/3498680
41. Lago, U.D., Gavazzo, F., Yoshimizu, A.: Differential Logical Relations, Part I: The Simply-Typed Case. In: 46th International Colloquium on Automata, Languages, and Programming (ICALP 2019). LIPIcs, vol. 132, pp. 111:1–111:14. Schloss Dagstuhl – Leibniz-Zentrum fuer Informatik (2019). https://doi.org/10.4230/LIPIcs.ICALP.2019.111
42. Levy, P., Power, J., Thielecke, H.: Modelling environments in call-by-value programming languages. Inf. Comput. **185**(2), 182–210 (2003)
43. Mac Lane, S.: Categories for the Working Mathematician, Graduate Texts in Mathematics, vol. 5. Springer, 2 edn. (1978), http://link.springer.com/10.1007/978-1-4757-4721-8
44. Milner, R.: A theory of type polymorphism in programming. Journal of Computer and System Sciences **17**(3), 348–375 (1978). https://doi.org/10.1016/0022-0000(78)90014-4
45. New, M.S., Bowman, W.J., Ahmed, A.: Fully abstract compilation via universal embedding. In: 21st ACM SIGPLAN International Conference on Functional Programming (ICFP 2016). pp. 103–116. ACM (2016). https://doi.org/10.1145/2951913.2951941
46. O'Hearn, P.W., Riecke, J.G.: Kripke logical relations and PCF. Inf. Comput. **120**(1), 107–116 (1995). https://doi.org/10.1006/inco.1995.1103
47. Ong, C.H.L.: The Lazy Lambda Calculus: An Investigation into the Foundations of Functional Programming. Ph.D. thesis, Imperial College London (1988), http://hdl.handle.net/10044/1/47211
48. Patrignani, M., Martin, E.M., Devriese, D.: On the semantic expressiveness of recursive types. In: 48th ACM SIGPLAN Symposium on Principles of Programming Languages (POPL 2021). Proc. ACM Program. Lang., vol. 5. ACM (2021). https://doi.org/10.1145/3434302
49. Pierce, B.C.: Types and programming languages. MIT Press (2002)
50. Pitts, A.M.: Reasoning about local variables with operationally-based logical relations. In: 11th Annual IEEE Symposium on Logic in Computer Science (LICS 1996). pp. 152–163. IEEE Computer Society (1996). https://doi.org/10.1109/LICS.1996.561314
51. Pitts, A.M.: Relational properties of domains. Information and Computation **127**(2), 66–90 (1996). https://doi.org/10.1006/inco.1996.0052

52. Pitts, A.M.: Parametric polymorphism and operational equivalence. Mathematical Structures in Computer Science **10**(3), 321–359 (2000). https://doi.org/10.1017/S0960129500003066
53. Pitts, A.M., Stark, I.D.B.: Observable properties of higher order functions that dynamically create local names, or: What's new? In: 8th International Symposium on Mathematical Foundations of Computer Science (MFCS 1993). LNCS, vol. 711, pp. 122–141. Springer (1993). https://doi.org/10.1007/3-540-57182-5_8
54. Pitts, A.M., Stark, I.D.B.: Operational reasoning for functions with local state. In: Gordon, A.D., Pitts, A.M. (eds.) Higher Order Operational Techniques in Semantics, pp. 227–274. Cambridge University Press, New York, NY, USA (1998)
55. Plotkin, G.D.: Lambda-definability and logical relations. Tech. rep., University of Edinburgh (1973)
56. Sieber, K.: Reasoning about sequential functions via logical relations. In: Fourman, M.P., Johnstone, P.T., Pitts, A.M. (eds.) Applications of Categories in Computer Science: Proceedings of the London Mathematical Society Symposium, Durham 1991. p. 258–269. London Mathematical Society Lecture Note Series, Cambridge University Press (1992). https://doi.org/10.1017/CBO9780511525902.015
57. Skorstengaard, L.: An Introduction to Logical Relations (2019). https://doi.org/10.48550/arXiv.1907.11133
58. Statman, R.: Logical relations and the typed lambda-calculus. Information and Control **65**(2), 85–97 (1985). https://doi.org/10.1016/S0019-9958(85)80001-2
59. Tait, W.W.: Intensional interpretations of functionals of finite type I. J. Symb. Log. **32**(2), 198–212 (1967). https://doi.org/10.2307/2271658
60. Timany, A., Stefanesco, L., Krogh-Jespersen, M., Birkedal, L.: A logical relation for monadic encapsulation of state: Proving contextual equivalences in the presence of runst. In: 44th ACM SIGPLAN Symposium on Principles of Programming Languages (POPL 2017). Proc. ACM Program. Lang., vol. 2. ACM (2017). https://doi.org/10.1145/3158152
61. Turi, D., Plotkin, G.D.: Towards a mathematical operational semantics. In: 12th Annual IEEE Symposium on Logic in Computer Science (LICS 1997). pp. 280–291 (1997). https://doi.org/10.1109/LICS.1997.614955
62. Urbat, H., Tsampas, S., Goncharov, S., Milius, S., Schröder, L.: Weak similarity in higher-order mathematical operational semantics. In: 38th Annual ACM/IEEE Symposium on Logic in Computer Science (LICS 2023). IEEE Computer Society Press (2023). https://doi.org/10.1109/LICS56636.2023.10175706
63. Wand, M., Culpepper, R., Giannakopoulos, T., Cobb, A.: Contextual equivalence for a probabilistic language with continuous random variables and recursion. In: 23rd ACM SIGPLAN International Conference on Functional Programming (ICFP 2018). Proc. ACM Program. Lang., vol. 2. ACM (2018). https://doi.org/10.1145/3236782

On Basic Feasible Functionals and the Interpretation Method*

Patrick Baillot[1], Ugo Dal Lago[2,3], Cynthia Kop[4], and Deivid Vale[4](✉)

[1] Univ. Lille, CNRS, Inria, Centrale Lille, UMR 9189 CRIStAL, F-59000 Lille, France
patrick.baillot@univ-lille.fr
[2] University of Bologna, Bologna, Italy
ugo.dallago@unibo.it
[3] INRIA Sophia Antipolis, Valbonne, France
[4] Radboud University Nijmegen, Nijmegen, The Netherlands
{c.kop,deividvale}@cs.ru.nl

Abstract. The class of basic feasible functionals (BFF) is the analog of FP (polynomial time functions) for type-2 functionals, that is, functionals that can take (first-order) functions as arguments. BFF can be defined through Oracle Turing machines with running time bounded by second-order polynomials. On the other hand, higher-order term rewriting provides an elegant formalism for expressing higher-order computation. We address the problem of characterizing BFF by higher-order term rewriting. Various kinds of interpretations for *first-order* term rewriting have been introduced in the literature for proving termination and characterizing (first-order) complexity classes. In this paper, we consider a recently introduced notion of cost–size interpretations for higher-order term rewriting and see definitions as ways of computing functionals. We then prove that the class of functionals represented by higher-order terms admitting a certain kind of cost–size interpretation is exactly BFF.

Keywords: Basic Feasible Functions · Higher-Order Term Rewriting · Tuple Interpretations · Computational Complexity

1 Introduction

Computational complexity classes, and in particular those relating to polynomial time and space [20,11] capture the concept of a feasible problem, and as such have been scrutinized with great care by the scientific community in the last fifty years. The fact that even apparently simple problems, such as nontrivial separation between those classes, remain open today has highlighted the need for a comprehensive study aimed at investigating the deep nature of computational complexity. The so-called implicit computational complexity [8,30,33,13,4] fits into this picture, and is concerned with characterizations of complexity classes based on tools from mathematical logic and the theory of programming languages.

* This work is supported by the NWO TOP project "Implicit Complexity through Higher-Order Rewriting", NWO 612.001.803/7571, the NWO VIDI project "Constrained Higher-Order Rewriting and Program Equivalence", NWO VI.Vidi.193.075, and the ERC CoG "Differential Program Semantics", GA 818616.

N. Kobayashi and J. Worrell (Eds.): FoSSaCS 2024, LNCS 14575, pp. 70–91, 2024.
https://doi.org/10.1007/978-3-031-57231-9_4

One of the areas involved in this investigation is certainly that of term rewriting [34], which has proved useful as a tool for the characterization of complexity classes. In particular, the class FP (i.e., of polytime first-order functions) has been characterized through variations of techniques originally introduced for *termination*, e.g., the interpretation method [31,29], path orders [15], or dependency pairs [16]. Some examples of such characterizations can be found in [7,9,10,1,3].

After the introduction of FP, it became clear that the study of computational complexity also applies to *higher-order functionals*, which are functions that take not only data but also other functions as inputs. The pioneering work of Constable [12], Mehlhorn [32], and Kapron and Cook [22] laid the foundations of the so-called higher-order complexity, which remains a prolific research area to this day. Some motivations for this line of work can be found e.g. in computable analysis [24], NP search problems [6], and programming language theory [14].

There have been several proposals for a class of type-two functionals that correctly generalizes FP. However, the most widely accepted one is the class BFF of *basic feasible functionals*. This class can be characterized based on function algebras, similar to Cobham-style, but it can also be described using Oracle Turing machines. The class BFF was then the object of study by the research community, which over the years has introduced a variety of characterizations, e.g., in terms of programming languages with restricted recursion schemes [21,14], typed imperative languages [17,18], and restricted forms of iteration in OTMs [23]. An investigation of higher-order complexity classes employing the higher-order interpretation method (in the context of a pure higher-order functional language) was also proposed in [19]. However, this paper does not provide a characterization of the standard BFF class. Instead, it characterizes a newly proposed class $\mathtt{SFF}_2$ (Safe Feasible Functionals) which is defined as the restriction of BFF to argument functions in FP (see Sect. 4.2 and the conclusion in [19]).

The studies cited above present structurally complex programming languages and logical systems, precisely due to the presence of higher-order functions. It is not currently known whether it is possible to give a characterization of BFF in terms of mainstream concepts of rewriting theory, although the latter has long been known to provide tools for the modeling and analysis of functional programs with higher-order functions [25].

This paper goes precisely in that direction by showing that the interpretation method in the form studied by Kop and Vale [27,26] provides the right tools to characterize BFF. More precisely, we consider a class of higher-order rewriting systems admitting cost–size tuple interpretations (with some mild upper-bound conditions on their cost and size components) and show that this class contains exactly the functionals in BFF. Such a characterization could not have been obtained employing classical integer interpretations as e.g. in [9] because BFF crucially relies on some conditions both on size and on time. This is the main contribution of our paper, formally stated in Theorem 2.

We believe that a benefit of this characterization is that it opens the way to effectively handling programs or executable specifications implementing BFF functions, in full generality. For instance, we expect that such a characterization could be integrated into rewriting-based tools for complexity analysis of term rewriting systems such as e.g. [2].

Our result is proved in two parts. We first prove that if any term rewriting system in this class computes a higher-order functional, then this functional has to be in BFF (*soundness*). Conversely, we prove that all functionals in BFF are computed by this class of rewriting systems (*completeness*). We argue that the key ingredient towards achieving this characterization is the ability to split the dual notions of cost and size given by the usage of tuple interpretations.

2 Preliminaries

2.1 Higher-Order Rewriting

We roughly follow the definition of *simply-typed term rewriting system* [28] (STRS): terms are applicative, and we limit our interest to *second-order* STRSs where all rules have base type. Reductions follow an innermost evaluation strategy.

Let $\mathbb{B}$ be a nonempty set whose elements are called *base types* and range over ι, κ, ν. The set $\mathbb{T}(\mathbb{B})$ of *simple types* over $\mathbb{B}$ is defined by the grammar $\mathbb{T}(\mathbb{B}) := \mathbb{B} \mid \mathbb{T}(\mathbb{B}) \Rightarrow \mathbb{T}(\mathbb{B})$. Types from $\mathbb{T}(\mathbb{B})$ are ranged over by σ, τ, ρ. The $\Rightarrow$ type constructor is right-associative, so we write $\sigma \Rightarrow \tau \Rightarrow \rho$ for $(\sigma \Rightarrow (\tau \Rightarrow \rho))$. Hence, every type σ can be written as $\sigma_1 \Rightarrow \cdots \Rightarrow \sigma_n \Rightarrow \iota$. We may write such types as $\vec{\sigma} \Rightarrow \iota$. The *order* of a type is: $\mathtt{ord}(\iota) = 0$ for $\iota \in \mathbb{B}$ and $\mathtt{ord}(\sigma \Rightarrow \tau) = \max(1 + \mathtt{ord}(\sigma), \mathtt{ord}(\tau))$. A *signature* $\mathbb{F}$ is a triple $(\mathbb{B}, \Sigma, \mathtt{typeOf})$ where $\mathbb{B}$ is a set of base types, Σ is a nonempty set of symbols, and $\mathtt{typeOf} : \Sigma \longrightarrow \mathbb{T}(\mathbb{B})$. For each type σ, we assume given a set $\mathbb{X}_\sigma$ of countably many variables and assume that $\mathbb{X}_\sigma \cap \mathbb{X}_\tau = \emptyset$ if $\sigma \neq \tau$. We let $\mathbb{X}$ denote $\cup_\sigma \mathbb{X}_\sigma$ and assume that $\Sigma \cap \mathbb{X} = \emptyset$.

The set $\mathsf{T}(\mathbb{F}, \mathbb{X})$ — of *terms* built from $\mathbb{F}$ and $\mathbb{X}$ — collects those expressions s for which a judgment $s : \sigma$ can be deduced using the following rules:

$$\text{(ax)}\ \frac{x \in \mathbb{X}_\sigma}{x : \sigma} \qquad \text{(f-ax)}\ \frac{\mathsf{f} \in \Sigma \qquad \mathtt{typeOf}(\mathsf{f}) = \sigma}{\mathsf{f} : \sigma} \qquad \text{(app)}\ \frac{s : \sigma \Rightarrow \tau \qquad t : \sigma}{(s\,t) : \tau}$$

As usual, application of terms is left-associative, so we write $s\,t\,u$ for $((s\,t)\,u)$. Let $\mathtt{vars}(s)$ be the set of variables occurring in s. A term s is *ground* if $\mathtt{vars}(s) = \emptyset$. The *head symbol* of a term $\mathsf{f}\, s_1 \cdots s_n$ is f. We say t is a *subterm* of s (written $s \trianglerighteq t$) if either (a) $s = t$, or (b) $s = s'\, s''$ and $s' \trianglerighteq t$ or $s'' \trianglerighteq t$. It is a *proper subterm* of s if $s \neq t$. For a term s, $\mathtt{pos}(s)$ is the set of *positions* in s: $\mathtt{pos}(x) = \mathtt{pos}(\mathsf{f}) = \{\sharp\}$ and $\mathtt{pos}(s\,t) = \{\sharp\} \cup \{1 \cdot u \mid u \in \mathtt{pos}(s)\} \cup \{2 \cdot u \mid u \in \mathtt{pos}(t)\}$. For $p \in \mathtt{pos}(s)$, the subterm $s|_p$ at position p is given by: $s|_\sharp = s$ and $(s_1\, s_2)|_{i \cdot p} = s_i|_p$.

In this paper, we require that for all $\mathsf{f} \in \Sigma$, $\mathtt{ord}(\mathtt{typeOf}(\mathsf{f})) \leq 2$, so w.l.o.g., $\mathsf{f} : (\vec{\iota_1} \Rightarrow \kappa_1) \Rightarrow \cdots \Rightarrow (\vec{\iota_k} \Rightarrow \kappa_k) \Rightarrow \nu_1 \Rightarrow \cdots \Rightarrow \nu_l \Rightarrow \iota$. Hence, in a fully applied term $\mathsf{f}\, s_1 \ldots s_k\, t_1 \ldots t_l$ we say the s_i are the arguments of type-1 and the t_j are the arguments of type-0 for f. A *substitution* γ is a type-preserving map from variables to terms such that $\{x \in \mathbb{X} \mid \gamma(x) \neq x\}$ is finite. We extend γ to terms as usual: $x\gamma = \gamma(x)$, $\mathsf{f}\gamma = \mathsf{f}$, and $(s\,t)\gamma = (s\gamma)\,(t\gamma)$. A *context* C is a term with a single occurrence of a variable $\Box$; the term $C[s]$ is obtained by replacing $\Box$ by s.

A *rewrite rule* $\ell \to r$ is a pair of terms of the same type such that $\ell = \mathsf{f}\,\ell_1 \cdots \ell_m$ and $\mathtt{vars}(\ell) \supseteq \mathtt{vars}(r)$. It is *left-linear* if no variable occurs more than once in ℓ. A *simply-typed term rewriting system* $(\mathbb{F}, \mathcal{R})$ is a set of rewrite rules $\mathcal{R}$ over $\mathsf{T}(\mathbb{F}, \mathbb{X})$. In this paper, we require that all rules have *base* type. An STRS is *innermost orthogonal* if all rules are left-linear, and for any two distinct rules $\ell_1 \to r_1, \ell_2 \to r_2$, there are no substitutions γ, δ such that $\ell_1\gamma = \ell_2\delta$. A *reducible expression* (redex) is a term of the form $\ell\gamma$ for a rule $\ell \to r$ and substitution γ. The *innermost rewrite relation* induced by $\mathcal{R}$ is defined as follows:

- $\ell\gamma \to_{\mathcal{R}} r\gamma$, if $\ell \to r \in \mathcal{R}$ and $\ell\gamma$ has no proper subterm that is a redex;
- $s\,t \to_{\mathcal{R}} u\,t$, if $s \to_{\mathcal{R}} u$ and $s\,t \to_{\mathcal{R}} s\,u$, if $t \to_{\mathcal{R}} u$.

We write $\to_{\mathcal{R}}^{+}$ for the transitive closure of $\to_{\mathcal{R}}$. An STRS $\mathcal{R}$ is *innermost terminating* if no infinite rewrite sequence $s \to_{\mathcal{R}} t \to_{\mathcal{R}} \ldots$ exists. It is *innermost confluent* if $s \to_{\mathcal{R}}^{+} t$ and $s \to_{\mathcal{R}}^{+} u$ implies that some v exists with $t \to_{\mathcal{R}}^{+} v$ and $u \to_{\mathcal{R}}^{+} v$. It is well-known that innermost orthogonality implies innermost confluence. In this paper, we will typically drop the "innermost" adjective and simply refer to terminating/orthogonal/confluent STRSs.

Example 1. Let $\mathbb{B} = \{\mathsf{nat}\}$ and $\mathsf{0} : \mathsf{nat}, \mathsf{s} : \mathsf{nat} \Rightarrow \mathsf{nat}, \mathsf{add}, \mathsf{mult} : \mathsf{nat} \Rightarrow \mathsf{nat} \Rightarrow \mathsf{nat}$, and $\mathsf{funcProd} : (\mathsf{nat} \Rightarrow \mathsf{nat}) \Rightarrow \mathsf{nat} \Rightarrow \mathsf{nat} \Rightarrow \mathsf{nat}$. We then let $\mathcal{R}$ be given by:

$$\begin{array}{rr} \mathsf{add}\,\mathsf{0}\,y \to y & \mathsf{add}\,(\mathsf{s}\,x)\,y \to \mathsf{s}\,(\mathsf{add}\,x\,y) \\ \mathsf{mult}\,\mathsf{0}\,y \to \mathsf{0} & \mathsf{mult}\,(\mathsf{s}\,x)\,y \to \mathsf{add}\,y\,(\mathsf{mult}\,x\,y) \\ \mathsf{funcProd}\,F\,\mathsf{0}\,y \to y & \mathsf{funcProd}\,F\,(\mathsf{s}\,x)\,y \to \mathsf{funcProd}\,F\,x\,(\mathsf{mult}\,y\,(F\,x)) \end{array}$$

Hereafter, we write $\ulcorner \mathsf{n} \urcorner$ for the term $\mathsf{s}\,(\mathsf{s}\,(\ldots\,\mathsf{0}\ldots))$ with n ss.

2.2 Cost–Size Interpretations

For sets A and B, we write $A \longrightarrow B$ for the set of functions from A to B. A *quasi-ordered set* $(A, \sqsupseteq)$ consists of a nonempty set A and a reflexive and transitive relation $\sqsupseteq$ on A. For quasi-ordered sets $(A_1, \sqsupseteq_1)$ and $(A_2, \sqsupseteq_2)$, we write $A_1 \Longrightarrow A_2$ for the set of functions $f \in A_1 \longrightarrow A_2$ such that $f(x) \sqsupseteq_2 f(y)$ whenever $x \sqsupseteq_1 y$, i.e., $A_1 \Longrightarrow A_2$ is the space of functions that preserve quasi-ordering.

For every $\iota \in \mathbb{B}$, let a quasi-ordered set $(\mathcal{S}_\iota, \sqsupseteq_\iota)$ be given. We extend this to $\mathbb{T}(\mathbb{B})$ by defining $\mathcal{S}_{\sigma \Rightarrow \tau} = (\mathcal{S}_\sigma \Longrightarrow \mathcal{S}_\tau, \sqsupseteq_{\sigma \Rightarrow \tau})$ where $f \sqsupseteq_{\sigma \Rightarrow \tau} g$ iff $f(x) \sqsupseteq_\tau f(x)$ for any $x \in \mathcal{S}_\sigma$. Given a function $\mathcal{J}^{\mathsf{s}}$ mapping $\mathsf{f} \in \Sigma$ to some $\mathcal{J}^{\mathsf{s}}_{\mathsf{f}} \in \mathcal{S}_{\mathtt{typeOf(f)}}$ and a valuation α mapping $x \in \mathbb{X}_\sigma$ to $\mathcal{S}_\sigma$, we can map each term $s : \sigma$ to an element of $\mathcal{S}_\sigma$ naturally as follows: (a) $[\![x]\!]^{\mathsf{s}}_\alpha = \alpha(x)$; (b) $[\![\mathsf{f}]\!]^{\mathsf{s}}_\alpha = \mathcal{J}^{\mathsf{s}}_{\mathsf{f}}$; (c) $[\![s\,t]\!]^{\mathsf{s}}_\alpha = [\![s]\!]^{\mathsf{s}}_\alpha([\![t]\!]^{\mathsf{s}}_\alpha)$.

For every type σ with $\mathtt{ord}(\sigma) \le 2$, we define $\mathcal{C}_\sigma$ as follows: (a) $\mathcal{C}_\kappa = \mathbb{N}$ for $\kappa \in \mathbb{B}$; (b) $\mathcal{C}_{\iota \Rightarrow \tau} = \mathcal{S}_\iota \Longrightarrow \mathcal{C}_\tau$ for $\iota \in \mathbb{B}$; and (c) $\mathcal{C}_{\sigma \Rightarrow \tau} = \mathcal{C}_\sigma \Longrightarrow \mathcal{S}_\sigma \Longrightarrow \mathcal{C}_\tau$ if $\mathtt{ord}(\sigma) = 1$. We want to interpret terms $s : \sigma$ where both σ and all variables occurring in s are of type order either 0 or 1, as is the case for the left- and right-hand side of rules. Thus, we let $\mathcal{J}^{\mathsf{c}}$ be a function mapping $\mathsf{f} \in \Sigma$ to some $\mathcal{J}^{\mathsf{c}}_{\mathsf{f}} \in \mathcal{C}_{\mathtt{typeOf(f)}}$ and assume given, for each type σ, *valuations* $\alpha : \mathbb{X}_\sigma \longrightarrow \mathcal{S}_\sigma$ and $\zeta : \mathbb{X}_\sigma \longrightarrow \mathcal{C}_\sigma$. We then define:

$$\begin{aligned} [\![x\,s_1 \cdots s_n]\!]^{\mathsf{c}}_{\alpha,\zeta} &= \zeta(x)([\![s_1]\!]^{\mathsf{s}}_\alpha, \ldots, [\![s_n]\!]^{\mathsf{s}}_\alpha) \\ [\![\mathsf{f}\,s_1 \cdots s_k\,t_1 \cdots t_n]\!]^{\mathsf{c}}_{\alpha,\zeta} &= \mathcal{J}^{\mathsf{c}}_{\mathsf{f}}([\![s_1]\!]^{\mathsf{c}}_{\alpha,\zeta}, [\![s_1]\!]^{\mathsf{s}}_\alpha, \ldots, [\![s_k]\!]^{\mathsf{c}}_{\alpha,\zeta}, [\![s_k]\!]^{\mathsf{s}}_\alpha, [\![t_1]\!]^{\mathsf{s}}_\alpha, \ldots, [\![t_n]\!]^{\mathsf{s}}_\alpha) \end{aligned}$$

We let $\mathsf{cost}(s)_{\alpha,\zeta} = \sum\{[\![t]\!]^{\mathsf{c}}_{\alpha,\zeta} \mid s \trianglerighteq t$ and t is a non-variable term of base type$\}$. This is all well-defined under our assumptions that all variables have a type of order 0 or 1, and $\mathsf{f} : (\vec{\iota_1} \Rightarrow \kappa_1) \Rightarrow \cdots \Rightarrow (\vec{\iota_k} \Rightarrow \kappa_k) \Rightarrow \nu_1 \Rightarrow \cdots \Rightarrow \nu_l \Rightarrow \iota$. We also define $\mathsf{cost}'(s)_{\alpha,\zeta} = \sum\{[\![t]\!]^{\mathsf{c}}_{\alpha,\zeta} \mid s \trianglerighteq t$ and $t \notin \mathbb{X}$ is of base type not in normal form$\}$.

A *cost–size interpretation* $\mathcal{F}$ for a second order signature $\mathbb{F} = (\mathbb{B}, \Sigma, \mathtt{typeOf})$ is a choice of a quasi-ordered set $\mathcal{S}_\iota$, for each $\iota \in \mathbb{B}$, along with cost- and size-interpretations $\mathcal{J}^{\mathsf{c}}$ and $\mathcal{J}^{\mathsf{s}}$ defined as above. Let $(\mathbb{F}, \mathcal{R})$ be an STRS over $\mathbb{F}$. We say $(\mathbb{F}, \mathcal{R})$ is *compatible* with a cost–size interpretation if for any valuations α and ζ, we have (a) $[\![\ell]\!]^{\mathsf{c}}_{\alpha,\zeta} > \mathsf{cost}(r)_{\alpha,\zeta}$ and (b) $[\![\ell]\!]^{\mathsf{s}}_{\alpha} \sqsupseteq [\![r]\!]^{\mathsf{s}}_{\alpha}$, for all rules $\ell \to r$ in $\mathcal{R}$. In this case we say such cost–size interpretation *orients* all rules in $\mathcal{R}$.

Theorem 1 (Innermost Compatibility). *Suppose $\mathcal{R}$ is an STRS compatible with a cost–size interpretation $\mathcal{F}$, then for any valuations α and ζ we have $\mathsf{cost}'(s)_{\alpha,\zeta} > \mathsf{cost}'(t)_{\alpha,\zeta}$ and $[\![s]\!]^{\mathsf{s}}_{\alpha} \sqsupseteq [\![t]\!]^{\mathsf{s}}_{\alpha}$ whenever $s \to_{\mathcal{R}} t$.*

From compatibility, we have that if $s_0 \to_{\mathcal{R}} \cdots \to_{\mathcal{R}} s_n$, then $n \leq \mathsf{cost}'(s_0)$. Hence, $\mathsf{cost}'(s)$ bounds the *derivation height* of s. This follows from [26, Corollary 34], although we significantly simplified the presentation: the limitation to second-order fully applied rules and the lack of abstraction terms allow us to avoid many of the complexities in [26]. We also adapted it to *innermost* rather than *call-by-value* evaluation. A correctness proof of this version is supplied in [5]. Since α and ζ are universally quantified, we typically omit them, and just write x instead of $\alpha(x)$ and F^{c} instead of $\zeta(F)$.

Example 2. We let $\mathcal{S}_{\mathsf{nat}} = (\mathbb{N}, \geq)$ and assign $\mathcal{J}^{\mathsf{s}}_{0} = 0$ and $\mathcal{J}^{\mathsf{s}}_{\mathsf{s}} = \boldsymbol{\lambda} x.x + 1$, as well as $\mathcal{J}^{\mathsf{c}}_{0} = 0$ and $\mathcal{J}^{\mathsf{c}}_{\mathsf{s}} = \boldsymbol{\lambda} x.0$. This gives us $[\![\ulcorner \mathsf{n} \urcorner]\!]^{\mathsf{s}} = n$ for all $n \in \mathbb{N}$, and $[\![\ulcorner \mathsf{n} \urcorner]\!]^{\mathsf{c}} = \mathsf{cost}(n) = 0$. Now, we let $\mathcal{J}^{\mathsf{s}}_{\mathsf{add}} = \boldsymbol{\lambda} xy.x + y$ and $\mathcal{J}^{\mathsf{s}}_{\mathsf{mult}} = \boldsymbol{\lambda} xy.x * y$; then indeed $[\![\ell]\!]^{\mathsf{s}} \geq [\![r]\!]^{\mathsf{s}}$ for the first four rules of Example 1 (e.g., $[\![\mathsf{mult}\,(\mathsf{s}\,x)\,y]\!]^{\mathsf{s}} = (x+1) * y \geq y + (x * y) = [\![\mathsf{add}\,y\,(\mathsf{mult}\,x\,y)]\!]^{\mathsf{s}}$). Moreover, let us choose $\mathcal{J}^{\mathsf{c}}_{\mathsf{add}} = \boldsymbol{\lambda} xy.x + 1$ and $\mathcal{J}^{\mathsf{c}}_{\mathsf{mult}} = \boldsymbol{\lambda} xy.x * y + x + 1$. Then also $[\![\ell]\!]^{\mathsf{c}} > \mathsf{cost}(r)$ for all rules; for example, $[\![\mathsf{mult}\,(\mathsf{s}\,x)\,y]\!]^{\mathsf{c}} = (x+1) * y + 2 * x + 3 > (y+1) + (x * y + 2 * x + 1) = [\![\mathsf{add}\,y\,(\mathsf{mult}\,x\,y)]\!]^{\mathsf{c}} + [\![\mathsf{mult}\,x\,y]\!]^{\mathsf{c}} = \mathsf{cost}(\mathsf{add}\,y\,(\mathsf{mult}\,x\,y))$. Regarding $\mathsf{funcProd}$, we can orient both rules by choosing $\mathcal{J}^{\mathsf{s}}_{\mathsf{funcProd}} = \boldsymbol{\lambda} Fxy.y * \max(F(x), 1)^{x}$ and $\mathcal{J}^{\mathsf{c}}_{\mathsf{funcProd}} = \boldsymbol{\lambda} FGxy.2 * x * y * \max(F(x), 1)^{x+1} + x * G(x) + 2 * x + 1$. This works due to the monotonicity assumption, which provides, e.g., $G(x+1) \geq G(x)$. (This function is not polynomial, but that is allowed in the general case.)

2.3 Basic Feasible Functionals

We assume familiarity with Turing machines. In this paper, we consider *deterministic multi-tape Turing machines*. Those are, conceptually, machines consisting of a finite set of states, one or more (but a fixed number of) right-infinite *tapes* divided into cells. Each tape is equipped with a tape head that scans the symbols on the tape's cells and may write on it. The head can move to the left or right. Let $W = \{0, 1\}^*$. A k-ary *Oracle Turing Machine* (OTM) is a deterministic multi-tape Turing machine with at least $2k+1$ tapes: one main tape for (input/output),

k designated *query* tapes, and k designated *answer* tapes. It also has k distinct *query states* q_i and k *answer states* a_i.

A computation with a k-ary OTM M requires k fixed *oracle functions* $f_1, \ldots, f_k : W \longrightarrow W$. We write $M_{\vec{f}}$ to denote a run of M with these functions. A run of $M_{\vec{f}}$ on w starts with w written in the main tape. It ends when the machine halts, and yields the word that is written in the main tape as output. As usual, we only consider machines that halt on all inputs. The computation proceeds as usual for non-query states. To query the value of f_i on w, the machine writes w on the corresponding query tape and enters the query state q_i. Then, *in one step*, the machine transitions to the answer state a_i as follows: (a) the query value w written in the query tape for f_i is read; (b) the contents of the answer tape for f_i are changed to $f_i(w)$; (c) the query value w is erased from the query tape; and (d) the head of the answer tape is moved to its first symbol. The *running time* of $M_{\vec{f}}$ on w is the number of steps used in the computation.

A *type-1 function* is a mapping in $W \longrightarrow W$. A *type-2 functional* of rank (k, l) is a mapping in $(W \longrightarrow W)^k \longrightarrow W^l \longrightarrow W$.

Definition 1. *We say an OTM M* **computes** *a type-2 functional Ψ of rank (k, l) iff for all type-1 functions $f_1, \ldots, f_k$ and $x_1, \ldots, x_l \in W$, whenever $M_{f_1,\ldots,f_k}$ is started with $x_1, \ldots, x_l$ written on its main tape (separated by blanks), it halts with $\Psi(f_1, \ldots, f_k, x_1, \ldots, x_l)$ written on its main tape.*

Definition 2. *Let $\{F_1, \ldots, F_k\}$ be a set of* type-1 variables *and $\{x_1, \ldots, x_l\}$ a set of* type-0 variables. *The set $\mathtt{Pol}^2_{\mathbb{N}}[F_1, \ldots, F_k; x_1, \ldots, x_l]$ of* ***second-order polynomials*** *over $\mathbb{N}$ with indeterminates $F_1, \ldots, F_k, x_1, \ldots, x_l$ is generated by:*

$$P, Q := n \mid x \mid P + Q \mid P * Q \mid F(Q)$$

where $n \in \mathbb{N}$, $x \in \{x_1, \ldots, x_l\}$, and $F \in \{F_1, \ldots, F_k\}$.

Notice that a polynomial expression can be viewed as a type-2 functional in the natural way, e.g., $P(F, x) = 3 * F(x) + x$ is a second-order polynomial functional. Given $w \in W$, we write $|w|$ for its length and define the length $|f|$ of $f : W \longrightarrow W$ as $|f| = \boldsymbol{\lambda} n. \max_{|y| \leq n} |f(y)|$. This allows us to define BFF as the class of functionals computable by OTMs with running time bounded by a second-order polynomial.

Definition 3. *A type-2 functional Ψ is in* BFF *iff there exist an OTM M and a second-order polynomial P such that M computes Ψ and for all $\vec{f}$ and $\vec{x}$: the running time of $M_{f_1,\ldots,f_k}$ on $x_1, \ldots, x_l$ is at most $P(|f_1|, \ldots, |f_k|, |x_1|, \ldots, |x_l|)$.*

3 Statement of the Main Result

The main result of this paper roughly states that BFF consists exactly of those type-2 functionals computed by an STRS compatible with a polynomially bounded cost–size tuple interpretation. To formally state this result, we must first define what it means for an STRS to compute a type-2 functional and define precisely the class of cost–size interpretations we are interested in.

Indeed, let us start by encoding words in W as terms. We let $\mathsf{bit}, \mathsf{word} \in \mathbb{B}$ and introduce symbols $\mathsf{o}, \mathsf{i} : \mathsf{bit}$ and $[] : \mathsf{word}$, $:: : \mathsf{bit} \Rightarrow \mathsf{word} \Rightarrow \mathsf{word}$. Then for instance 001 is encoded as the term $::\mathsf{o}\,(::\mathsf{o}\,(::\mathsf{i}\,[]))$. We use the cleaner list-like notation $[\mathsf{o}; \mathsf{o}; \mathsf{i}]$ in practice. Let $\underline{\mathsf{w}}$ denote the term encoding of a word w. Next, we encode type-1 functions as a possibly infinite set of one-step rewrite rules.

Definition 4. *Consider a type-1 function $f : W \longrightarrow W$ and let $\mathsf{S}_f : \mathsf{word} \Rightarrow \mathsf{word}$ be a fresh function symbol. A set of rules $\mathcal{R}_f$* ***defines*** f ***by way of*** S_f *if for each $w \in W$ there is exactly one rule of the form $\mathsf{S}_f\,\underline{\mathsf{w}} \rightarrow \underline{\mathsf{f(w)}}$ in $\mathcal{R}_f$.*

Henceforth, we assume given that our STRS $(\mathbb{F}, \mathcal{R})$ at hand is such that $\mathbb{F}$ contains $\mathsf{o}, \mathsf{i}, [], ::$ typed as above and a distinguished symbol $\mathsf{F} : (\mathsf{word} \Rightarrow \mathsf{word})^k \Rightarrow \mathsf{word}^l \Rightarrow \mathsf{word}$. Given type-1 functions $f_1, \ldots, f_k$, we write $\mathbb{F}_{\vec{f}}$ for $\mathbb{F}$ extended with function symbols $\mathsf{S}_{f_i} : \mathsf{word} \Rightarrow \mathsf{word}$, with $1 \leq i \leq k$, and let $\mathcal{R}_{+\vec{f}} = \mathcal{R} \cup \bigcup_{i=1}^{k} \mathcal{R}_f$. Now we can define the notion of type-2 computability for such STRSs.

Definition 5. *Let $(\mathbb{F}, \mathcal{R})$ be an STRS. We say that* F ***computes*** *the type-2 functional Ψ in $(\mathbb{F}, \mathcal{R})$ iff for all type-1 functions $f_1, \ldots, f_k$ and all $w_1, \ldots, w_l \in W$, $\mathsf{F}\,\mathsf{S}_{f_1} \cdots \mathsf{S}_{f_k}\,\underline{\mathsf{w_1}} \cdots \underline{\mathsf{w_l}} \rightarrow^{+}_{\mathcal{R}_{+\vec{f}}} \underline{\mathsf{u}}$, where $u = \Psi(f_1, \ldots, f_k, w_1, \ldots, w_l)$.*

Next, we define what we mean by polynomially bounded interpretation.

Definition 6. *We say an STRS $(\mathbb{F}, \mathcal{R})$* ***admits*** *a polynomially bounded interpretation iff $(\mathbb{F}, \mathcal{R})$ is compatible with a cost–size interpretation such that:*

- $\mathcal{S}_{\mathsf{word}} = (\mathbb{N}, \geq)$;
- $\mathcal{J}^{\mathsf{c}}_{\mathsf{o}} = \mathcal{J}^{\mathsf{c}}_{\mathsf{i}} = \mathcal{J}^{\mathsf{c}}_{[]} = 0$, $\mathcal{J}^{\mathsf{c}}_{::} = \boldsymbol{\lambda} xy.0$, *and* $\mathcal{J}^{\mathsf{s}}_{::} = \boldsymbol{\lambda} xy.x + y + c$ *for some* $c \geq 1$;
- $\mathcal{J}^{\mathsf{c}}_{\mathsf{F}}$ *is bounded by a polynomial in* $\mathtt{Pol}^2_{\mathbb{N}}[F^{\mathsf{c}}_1, F^{\mathsf{s}}_1, \ldots, F^{\mathsf{c}}_k, F^{\mathsf{s}}_k; x_1, \ldots, x_l]$.

Finally, we can formally state our main result.

Theorem 2. *A type-2 functional Ψ is in* BFF *if and only if there exists a finite orthogonal STRS $(\mathbb{F}, \mathcal{R})$ such that the distinguished symbol* F *computes Ψ in $(\mathbb{F}, \mathcal{R})$ and $\mathcal{R}$ admits a polynomially bounded cost–size interpretation.*

We prove this result in two parts. First, we prove soundness in Section 4 which states that every type-2 functional computed by an STRS as above is in BFF. Then in Section 5 we prove completeness which states that every functional in BFF can be computed by such an STRS. In order to simplify proofs, we only consider type-2 functions of rank (1,1). We claim that the results can be easily generalized, but the proofs become more tedious when handling multiple arguments.

Example 3. Let us consider the type-2 functional defined by $\Psi := \boldsymbol{\lambda} fx. \sum_{i<|x|} f(i)$. Notice that Ψ adds all $f(i)$ over each word $i \in W$ whose value (as a natural number) is smaller than the length of x. This functional was proved to lie in BFF in [21], where the authors utilized an encoding of Ψ as a BTLP_2 program. We can encode Ψ as an STRS as follows. Let us consider ancillary symbols $\mathsf{lengthOf} : \mathsf{word} \Rightarrow \mathsf{nat}$ and $\mathsf{toBin} : \mathsf{nat} \Rightarrow \mathsf{word}$. The former computes the length of a given word and the latter converts a number from unary to binary representation. We also consider rules for addition on binary words, i.e., $+_{\mathsf{B}} : \mathsf{word} \Rightarrow \mathsf{word} \Rightarrow \mathsf{word}$, which we use in infix notation below.

$$\begin{aligned} \mathsf{compute}\, F\, x\, \mathsf{0}\, acc &\to acc \\ \mathsf{compute}\, F\, x\, (\mathsf{s}\, i)\, acc &\to \mathsf{compute}\, F\, x\, i\, (acc +_{\mathsf{B}} F(\mathsf{toBin}\, i)) \\ \mathsf{start}\, F\, x &\to \mathsf{compute}\, F\, x\, (\mathsf{lengthOf}\, x)\, [] \end{aligned}$$

Now, if we want to compute $\Psi(f, x)$ we simply reduce the term $\mathsf{start}\,\mathsf{S}_f\,\underline{\mathsf{x}}$ to normal form. To show that this system is in BFF via our rewriting formalism, we need to exhibit a cost–size tuple interpretation for it that satisfies Definition 6, see [5, Example 3].

4 Soundness

In order to prove soundness, let us consider a fixed finite orthogonal STRS $\mathcal{R}$ admitting a polynomially bounded cost–size interpretation such that it computes a type-2 functional Ψ. We proceed to show that Ψ is in BFF roughly as follows:

1. Since $\mathcal{R}$ computes Ψ and admits a polynomially bounded interpretation, we show that so does the extended system $\mathcal{R}_{+f}$ (Definition 5). The restriction on $\mathcal{J}^{\mathsf{s}}_{::}$ (Definition 6) implies that $[\![\mathsf{F}\,\mathsf{S}_f\,\underline{\mathsf{w}}]\!]^{\mathsf{c}}$ is bounded by a second-order polynomial over $|f|, |w|$. We show this in Lemma 1. By compatibility (Theorem 1), we can do at most polynomially many steps when reducing $\mathsf{F}\,\mathsf{S}_f\,\underline{\mathsf{w}}$.
2. The cost polynomial restricts the size of any input that the function variable F is applied to (e.g., a cost bound of $3 + F^{\mathsf{c}}(m)$ implies that F is never called on a term with size interpretation $> m$). This is the subject of Lemma 3.
3. Using the observations above, we then show that by graph rewriting we can simulate $\mathcal{R}_{+f}$ and compute each $\mathcal{R}_{+f}$-reduction step in polynomial time on an OTM. This guarantees that Ψ is in BFF, Theorem 3.

4.1 Interpreting The Extended STRS, Polynomially

Our first goal is to provide a polynomially bounded cost–size interpretation to the extended system $\mathcal{R}_{+f}$. We start with the observation that the size interpretation of words in W is proportional to their length. Indeed, since $\mathcal{J}^{\mathsf{s}}_{::} = \boldsymbol{\lambda} xy.x + y + c$ (Definition 6) let $\mu := \max(\mathcal{J}^{\mathsf{s}}_{\mathsf{o}}, \mathcal{J}^{\mathsf{s}}_{\mathsf{i}}) + c$ and $\nu := \mathcal{J}^{\mathsf{s}}_{[]}$. Consequently, for all $w \in W$:

$$|w| \leq [\![\underline{\mathsf{w}}]\!]^{\mathsf{s}} \leq \mu * |w| + \nu \tag{1}$$

Recall that by Definition 4 the extended system $\mathcal{R}_{+f}$ has possibly infinitely many rules of the form $\mathsf{S}_f\underline{\mathsf{w}} \to \underline{\mathsf{f(w)}}$. Such rules S_f represent calls for an oracle to compute f in a single step. Thus, we set their cost to 1. The size should be given by the length of the oracle output, taking the overhead of interpretation into account. Hence, we obtain:

$$\mathcal{J}^{\mathsf{c}}_{\mathsf{S}_f} = \boldsymbol{\lambda} x.1 \qquad \mathcal{J}^{\mathsf{s}}_{\mathsf{S}_f} = \boldsymbol{\lambda} x.\mu * |f|(x) + \nu$$

This is weakly monotonic because $|f|$ is. It orients the rules in $\mathcal{R}_f$ because $[\![\mathsf{S}_f\,\underline{\mathsf{w}}]\!]^{\mathsf{c}} = 1 > 0 = \mathsf{cost}(\underline{\mathsf{f}(\mathsf{w})})$, and $[\![\mathsf{S}_f\,\underline{\mathsf{w}}]\!]^{\mathsf{s}} = \mu * |f|([\![\underline{\mathsf{w}}]\!]^{\mathsf{s}}) + \nu \geq \mu * |f|(|w|) + \nu \geq \mu * |f(w)| + \nu$ by definition of $|f|$, which is superior or equal to $[\![\underline{\mathsf{f}(\mathsf{w})}]\!]^{\mathsf{s}}$.

As $\mathcal{J}^{\mathsf{c}}_{\mathsf{F}}$ is bounded by a second-order polynomial $\boldsymbol{\lambda} F^{\mathsf{c}} F^{s} x.P$, we can let $D(F, n) := P(\boldsymbol{\lambda} x.1, \boldsymbol{\lambda} x.\mu{*}F(x){+}\nu, \mu{*}n{+}\nu)$. Then D is a second-order polynomial, and $D(|f|, |w|) \geq \mathcal{J}^{\mathsf{c}}_{\mathsf{F}}(\mathcal{J}^{\mathsf{c}}_{\mathsf{S}_f}, \mathcal{J}^{\mathsf{s}}_{\mathsf{S}_f}, [\![\underline{\mathsf{w}}]\!]^{\mathsf{s}}) = \mathsf{cost}(\mathsf{F}\,\mathsf{S}_f\,\underline{\mathsf{w}})$. By Theorem 1 we see:

Lemma 1. *There exists a second-order polynomial D so that $D(|f|, |w|)$ bounds the derivation height of $\mathsf{F}\,\mathsf{S}_f\,\underline{\mathsf{w}}$ for any $f \in W \longrightarrow W$ and $w \in W$.*

Notice that this lemma does not imply that Ψ is in BFF. It only guarantees that there is a polynomial bound to the number of rewriting steps for such systems. However, it does not immediately follow that this number is a reasonable bound for the actual computational cost of simulating a reduction on an OTM. Consider for example a rule $\mathsf{f}\,(\mathsf{s}\,n)\,t \to \mathsf{f}\,n\,(\mathsf{c}\,t\,t)$. Every step doubles the size of the term. A naive implementation – which copies the duplicated term in each step – would take exponential time. Moreover, a single step using the oracle can create a very large output, which is not considered part of the cost of the reduction, even though an OTM would be unable to use it without first fully reading it.

Therefore, in order to prove soundness, we show how to realize a reasonable implementation of rewriting w.r.t. OTMs. In essence, we will show that (1) oracle calls are not problematic in the presence of polynomially bounded interpretations, and (2) we can handle duplication with an appropriate representation of rewriting.

4.2 Bounding The Oracle Input

We first deal with the reasonability of oracle calls. We will show that there exists a second-order polynomial B such that if an oracle call $\mathsf{S}_f\,\underline{\mathsf{x}}$ occurs anywhere along the reduction $\mathsf{F}\,\mathsf{S}_f\,\underline{\mathsf{w}} \to^{+}_{\mathcal{R}} \underline{\mathsf{v}}$, then $|x| \leq B(|f|, |w|)$. From this, we know that the growth of the overall term size during an oracle call is at most $|f|(B(|f|, |w|))$.

Let P again be the polynomial bounding $\mathcal{J}^{\mathsf{c}}_{\mathsf{F}}$. Since P is a second-order polynomial, each occurrence of a sub-expression $F^{\mathsf{c}}(E)$ in P is a second-order polynomial, and so is E. Let us enumerate these arguments as $E_1, \ldots, E_n$. We can then form the new polynomial Q defined as

$$Q := \sum_i E_i \quad \text{where occurrences of } F^{\mathsf{c}}(E'_j) \text{ inside } E_i \text{ are replaced by } 1$$

We let $B(G, y) := Q(\boldsymbol{\lambda} z.\mu * G(z) + \nu, \mu * y + \nu)$.

Example 4. If $P = \boldsymbol{\lambda} F^{\mathsf{c}} F^{s} x.x * F^{\mathsf{c}}(3 + F^{\mathsf{s}}(9 * x)) + F^{\mathsf{c}}(12) * F^{\mathsf{c}}(3 + x * F^{\mathsf{c}}(2)) + 5$, then $Q = 3 + F^{\mathsf{s}}(9 * x) + 12 + 3 + x * 1 + 2 = 20 + F^{\mathsf{s}}(9 * x) + x$. We have $B(G, x) = 20{+}\mu{*}G(9{*}(\mu{*}x{+}\nu)){+}\nu{+}(\mu{*}x{+}\nu) = 20{+}2{*}\nu{+}G(9{*}\mu{*}x{+}9{*}\nu){+}\mu{*}x$.

Now B gives an upper bound to the argument values for F^{c} that are considered: if a function differs from $\mathcal{J}^{\mathsf{c}}_{\mathsf{S}_f}$ only on argument values greater than $B(|f|, |w|)$, then we can use it in P and obtain the same result. Formally:

Lemma 2. *Fix f, w. Let $G \in \mathbb{N} \longrightarrow \mathbb{N}$ with $G(z) = 1$ if $z \leq B(|f|, |w|)$. Then $P(G, \mathcal{J}^{\mathsf{s}}_{\mathsf{S}_f}, [\![\underline{\mathsf{w}}]\!]^{\mathsf{s}}) = P(\mathcal{J}^{\mathsf{c}}_{\mathsf{S}_f}, \mathcal{J}^{\mathsf{s}}_{\mathsf{S}_f}, [\![\underline{\mathsf{w}}]\!]^{\mathsf{s}})$.*

This is proved by induction on the form of P, using that G is never applied on arguments larger than $B(|f|, |w|)$. Lemma 2 is used in the following key result:

Lemma 3 (Oracle Subterm Lemma). *Let $f : W \longrightarrow W$ be a type-1 function and $w \in W$. If $\mathsf{F}\,\mathsf{S}_f\,\underline{\mathsf{w}} \rightarrow^*_{\mathcal{R}_{+f}} C[\mathsf{S}_f\,\underline{\mathsf{x}}]$ for some context C, then $|x| \leq B(|f|, |w|)$.*

Proof. In view of a contradiction, suppose there exist f, w, and x such that $\mathsf{F}\,\mathsf{S}_f\,\underline{\mathsf{w}} \rightarrow^*_{\mathcal{R}_{+f}} C[\mathsf{S}_f\,\underline{\mathsf{x}}]$ for some context C, and $|x| > B(|f|, |w|)$. Let us now construct an alternative oracle: let $\mathsf{0} : \mathsf{nat}, \mathsf{s} : \mathsf{nat} \Rightarrow \mathsf{nat}, \mathsf{S}'_f : \mathsf{word} \Rightarrow \mathsf{word}$ and $\mathsf{helper} : \mathsf{nat} \Rightarrow \mathsf{nat} \Rightarrow \mathsf{nat}$, and for $N := D(|f|, |w|)$, let $\mathcal{R}'_{f,w}$ be given by:

$$\begin{array}{lll} \mathsf{S}'_f\,\underline{\mathsf{x}} \rightarrow \underline{\mathsf{f(x)}} & \text{if } |x| \leq B(|f|, |w|) & \mathsf{helper}\,\mathsf{0}\,y \rightarrow y \\ \mathsf{S}'_f\,\underline{\mathsf{x}} \rightarrow \mathsf{helper}\,\ulcorner\mathsf{N}\urcorner\,\underline{\mathsf{f(x)}} & \text{otherwise} & \mathsf{helper}\,(\mathsf{s}\,x)\,y \rightarrow \mathsf{helper}\,x\,y \end{array}$$

Where $\ulcorner\mathsf{N}\urcorner$ is the unary number encoding of N as introduced in Section 2.1. Notice that by definition, the rules for S'_f will produce $\underline{\mathsf{f(x)}}$ in one step if $|x| \leq B(|f|, |w|)$, but they will take $N + 2$ steps otherwise. Also observe that S_f and S'_f behave the same; that is, $\mathsf{S}_f\,\underline{\mathsf{x}}$ and $\mathsf{S}'_f\,\underline{\mathsf{x}}$ have the same normal form on any input $\underline{\mathsf{x}}$. We extend the interpretation function of the original signature with:

$$\mathcal{J}^{\mathsf{c}}_{\mathsf{S}'_f} = \boldsymbol{\lambda} x. \begin{cases} 1 & \text{if } x \leq B(|f|, |n|) \\ N + 2 & \text{if } x > B(|f|, |n|) \end{cases} \qquad \mathcal{J}^{\mathsf{s}}_{\mathsf{S}'_f} = \mathcal{J}^{\mathsf{s}}_{\mathsf{S}_f}(y)$$

$$\mathcal{J}^{\mathsf{c}}_{\mathsf{helper}} = \boldsymbol{\lambda} xy.x + 1 \qquad \mathcal{J}^{\mathsf{s}}_{\mathsf{helper}} = \boldsymbol{\lambda} xy.y \qquad \mathcal{J}^{\mathsf{s}}_{\mathsf{0}} = 0 \qquad \mathcal{J}^{\mathsf{s}}_{\mathsf{s}} = \boldsymbol{\lambda} x.x + 1$$

We easily see that this orients all rules in $\mathcal{R}_{f,w}$. Then, by Lemma 2, $\mathsf{cost}(\mathsf{F}\,\mathsf{S}'_f\,\underline{\mathsf{w}}) \leq P(\mathcal{J}^{\mathsf{c}}_{\mathsf{S}'_f}, \mathcal{J}^{\mathsf{s}}_{\mathsf{S}'_f}, [\![\underline{\mathsf{w}}]\!]^{\mathsf{s}}) = P(\mathcal{J}^{\mathsf{c}}_{\mathsf{S}_f}, \mathcal{J}^{\mathsf{s}}_{\mathsf{S}_f}, [\![\underline{\mathsf{w}}]\!]^{\mathsf{s}}) \leq D(|f|, |w|) = N$. Yet, as we have $\mathsf{F}\,\mathsf{S}_f\,\underline{\mathsf{w}} \rightarrow^*_{\mathcal{R}_{+f}} C[\mathsf{S}_f\,\underline{\mathsf{x}}]$, we also have $\mathsf{F}\,\mathsf{S}_f\,\underline{\mathsf{w}} \rightarrow_{\mathcal{R} \cup \mathcal{R}'_{f,w}} C'[\mathsf{S}'_f\,\underline{\mathsf{x}}]$, where C' is obtained from C by replacing all occurrences of S_f by S'_f. Since $|x| > B(|f|, |w|)$ by assumption, the reduction $\mathsf{F}\,\mathsf{S}'_f\,\underline{\mathsf{w}} \rightarrow^*_{\mathcal{R} \cup \mathcal{R}'_{f,w}} C[\mathsf{S}'_f\,\underline{\mathsf{w}}] \rightarrow^*_{\mathcal{R} \cup \mathcal{R}_{f,w'}} C[\underline{\mathsf{f(x)}}]$ takes strictly more than N steps, contradicting Theorem 1. □

4.3 Graph Rewriting

Lemma 1 guarantees that if $\mathcal{R}$ is compatible with a suitable interpretation, then at most polynomially many $\mathcal{R}_{+f}$-steps can be performed starting in $\mathsf{F}\,\mathsf{S}_f\,\underline{\mathsf{w}}$. However, as observed in Section 4.1, this does not yet imply that a type-2 functional computed by an STRS with such an interpretation is in BFF. To simulate a reduction on an OTM, we must find a representation whose size does not increase too much in any given step. The answer is *graph rewriting*.

Definition 7. *A **term graph** for a signature Σ is a tuple $(V, \mathtt{label}, \mathtt{succ}, \Lambda)$ with V a finite nonempty set of vertices; $\Lambda \in V$ a designated vertex called the* root*; $\mathtt{label} : V \longrightarrow \Sigma \cup \{@\}$ a partial function with $@$ fresh; and $\mathtt{succ} : V \longrightarrow V^*$*

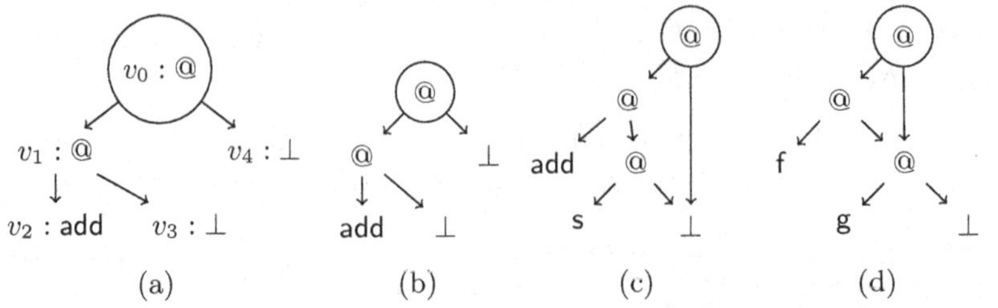

Fig. 1: A term graph, its simplified version, and two graphs with sharing

a total function such that $\texttt{succ}(v) = v_1v_2$ *when* $\texttt{label}(v) = @$ *and* $\texttt{succ}(v) = \varepsilon$ *otherwise. We view this as a directed graph, with an edge from v to v' if $v' \in$* $\texttt{succ}(v)$*, and require that this graph is* acyclic *(i.e., there is no path from any v to itself). Given term graph G, we will often directly refer to* $V_G, \texttt{label}_G$*, etc.*

Term graphs can be denoted visually in an intuitive way. For example, using Σ from Example 1, the graph with $V = \{v_0, \dots, v_4\}$, $\texttt{label} = \{v_0, v_1 \mapsto @,\ v_2 \mapsto \textsf{add}\}$, $\texttt{succ} = \{v_0 \mapsto v_1v_4,\ v_1 \mapsto v_2v_3,\ v_3, v_4, v_5 \mapsto \varepsilon\}$ and $\Lambda = v_0$ is pictured in Figure 1a. We use $\bot$ to indicate unlabeled vertices and a circle for Λ. We will typically omit vertex names, as done in Figure 1b. Note that the definition allows multiple vertices to have the same vertex as successor; these successor vertices with in-degree > 1 are *shared.* Two examples are denoted in Figures 1c and 1d.

Each term has a natural representation as a tree. Formally, for a term s we let $[s]_{\mathbb{G}} = (\texttt{pos}(s), \texttt{label}, \texttt{succ}, \sharp)$ where $\texttt{label}(p) = @$ if $s|_p = s_1s_2$ and $\texttt{label}(p) = \mathsf{f}$ if $s|_p = \mathsf{f}$; $\texttt{label}(p)$ is not defined if $s|_p$ is a variable; and $\texttt{succ}(p) = (1 \cdot p)(2 \cdot p)$ if $s|_p = s_1\, s_2$ and $\texttt{succ}(p) = \varepsilon$ otherwise. Essentially, $[s]_{\mathbb{G}}$ maintains the positioning structure of s and forgets variable names. For example, Figure 1b denotes both $[\textsf{add}\, x\, y]_{\mathbb{G}}$ and $[\textsf{add}\, x\, x]_{\mathbb{G}}$.

Our next step is to *reduce* term graphs using rules. We limit interest to *left-linear* rules, which includes all rules in $\mathcal{R}_{+f}$ (as $\mathcal{R}$ is orthogonal, and the rules in $\mathcal{R}_f$ are ground). To define reduction, we will need some helper definitions.

Definition 8. *Let* $G = (V, \texttt{label}, \texttt{succ}, \Lambda), v \in V$*. The* ***subgraph*** $\textsf{reach}(G, v)$ *of G rooted at v is the term graph* $(V', \texttt{label}', \texttt{succ}', v)$ *where V' contains those $v' \in V$ such that a path from v to v' exists, and* $\texttt{label}', \texttt{succ}'$ *are respectively the limitations of* $\texttt{label}$ *and* $\texttt{succ}$ *to V'.*

Definition 9. *A* ***homomorphism*** *between two term graphs G and H is a function* $\phi : V_G \longrightarrow V_H$ *with* $\phi(\Lambda_G) = \Lambda_H$*, and for $v \in V_G$ such that* $\texttt{label}_G(v)$ *is defined,* $\texttt{label}_H(\phi(v)) = \texttt{label}_G(v)$ *and* $\texttt{succ}_H(\phi(v)) = \phi(v_1) \dots \phi(v_k)$ *when* $\texttt{succ}_G(v) = v_1 \dots v_k$*. (If* $\texttt{label}_G(v)$ *is undefined,* $\texttt{succ}_H(\phi(v))$ *may be anything.)*

Definition 10. *A* ***redex*** *in G is a triple (ρ, v, ϕ) consisting of some rule* $\rho = \ell \to r \in \mathcal{R}_{+f}$*, a vertex v in V_G, and a homomorphism* $\phi : [\ell]_{\mathbb{G}} \longrightarrow \textsf{reach}(G, v)$*.*

Definition 11. *Let G be a term graph and v_1, v_2 vertices in G. The* **redirection** *of v_1 to v_2 is the term graph $G[v_1 \gg v_2] \equiv (V_G, \mathtt{label}_G, \mathtt{succ}_{G'}, \Lambda'_G)$ with*

$$\mathtt{succ}_{G'}(v)_i = \begin{cases} v_2, & \textit{if } \mathtt{succ}_G(v)_i = v_1 \\ \mathtt{succ}_G(v)_i, & \textit{otherwise} \end{cases} \qquad \Lambda'_G = \begin{cases} v_2 & \textit{if } \Lambda_G = v_1 \\ \Lambda_G & \textit{otherwise} \end{cases}$$

That is, we replace every reference to v_1 by a reference to v_2. With these definitions in hand, we can define *contraction* of term graphs:

Definition 12. *Let G be a term graph, and (ρ, v, ϕ) a redex in G with $\rho \in \mathcal{R}_{+f}$, such that no other vertex v' in $\mathsf{reach}(G, v)$ admits a redex (so v is an* innermost redex position*). Denote a_x for the position of variable x in ℓ, and recall that a_x is a vertex in $[\ell]_{\mathbb{G}}$. By left-linearity, a_x is unique for $x \in \mathtt{vars}(\ell)$. The* **contraction** *of (ρ, v, ϕ) in G is the term graph J produced after the following steps: H (building), I (redirection), and J (garbage collection).*

(building) *Let $H = (V_H, \mathtt{label}_H, \mathtt{succ}_H, \Lambda_G)$ where:*
- *$V_H = V_G \uplus \{\overline{p} \in \mathtt{pos}(r) \mid r|_p$ is not a variable$\}$ ($\uplus$ means disjoint union);*
- *for $v \in V_G$: $\mathtt{label}_H(v) = \mathtt{label}_G(v)$ and $\mathtt{succ}_H(v) = \mathtt{succ}_G(v)$*
- *for $p \in V_H$ with $r|_p$ not a variable:*
  - *$\mathtt{label}_H(\overline{p}) = \mathsf{f}$ if $r|_p = \mathsf{f}$ and $\mathtt{label}_H(\overline{p}) = @$ otherwise*
  - *$\mathtt{succ}_H(\overline{p}) = \varepsilon$ if $r|_p = \mathsf{f}$; otherwise, $\mathtt{succ}_H(\overline{p}) = \psi(1 \cdot p)\psi(2 \cdot p)$*
    Here, $\psi(q) = \overline{q}$ if $r|_q$ is not a variable; if $r|_q = x$ then $\psi(q) = \phi(a_x)$.

(redirection) *If r is a variable x (so $H = G$), then let $I = G[v \gg \phi(a_x)]$. Otherwise, let $I = H[v \gg \overline{\sharp}]$, so with all references to v redirected to the root vertex for r.*

(garbage collection) *Let $J := \mathsf{reach}(I, \Lambda_I)$ (so remove unreachable vertices).*

We then write $G \rightsquigarrow J$ in one step, and $G \rightsquigarrow^n J$ for the n-step reduction.

We illustrate this with two examples. First, we aim to rewrite the graph of Figure 2a with a rule $\mathsf{add}\,0\,y \to y$ at vertex v. Since the right-hand side is a variable, the building phase does nothing. The result of the redirection phase is given in Figure 2b, and the result of the garbage collection in Figure 2c.

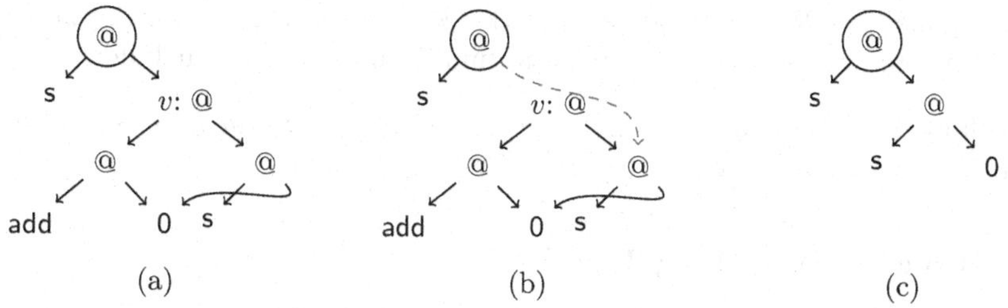

Fig. 2: Reducing a graph with the rule $\mathsf{add}\,0\,y \to y$

Second, we consider a reduction by $\mathsf{mult}\,(\mathsf{s}\,x)\,y \to \mathsf{add}\,y\,(\mathsf{mult}\,x\,y)$. Figure 3a shows the result of the building phase, with the vertices and edges added during this phase in red. Redirection sets the root to the squared node (the root of the right-hand side), and the result after garbage collection is in Figure 3b.

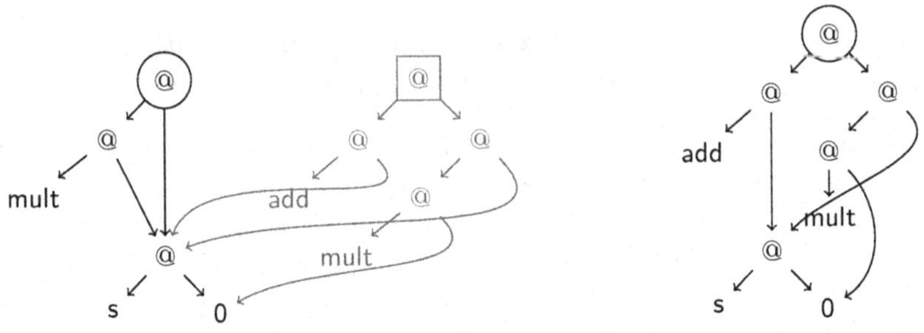

(a) (b)

Fig. 3: Reducing a term graph with substantial sharing

Note that, even when a term graph G is not a tree, we can find a corresponding term: we assign a variable $var(v)$ to each unlabeled vertex v in G, and let:

$$\theta(v) = \begin{cases} \theta(v_1)\,\theta(v_2) & \text{if } \mathtt{label}(v) = @ \text{ and } \mathtt{succ}(v) = v_1 v_2 \\ \mathsf{f} & \text{if } \mathtt{label}(v) = \mathsf{f} \\ var(v) & \text{if } \mathtt{label}(v) \text{ is undefined} \end{cases}$$

Then we may define $[G]_{\mathbb{G}}^{-1} = \theta(\Lambda_G)$. For a linear term, clearly $[[s]_{\mathbb{G}}]_{\mathbb{G}}^{-1} = s$ (modulo variable renaming). We make the following observation:

Lemma 4. *Assume given a term graph G such that there is a path from Λ_G to every vertex in V_G, and let $[G]_{\mathbb{G}}^{-1} = s$. If $G \leadsto H$ then $[G]_{\mathbb{G}}^{-1} \to_{\mathcal{R}_{+f}}^{+} [H]_{\mathbb{G}}^{-1}$. Moreover, if $s \to_{\mathcal{R}_{+f}} t$ for some t, then there exists H such that $G \leadsto H$.*

Consequently, if $\to_{\mathcal{R}_{+f}}$ is terminating, then so is $\leadsto$; and if $[s]_{\mathbb{G}} \leadsto^n G$ for some ground term s then $s \to_{\mathcal{R}_{+f}}^{*} [G]_{\mathbb{G}}^{-1}$ in at least n steps. Notice that if G does not admit any redex, then $[G]_{\mathbb{G}}^{-1}$ is in normal form. Moreover, since $\mathcal{R}_{+f} = \mathcal{R} \cup \mathcal{R}_f$ is orthogonal (as $\mathcal{R}$ is orthogonal and the $\mathcal{R}_f$ rules are non-overlapping) and therefore confluent, this is the *unique* normal form of s. We conclude:

Corollary 1. *If $[\mathsf{F}\,\mathsf{S}_f\,\underline{\mathsf{w}}]_{\mathbb{G}} \leadsto^n G$, then $n \le D(|f|, |w|)$; and if G is in normal form, then $[G]_{\mathbb{G}}^{-1} = \underline{\Psi(\mathsf{f}, \mathsf{w})}$.*

4.4 Bringing Everything Together

We are now ready to complete the soundness proof following the recipe at the start of the section. Towards the third bullet point, we make the following observation.

Lemma 5. *There is a constant a such that, whenever $G \rightsquigarrow H$ by a rule in $\mathcal{R}$, then $|H| \leq |G| + a$, where $|G|$ denotes the total number of nodes in the graph G.*

Proof. In a step using a rule $\ell \to r$, the number of nodes in the graph can be increased at most by $|[r]_{\mathbb{G}}|$. As there are only finitely many rules in $\mathcal{R}$, we can let a be the number of nodes in the largest graph for a right-hand side r. □

To see that graph rewriting with S_f can be implemented in an efficient way, we observe that the size of any intermediary graph in the reduction $[\mathsf{G}\,\underline{\mathsf{w}}]_{\mathbb{G}} \to^{!}_{\mathcal{R}} [q]_{\mathbb{G}}$ is polynomially bounded by a second-order polynomial over $|f|, |w|$:

Lemma 6. *There is a second-order polynomial Q such that if $[\mathsf{F}\,\mathsf{S}_f\,\underline{\mathsf{w}}]_{\mathbb{G}} \rightsquigarrow^* H$, then $|H| \leq Q(|f|, |w|)$.*

Proof. Let $Q(F, x) := x + D(F, x) * (a + F(B(F, x)))$, where D is the polynomial from Lemma 1, a is the constant from Lemma 5, and B is the polynomial from Section 4.2. This suffices, because there are at most $D(|f|, |w|)$ steps (Lemma 1, Corollary corollary 1), each of which increases the graph size by at most $\max(a, |f|(B(|f|, |w|)))$. □

All in all, we are finally ready to prove the *soundness* side of the main theorem:

Theorem 3. *Let $\mathcal{R}$ be a finite orthogonal STRS admitting a polynomially bounded interpretation. If F computes a type-2 functional Ψ, then $\Psi \in$ BFF.*

Proof. Given $(\mathbb{F}, \mathcal{R})$, we can construct an OTM M so that for a given $f \in W \longrightarrow W$, the machine M_f executed on $w \in W$ computes the normal form of $\mathsf{F}\,\mathsf{S}_f\,\underline{\mathsf{w}}$ under $\to_{\mathcal{R}_{+f}}$ using graph rewriting. We omit the exact construction, but observe:

- that we can represent each graph in polynomial space in the size of the graph;
- that we can do a rewriting step that does not call the oracle (so using a rule in $\mathcal{R}$) following the contraction algorithm we defined in Definition 12, which is clearly feasible to do in polynomial time in the size of the graph;
- and that each oracle call (implemented in rewriting by a $\mathcal{R}_f$-step $\mathsf{S}_f\,\underline{\mathsf{x}} \to \underline{\mathsf{y}}$) is resolved by copying $\underline{\mathsf{x}}$ to the query tape, transitioning to the query state, and from the answer state copying $\underline{\mathsf{y}}$ from the answer tape to the main tape. By Lemma 3 this is doable in polynomial time in $|f|, |w|$ and the graph size.

By Lemma 6, graph sizes are bounded by a polynomial over $|f|, |w|$, so using the above reasoning, the same holds for the cost of each reduction step. In summary: the total cost of M_f running on w is bounded by a second-order polynomial in terms of $|f|$ and $|w|$. As M_f simulates $\mathcal{R}_{+f}$ via graph rewriting and $\mathcal{R}_{+f}$ computes Ψ, M also computes Ψ. By Definition 3, Ψ is in BFF. □

5 Completeness

Recall from Section 3 that to prove completeness we have to show the following: if a given type-2 functional Ψ is in BFF, then there exists an orthogonal STRS that computes Ψ and admits a polynomially bounded interpretation. We prove this by providing an encoding of OTMs as STRSs that admit a polynomially bounded interpretation.

The encoding is divided into three steps. In Section 5.1, we will define the function symbols that will allow us to encode any possible machine configuration as terms. In Section 5.2, we will encode transitions as reduction rules that rewrite configuration terms. Lastly, we will design an STRS to simulate a complete execution of an OTM in polynomially many steps. Achieving this polynomial bound is non-trivial and is done in Sections 5.3–5.4.

Henceforth, we assume given a fixed OTM M, and a second-order polynomial P_M, such that M operates in time P_M. For simplicity, we assume the machine has only three tapes (one input/output tape, one query tape, one answer tape); that each non-oracle transition only operates on one tape (i.e., reading/writing and moving the tape head); and that we only have tape symbols $\{0, 1, \mathtt{B}\}$.

5.1 Representing Configurations

Following 3, we have $\mathsf{o}, \mathsf{i} : \mathsf{bit}$, $:: : \mathsf{bit} \Rightarrow \mathsf{word} \Rightarrow \mathsf{word}$ and $[] : \mathsf{word}$. To represent a (partial) tape, we also introduce $\mathsf{b} : \mathsf{bit}$ for the blank symbol. Now for instance a tape with content $011\mathtt{B}01\mathtt{BB}\cdots$ (followed by infinitely many blanks) may be represented as the list $[\mathsf{o}; \mathsf{i}; \mathsf{i}; \mathsf{b}; \mathsf{o}; \mathsf{i}]$ of type word. We may also add an arbitrary number of blanks at the end of the representation; e.g., $[\mathsf{o}; \mathsf{i}; \mathsf{i}; \mathsf{b}; \mathsf{o}; \mathsf{i}; \mathsf{b}; \mathsf{b}]$.

We can think of a *tape configuration* — the combination of a tape and the position of the tape head — as a finite word $w_1 \dots w_{p-1} \# w_p w_{p+1} \dots w_k$ (followed by infinitely many blanks). Here, the tape's head is reading the symbol w_p. We can split this tape into two components: the *left* word $w_1 \dots w_{p-1}$, and the *right* word $w_p \dots w_k$. To represent a tape configuration, we introduce three symbols:

$$\mathsf{L} : \mathsf{word} \Rightarrow \mathsf{left} \qquad \mathsf{R} : \mathsf{word} \Rightarrow \mathsf{right} \qquad \mathsf{split} : \mathsf{left} \Rightarrow \mathsf{right} \Rightarrow \mathsf{tape}$$

Here, L, R hold the content of the left and right split of the tape, respectively. While we technically do not need these two constructors (we could have $\mathsf{split} : \mathsf{word} \Rightarrow \mathsf{word} \Rightarrow \mathsf{tape}$), they serve to make configurations more human-readable. For convenience in rewriting transitions, later on, we will encode the left side of the split in reverse order. Specifically, we encode $w_1 \dots w_{p-1} \# w_p w_{p+1} \dots w_k$ as

$$\mathsf{split}\,(\mathsf{L}\,[w_{p-1}; \dots; w_2; w_1])\,(\mathsf{R}\,[w_p; \dots; w_{k-1}; w_k])$$

The symbol currently being read is the first element of the list below R; in case of $\mathsf{R}\,[]$, this symbol is $\mathtt{B}$. For a concrete example, a tape configuration $1\mathtt{B}0\#10$ is represented by: $\mathsf{split}\,(\mathsf{L}\,[\mathsf{o}; \mathsf{b}; \mathsf{i}])\,(\mathsf{R}\,[\mathsf{i}; \mathsf{o}])$. Since we have assumed an OTM with three tapes, a configuration of the machine at any moment is a tuple (q, t_1, t_2, t_3), with q a state and t_1, t_2, t_3 tape configurations. To represent machine configurations, we introduce, for each state q, a symbol $\mathsf{q} : \mathsf{tape} \Rightarrow \mathsf{tape} \Rightarrow \mathsf{tape} \Rightarrow \mathsf{config}$. Thus, a configuration (q, t_1, t_2, t_3) is represented by a term $\mathsf{q}\,T_1\,T_2\,T_3$.

Example 5. The initial configuration for a machine M_f on input w is a tuple $(q_0, \#w, \#\mathtt{B}, \#\mathtt{B})$. This is represented by the term

$$\mathsf{initial}(w) := \mathsf{q_0}\,(\mathsf{split}\,(\mathsf{L}\,[])\,(\mathsf{R}\,\underline{\mathsf{w}}))\,(\mathsf{split}\,(\mathsf{L}\,[])\,(\mathsf{R}\,[]))\,(\mathsf{split}\,(\mathsf{L}\,[])\,(\mathsf{R}\,[]))$$

To interpret the symbols from this section, we let $(\mathcal{S}_\iota, \sqsupset_\iota) := (\mathbb{N}, \geq)$ for all ι, let $\mathcal{J}^{\mathsf{c}}_{\mathsf{f}} = \boldsymbol{\lambda} x_1 \dots x_m.0$ whenever f takes m arguments, and for the sizes:

$$\begin{array}{lllll} \mathcal{J}^{\mathsf{s}}_{\mathsf{o}} = 0 & \mathcal{J}^{\mathsf{s}}_{\mathsf{b}} = 0 & \mathcal{J}^{\mathsf{s}}_{\mathsf{L}} = \boldsymbol{\lambda} x.x & \mathcal{J}^{\mathsf{s}}_{::} = \boldsymbol{\lambda} xy.x + y + 1 & \mathcal{J}^{\mathsf{s}}_{\mathsf{q}} = \boldsymbol{\lambda} xyz.x + y \\ \mathcal{J}^{\mathsf{s}}_{\mathsf{i}} = 0 & \mathcal{J}^{\mathsf{s}}_{[]} = 0 & \mathcal{J}^{\mathsf{s}}_{\mathsf{R}} = \boldsymbol{\lambda} x.x & \mathcal{J}^{\mathsf{s}}_{\mathsf{split}} = \boldsymbol{\lambda} x.xy.x + y & \text{(for all states } q) \end{array}$$

Hence, $[\![\underline{\mathsf{w}}]\!]^{\mathsf{s}} = |w|$, which satisfies the requirements of Theorem 2; the size of a tape configuration $w_1 \dots w_{p-1}\#w_p \dots w_k$ is k, and the size of a configuration is the size of its first and second tapes combined. We do *not* include the third tape, as it does not directly affect either the result yielded by the final configuration (this is read from the first tape), nor the size of a word the oracle f is applied on.

5.2 Executing The Machine

A single step in an OTM can either be an oracle call (a transition from the `query` state to the `answer` state), or a traditional step: we assume that an OTM M has a fixed set $\mathcal{T}$ of *transitions* $q \xrightarrow[t]{r/i,\ d} l$ where q is the *input state*, l the *output state*, $t \in \{1, 2, 3\}$ the tape considered (recall that we have assumed that a non-oracle transition only operates on one tape), $r, i \in \{0, 1, \mathtt{B}\}$ respectively the symbol being read and the symbol being written, and $d \in \{L, R\}$ the direction for the read head of tape t to move. We will model the computation of M as rules that simulate the small step semantics for the machine.

To encode a single transition, let $\mathsf{step} : (\mathsf{word} \Rightarrow \mathsf{word}) \Rightarrow \mathsf{config} \Rightarrow \mathsf{config}$. For any transition of the form $q \xrightarrow[1]{r/i,\ L} l$ (so a transition operating on tape 1, and moving left), we introduce a rule (where we write $\underline{0} = \mathsf{o}$, $\underline{1} = \mathsf{i}$, $\underline{\mathtt{B}} = \mathsf{b}$):

$$\mathsf{step}\,F\,(\mathsf{q}\,(\mathsf{split}\,(\mathsf{L}\,(x{::}y))\,(\mathsf{R}\,(\underline{\mathsf{r}}{::}z)))\,u\,v) \to \mathsf{l}\,(\mathsf{split}\,(\mathsf{L}\,y)\,(\mathsf{R}\,(x{::}\underline{\mathsf{i}}{::}z)))\,u\,v$$

Moreover, for transitions $q \xrightarrow[1]{\mathtt{B}/w,\ L} l$ (so where $\mathtt{B}$ is read), we add a rule:

$$\mathsf{step}\,F\,(\mathsf{q}\,(\mathsf{split}\,(\mathsf{L}\,(x{::}y))\,(\mathsf{R}\,[]))\,u\,v) \to \mathsf{l}\,(\mathsf{split}\,(\mathsf{L}\,y)\,(\mathsf{R}\,(x{::}\underline{\mathsf{i}}{::}[])))\,u\,v$$

These rules respectively handle the steps where a tape configuration is changed from $u_1 \dots u_{p-1}u_p\#ru_{p+2} \dots u_k$ to $u_1 \dots u_{p-1}\#u_p i u_{p+2} \dots u_k$, and where a tape configuration is changed from $u_1 \dots u_k\#$ to $u_1 \dots \#u_k i$.

Transitions where $d = R$, or on the other two tapes, are encoded similarly.

Next, we encode oracle calls. Recall that, to query the machine for the value of f at u, we write u on the second tape, move its head to the leftmost position, and enter the query state. Then, the content of this tape is erased and the image of f over u is written in the third tape. Visually, this step is represented as:

$$(\mathtt{query}, \langle \mathrm{tape}_1 \rangle, v_1 \dots v_p\#\underline{u}\mathtt{B} \dots, \langle \mathrm{tape}_3 \rangle) \rightsquigarrow (\mathtt{answer}, \langle \mathrm{tape}_1 \rangle, \#\mathtt{B}, \#\underline{\mathsf{f}(\mathsf{u})})$$

This is implemented by the following rules:

$$\mathsf{step}\,F\,(\mathtt{query}\,t_1\,(\mathsf{split}\,x\,(\mathsf{R}\,y))\,t_3) \to \mathtt{answer}\,t_1\,(\mathsf{split}\,(\mathsf{L}\,[])\,(\mathsf{R}\,[]))\\ (\mathsf{split}\,(\mathsf{L}\,[])\,(\mathsf{R}\,(F\,(\mathsf{clean}\,y))))$$

$$\begin{array}{ll} \mathsf{clean}\,(\mathsf{o}{::}x) \to \mathsf{o}{::}(\mathsf{clean}\,x) & \mathsf{clean}\,(\mathsf{b}{::}x) \to [] \\ \mathsf{clean}\,(\mathsf{i}{::}x) \to \mathsf{i}{::}(\mathsf{clean}\,x) & \mathsf{clean}\,[] \to [] \end{array}$$

Here, clean : word $\Rightarrow$ word turns a word that may have blanks in it into a bitstring, by reading until the next blank; for instance replacing [o; i; b; i] by [o; i].

The various step rules, as well as the clean rules, are non-overlapping because we consider *deterministic* OTMs. They are also left-linear, and are oriented using:

$$\begin{array}{ll} \mathcal{J}^{\mathsf{s}}_{\mathsf{clean}} = \boldsymbol{\lambda} x.x & \mathcal{J}^{\mathsf{c}}_{\mathsf{clean}} = \boldsymbol{\lambda} x.x+1 \\ \mathcal{J}^{\mathsf{s}}_{\mathsf{step}} = \boldsymbol{\lambda} Fx.x+1 & \mathcal{J}^{\mathsf{c}}_{\mathsf{step}} = \boldsymbol{\lambda} F^c F^s x.F^c(x)+x+2 \end{array}$$

(Note that $\mathcal{J}^{\mathsf{s}}_{\mathsf{step}}$ is so simple because the size of a configuration does not include the size of the answer tape.) From the rules, the following result is obvious:

Lemma 7. *Let M_f be an OTM and C, C' be machine configurations of M_f such that $C \leadsto C'$. Then* step $\mathsf{S}_f\,[C] \to^+_{\mathcal{R}} [C']$, *where $[C]$ is the term encoding of C.*

5.3 A Bound on the Number of Steps

To generalize from performing a single step of the machine to tracing a full computation on the machine level, the natural idea would be to define rules such as:

$$\begin{array}{r} \mathsf{execute}\,F\,(\mathsf{q}\,x\,y\,z) \to \mathsf{execute}\,F\,(\mathsf{step}(\mathsf{q}\,x\,y\,z)) \text{ for } \mathsf{q} \neq \mathsf{end} \\ \mathsf{execute}\,F\,(\mathsf{end}\,(\mathsf{split}\,(\mathsf{L}\,x)\,(\mathsf{R}\,w))\,y\,z) \to \mathsf{clean}\,w \qquad\qquad\qquad \end{array}$$

Then, reducing execute S_f initial(w) to normal form simulates a full OTM execution of M_f on input w. Unfortunately, this rule does not admit an interpretation, as it may be non-terminating. A solution could be to give execute an additional argument $\ulcorner \mathsf{N} \urcorner$ suggesting an execution in at most N steps; this argument would ensure termination, and could be used to find an interpretation.

The challenge, however, is to compute a bound on the number of steps in the OTM: the obvious thought is to compute $P_M(|f|, |w|)$, but this cannot in general be done in polynomial time because the STRS does not have access to $|f|$: since $|f|(i) = \max\{x \in \mathbb{N} \mid |x| \leq i\}$, there are exponentially many choices for x.

To solve this, and following [22, Proposition 2.3], we observe that it suffices to know a bound for $f(x)$ for only those x on which the oracle is actually questioned. That is, for $A \subseteq W$, let $|f|_A = \boldsymbol{\lambda} n. \max\{|f(x)| \mid x \in A \wedge |x| \leq n\}$. Then:

Lemma 8. *Suppose an OTM M_f runs in time bounded by $P_M(|f|, |w|)$ on input w. If M_f transitions in N steps from its initial state to some configuration C, calling the oracle only on words in $A \subseteq W$, then $N \leq P_M(|f|_A, |w|)$.*

Proof (Sketch). We construct f' with $f'(x) = 0$ if $x \notin A$ and $f'(x) = f(x)$ if $x \in A$. Then $|f'| = |f|_A$, and $M_{f'}$ runs the same on input w as M_f does. □

Now, for A encoded as a term A (using symbols $\emptyset : \mathsf{set}$, $\mathsf{setcons} : \mathsf{word} \Rightarrow \mathsf{set} \Rightarrow \mathsf{set}$), we can compute $|f|_A$ using the rules below, where we use unary integers as in Example 1 ($0 : \mathsf{nat}, \mathsf{s} : \mathsf{nat} \Rightarrow \mathsf{nat}$), and defined symbols $\mathsf{len} : \mathsf{word} \Rightarrow \mathsf{nat}$, $\mathsf{max} : \mathsf{nat} \Rightarrow \mathsf{nat} \Rightarrow \mathsf{nat}$, $\mathsf{limit} : \mathsf{word} \Rightarrow \mathsf{nat} \Rightarrow \mathsf{word}$, $\mathsf{retif} : \mathsf{word} \Rightarrow \mathsf{nat} \Rightarrow \mathsf{word} \Rightarrow \mathsf{word}$, $\mathsf{tryapply} : (\mathsf{word} \Rightarrow \mathsf{word}) \Rightarrow \mathsf{word} \Rightarrow \mathsf{nat} \Rightarrow \mathsf{nat}$, $\mathsf{tryall} : (\mathsf{word} \Rightarrow \mathsf{word}) \Rightarrow \mathsf{set} \Rightarrow \mathsf{nat} \Rightarrow \mathsf{nat}$. By design, $\mathsf{retif}\, \underline{x}\, \ulcorner \mathsf{n} \urcorner\, \underline{\mathsf{y}}$ reduces to $\underline{\mathsf{y}}$ if $|x| \leq n$ and to $[]$ otherwise; $\mathsf{tryapply}\, \mathsf{S}_f\, \underline{\mathsf{x}}\, \ulcorner \mathsf{n} \urcorner$ reduces to the unary encoding of $|F|_{\{x\}}(n)$ and $\mathsf{tryall}\, \mathsf{a}\, \underline{\mathsf{x}}\, \ulcorner \mathsf{n} \urcorner$ yields $|F|_A(n)$.

$$\begin{array}{lll} \mathsf{len}\,[] \to 0 & \mathsf{len}\,(x{::}y) \to \mathsf{s}\,(\mathsf{len}\,y) & \\ \mathsf{max}\,0\,m \to m & \mathsf{max}\,(\mathsf{s}\,n)\,0 \to \mathsf{s}\,n & \mathsf{max}\,(\mathsf{s}\,n)\,(\mathsf{s}\,m) \to \mathsf{s}\,(\mathsf{max}\,n\,m) \\ \mathsf{limit}\,[]\,n \to [] & \mathsf{limit}\,(x{::}y)\,0 \to [] & \mathsf{limit}\,(x{::}y)\,(\mathsf{s}\,n) \to x{::}(\mathsf{limit}\,y\,n) \\ \mathsf{retif}\,[]\,n\,z \to z & \mathsf{retif}\,(x{::}y)\,0\,z \to [] & \mathsf{retif}\,(x{::}y)\,(\mathsf{s}\,n)\,z \to \mathsf{retif}\,y\,n\,z \end{array}$$

$$\begin{array}{ll} & \mathsf{tryapply}\,F\,a\,n \to \mathsf{len}\,(\mathsf{retif}\,a\,n\,(F\,(\mathsf{limit}\,a\,n))) \\ \mathsf{tryall}\,F\,\emptyset\,n \to 0 & \mathsf{tryall}\,F\,(\mathsf{setcons}\,a\,tl)\,n \to \mathsf{max}\,(\mathsf{tryapply}\,F\,a\,n)\,(\mathsf{tryall}\,F\,tl\,n) \end{array}$$

An interpretation is provided in [5]. Importantly, the limit function ensures that, in $\mathsf{tryall}\, F\, n$ we never apply F to a word w with $|w| > n$. Therefore we can let $[\![\mathsf{A}]\!]^{\mathsf{s}} = |A|$, the number of words in A, and have $\mathcal{J}^{\mathsf{s}}_{\mathsf{tryall}} = \boldsymbol{\lambda} Fan.F(n)$ and $\mathcal{J}^{\mathsf{c}}_{\mathsf{tryall}} = \boldsymbol{\lambda} F^c F^s an.1 + a + F^c(n) + 2 * F^s(n) + 2 * n + 6$.

Now, for a given second-order polynomial P, fixed f, n, and a term A encoding a set $A \subseteq W$, we can construct a term $\Theta^P_{\mathsf{S}_f;\ulcorner \mathsf{n} \urcorner;\mathsf{A}}$ that computes $P(|f|_A, n)$ using tryall and the functions $\mathsf{add}, \mathsf{mult}$ from Example 1. By induction on P, we have $[\![\Theta^P_{\mathsf{S}_f;\ulcorner \mathsf{n} \urcorner;\mathsf{A}}]\!]^{\mathsf{s}} = P(|f|, n)$, while its cost is bounded by a polynomial over $|f|, n, |A|$.

5.4 Finalising Execution

Now, we can define execution in a way that can be bounded by a polynomial interpretation. We let $\mathsf{execute} : (\mathsf{word} \Rightarrow \mathsf{word}) \Rightarrow \mathsf{nat} \Rightarrow \mathsf{nnat} \Rightarrow \mathsf{nat} \Rightarrow \mathsf{set} \Rightarrow \mathsf{config} \Rightarrow \mathsf{word}$ and will define rules to reduce expressions $\mathsf{execute}\, F\, n\, m\, z\, a\, c$ where

- F is the function to be used in oracle calls.
- $n - 1$ is a bound on the number of steps that can be done before the next oracle call (or until the machine completes execution).
- m is essentially a natural number that represents the number of steps that have been done so far. We use a new sort nnat with function symbols $\mathsf{o} : \mathsf{nnat}$ and $\mathsf{n} : \mathsf{nnat} \Rightarrow \mathsf{nnat}$ because we will let $\mathcal{S}_{\mathsf{nnat}} = (\mathbb{N}, \leq)$, so ordered in the other direction. This will be essential to find an interpretation for $\mathsf{execute}$.
- z is a unary representation of $|w|$, where w is the input to the OTM.
- c is the current configuration.

Using helper symbols $\mathsf{F}' : (\mathsf{word} \Rightarrow \mathsf{word}) \Rightarrow \mathsf{nat} \Rightarrow \mathsf{config} \Rightarrow \mathsf{word}$, $\mathsf{execute}' : (\mathsf{word} \Rightarrow \mathsf{word}) \Rightarrow \mathsf{nat} \Rightarrow \mathsf{nnat} \Rightarrow \mathsf{nat} \Rightarrow \mathsf{set} \Rightarrow \mathsf{config} \Rightarrow \mathsf{word}$, $\mathsf{extract} : \mathsf{tape} \Rightarrow \mathsf{word}$ and $\mathsf{minus} : \mathsf{nat} \Rightarrow \mathsf{nnat} \Rightarrow \mathsf{nat}$, we introduce the rules:

$$\mathsf{F}\,F\,w \to \mathsf{F}'\,F\,(\mathsf{len}\,w)\,(\mathsf{q_0}\,(\mathsf{split}(\mathsf{L}\,[])\,(\mathsf{R}\,w))\,(\mathsf{split}(\mathsf{L}\,[])\,(\mathsf{R}\,[]))\,(\mathsf{split}(\mathsf{L}\,[])\,(\mathsf{R}\,[])))$$
$$\mathsf{F}'\,F\,z\,c \to \mathsf{execute}\,F\,\Theta^{P_M+1}_{F;z;\emptyset}\,\mathsf{o}\,z\,\emptyset\,c$$

$$\mathsf{execute}\,F\,(\mathsf{s}\,n)\,m\,z\,a\,(\mathsf{q}\,t_1\,t_2\,t_3) \to$$
$$\mathsf{execute}\,F\,n\,(\mathsf{n}\,m)\,z\,(\mathsf{step}\,F\,(\mathsf{q}\,t_1\,t_2\,t_3)) \text{ for } \mathsf{q} \notin \{\mathtt{query}, \mathtt{end}\}$$
$$\mathsf{execute}\,F\,(\mathsf{s}\,n)\,m\,z\,a\,(\mathtt{query}\,t_1\,t_2\,t_3) \to$$
$$\mathsf{execute}'\,F\,n\,(\mathsf{n}\,m)\,z\,(\mathsf{setcons}\,(\mathsf{extract}\,t_2)\,a)\,(\mathtt{query}\,t_1\,t_2\,t_3)$$
$$\mathsf{execute}'\,F\,n\,m\,z\,a\,c \to \mathsf{execute}\,F\,(\mathsf{minus}\,\Theta^{P_M+1}_{F;z;a}\,m)\,m\,z\,a\,(\mathsf{step}\,F\,c)$$
$$\mathsf{execute}\,F\,n\,m\,z\,a\,(\mathtt{end}\,t_1\,t_2\,t_3) \to \mathsf{extract}\,t_1$$
$$\mathsf{extract}\,(\mathsf{split}\,(\mathsf{L}\,x)\,(\mathsf{R}\,y)) \to \mathsf{clean}\,y$$
$$\mathsf{minus}\,x\,\mathsf{o} \to x \qquad \mathsf{minus}\,0\,(\mathsf{n}\,y) \to \mathsf{o} \qquad \mathsf{minus}\,(\mathsf{s}\,x)\,(\mathsf{n}\,y) \to \mathsf{minus}\,x\,y$$

That is, an execution on $\mathsf{FS}_f\,\underline{\mathsf{w}}$ starts by computing the length of w and $P_M(|f|_\emptyset, |w|)$, and uses these as arguments to execute. Each normal transition lowers the number n of steps we are allowed to do and increases the number n of steps we have done. Each oracle transition updates A, and either lowers n by one, or updates it to the new value $P_M(|f|_A, |w|) - m$, since we have already done m steps. Once we read the final state, the answer is read off the first tape.

For the interpretation, note that the unusual size set of nnat allows us to choose $\mathcal{J}^{\mathsf{s}}_{\mathsf{minus}} = \boldsymbol{\lambda} xy.\max(x - y, 0)$ without losing monotonicity. Hence, in every step execute $F\,n\,m\,z\,a\,c$, the value $\max(P_M([\![F]\!]^{\mathsf{s}}, [\![z]\!]^{\mathsf{s}}) + 1 - [\![m]\!]^{\mathsf{s}}, [\![n]\!]^{\mathsf{s}})$ decreases by at least one. Since $[\![\Theta^{P_M+1}F; z; a]\!]^{\mathsf{s}} = P_M([\![F]\!]^{\mathsf{s}}, [\![z]\!]^{\mathsf{s}})$ regardless of a, we can use this component as part of the interpretation. The full interpretation functions for execute and F are long and complex, so we will not supply them here. They can be found in [5]. We will only conclude the other side of Theorem 2:

Theorem 4. *If $\Psi \in$ BFF, then there exists a finite orthogonal STRS $\mathcal{R}$ such that* F *computes Ψ in $\mathcal{R}$ and $\mathcal{R}$ admits a polynomially bounded interpretation.*

6 Conclusions and Future Work

In this paper, we have shown that BFF can be characterized through second-order term rewriting systems admitting polynomially bounded cost–size interpretations. This is arguably the first characterization of the basic feasible functionals purely in terms of rewriting theoretic concepts.

For the purpose of presentation, we have imposed some mild restrictions that we believe are not essential in practice. In future extensions, we can eliminate these restrictions, such as allowing lambda-abstraction, non-base type rules, and higher-order functions (assuming that F is still second-order). We can also allow arbitrary inductive data structures as input.

Another direction we definitely wish to explore is the characterization of polynomial time complexity for functionals of order strictly higher than two. It is well known that the underlying theory in this case becomes less robust than in type-2 complexity. As such, it is not clear which of the existing proposals for complexity classes of higher-order polytime complexity we can hope to capture within our framework.

References

1. Avanzini, M., Moser, G.: Polynomial path orders. Log. Methods Comput. Sci. **9**(4) (2013). https://doi.org/10.2168/LMCS-9(4:9)2013
2. Avanzini, M., Moser, G., Schaper, M.: Tct: Tyrolean complexity tool. In: Chechik, M., Raskin, J. (eds.) Proceedings of TACAS 2016 conference. Lecture Notes in Computer Science, vol. 9636, pp. 407–423. Springer (2016). https://doi.org/10.1007/978-3-662-49674-9_24
3. Baillot, P., Dal Lago, U.: Higher-order interpretations and program complexity. In: Proceedings of CSL 2012. LIPIcs, vol. 16, pp. 62–76. Schloss Dagstuhl - Leibniz-Zentrum für Informatik (2012). https://doi.org/10.4230/LIPICS.CSL.2012.62, A journal version in Information and Computation (248), 2016
4. Baillot, P., De Benedetti, E., Ronchi Della Rocca, S.: Characterizing polynomial and exponential complexity classes in elementary lambda-calculus. Inf. Comput. **261**, 55–77 (2018). https://doi.org/10.1016/J.IC.2018.05.005
5. Baillot, P., Dal Lago, U., Kop, C., Vale, D.: On basic feasible functionals and the interpretation method (2024), https://arxiv.org/abs/2401.12385
6. Beame, P., Cook, S.A., Edmonds, J., Impagliazzo, R., Pitassi, T.: The relative complexity of NP search problems. J. Comput. Syst. Sci. **57**(1), 3–19 (1998). https://doi.org/10.1006/JCSS.1998.1575
7. Beckmann, A., Weiermann, A.: A term rewriting characterization of the polytime functions and related complexity classes. Arch. Math. Log. **36**(1), 11–30 (1996). https://doi.org/10.1007/s001530050054
8. Bellantoni, S.J., Cook, S.A.: A new recursion-theoretic characterization of the polytime functions. Comput. Complex. **2**, 97–110 (1992). https://doi.org/10.1007/BF01201998
9. Bonfante, G., Cichon, A., Marion, J., Touzet, H.: Algorithms with polynomial interpretation termination proof. J. Funct. Program. **11**(1), 33–53 (2001). https://doi.org/10.1017/S0956796800003877
10. Bonfante, G., Marion, J., Moyen, J.: Quasi-interpretations a way to control resources. Theor. Comput. Sci. **412**(25), 2776–2796 (2011). https://doi.org/10.1016/j.tcs.2011.02.007
11. Cobham, A.: The intrinsic computational difficulty of functions. In: Bar-Hillel, Y. (ed.) Logic, Methodology and Philosophy of Science: Proceedings of the 1964 International Congress (Studies in Logic and the Foundations of Mathematics), pp. 24–30. North-Holland Publishing (1965)
12. Constable, R.L.: Type two computational complexity. In: Aho, A.V., Borodin, A., Constable, R.L., Floyd, R.W., Harrison, M.A., Karp, R.M., Strong, H.R. (eds.) Proceedings of the 5th Annual ACM Symposium on Theory of Computing, April 30 - May 2, 1973, Austin, Texas, USA. pp. 108–121. ACM (1973). https://doi.org/10.1145/800125.804041
13. Dal Lago, U., Hofmann, M.: Realizability models and implicit complexity. Theor. Comput. Sci. **412**(20), 2029–2047 (2011). https://doi.org/10.1016/J.TCS.2010.12.025
14. Danner, N., Royer, J.S.: Adventures in time and space. In: Morrisett, J.G., Jones, S.L.P. (eds.) Proceedings of the 33rd ACM SIGPLAN-SIGACT Symposium on Principles of Programming Languages, POPL 2006, Charleston, South Carolina, USA, January 11-13, 2006. pp. 168–179. ACM (2006). https://doi.org/10.1145/1111037.1111053

15. Dershowitz, N.: Orderings for term-rewriting systems. Theor. Comput. Sci. **17**, 279–301 (1982). https://doi.org/10.1016/0304-3975(82)90026-3
16. Giesl, J., Thiemann, R., Schneider-Kamp, P.: The dependency pair framework: Combining techniques for automated termination proofs. In: Baader, F., Voronkov, A. (eds.) Logic for Programming, Artificial Intelligence, and Reasoning, 11th International Conference, LPAR 2004, Montevideo, Uruguay, March 14-18, 2005, Proceedings. Lecture Notes in Computer Science, vol. 3452, pp. 301–331. Springer (2004). https://doi.org/10.1007/978-3-540-32275-7_21
17. Hainry, E., Kapron, B.M., Marion, J., Péchoux, R.: A tier-based typed programming language characterizing feasible functionals. In: Hermanns, H., Zhang, L., Kobayashi, N., Miller, D. (eds.) LICS '20: 35th Annual ACM/IEEE Symposium on Logic in Computer Science, Saarbrücken, Germany, July 8-11, 2020. pp. 535–549. ACM (2020). https://doi.org/10.1145/3373718.3394768
18. Hainry, E., Kapron, B.M., Marion, J., Péchoux, R.: Complete and tractable machine-independent characterizations of second-order polytime. In: Bouyer, P., Schröder, L. (eds.) Foundations of Software Science and Computation Structures - 25th International Conference, FOSSACS 2022, Held as Part of the European Joint Conferences on Theory and Practice of Software, ETAPS 2022, Munich, Germany, April 2-7, 2022, Proceedings. Lecture Notes in Computer Science, vol. 13242, pp. 368–388. Springer (2022). https://doi.org/10.1007/978-3-030-99253-8_19
19. Hainry, E., Péchoux, R.: Theory of higher order interpretations and application to basic feasible functions. Log. Methods Comput. Sci. **16**(4) (2020), https://lmcs.episciences.org/6973
20. Hartmanis, J., Stearns, R.E.: Automata-based computational complexity. Inf. Sci. **1**(2), 173–184 (1969). https://doi.org/10.1016/0020-0255(69)90014-0
21. Irwin, R.J., Royer, J.S., Kapron, B.M.: On characterizations of the basic feasible functionals (part i). J. Funct. Program. **11**(1), 117–153 (2001). https://doi.org/10.1017/s0956796800003841
22. Kapron, B.M., Cook, S.A.: A new characterization of type-2 feasibility. SIAM J. Comput. **25**(1), 117–132 (1996). https://doi.org/10.1137/S0097539794263452
23. Kapron, B.M., Steinberg, F.: Type-two polynomial-time and restricted lookahead. In: Dawar, A., Grädel, E. (eds.) Proceedings of the 33rd Annual ACM/IEEE Symposium on Logic in Computer Science, LICS 2018, Oxford, UK, July 09-12, 2018. pp. 579–588. ACM (2018). https://doi.org/10.1145/3209108.3209124
24. Kawamura, A., Cook, S.A.: Complexity theory for operators in analysis. ACM Trans. Comput. Theory **4**(2), 5:1–5:24 (2012). https://doi.org/10.1145/2189778.2189780
25. Klop, J.W., van Oostrom, V., van Raamsdonk, F.: Combinatory reduction systems: Introduction and survey. Theor. Comput. Sci. **121**(1&2), 279–308 (1993). https://doi.org/10.1016/0304-3975(93)90091-7
26. Kop, C., Vale, D.: Cost-size semantics for call-by-value higher-order rewriting. In: Proc. FSCD. LIPIcs, vol. 260, pp. 15:1–15:19 (2023). https://doi.org/10.4230/LIPIcs.FSCD.2023.15
27. Kop, C., Vale, D.: Tuple interpretations for higher-order complexity. In: Kobayashi, N. (ed.) 6th International Conference on Formal Structures for Computation and Deduction, FSCD 2021, July 17-24, 2021, Buenos Aires, Argentina (Virtual Conference). LIPIcs, vol. 195, pp. 31:1–31:22. Schloss Dagstuhl - Leibniz-Zentrum für Informatik (2021). https://doi.org/10.4230/LIPIcs.FSCD.2021.31
28. Kusakari, K.: On proving termination of term rewriting systems with higher-order variables. IPSJ Transactions on Programming **42**(SIG 7 (PRO 11)), 35–45 (2001), http://id.nii.ac.jp/1001/00016864/

29. Lankford, D.S.: On proving term rewriting systems are noetherian. Memo MTP-3 (1979), https://www.ens-lyon.fr/LIP/REWRITING/TERMINATION/Lankford_Poly_Term.pdf
30. Leivant, D.: A foundational delineation of computational feasiblity. In: Proceedings of the Sixth Annual Symposium on Logic in Computer Science (LICS '91), Amsterdam, The Netherlands, July 15-18, 1991. pp. 2–11. IEEE Computer Society (1991). https://doi.org/10.1109/LICS.1991.151625
31. Manna, Z., Ness, S.: On the termination of Markov algorithms. In: Proceedings of the Third Hawaii International Conference on System Science. pp. 789–792 (1970)
32. Mehlhorn, K.: Polynomial and abstract subrecursive classes. J. Comput. Syst. Sci. **12**(2), 147–178 (1976). https://doi.org/10.1016/S0022-0000(76)80035-9
33. Oitavem, I.: Implicit characterizations of pspace. In: Kahle, R., Schroeder-Heister, P., Stärk, R.F. (eds.) Proof Theory in Computer Science, International Seminar, PTCS 2001, Dagstuhl Castle, Germany, October 7-12, 2001, Proceedings. Lecture Notes in Computer Science, vol. 2183, pp. 170–190. Springer (2001). https://doi.org/10.1007/3-540-45504-3_11
34. Terese: Term rewriting systems, Cambridge tracts in theoretical computer science, vol. 55. Cambridge University Press (2003)

Logic and Proofs

Succinctness of Cosafety Fragments of LTL via Combinatorial Proof Systems

Luca Geatti[1(✉)], Alessio Mansutti[2(✉)], and Angelo Montanari[1(✉)]

[1] University of Udine, Udine, Italy
{luca.geatti,angelo.montanari}@uniud.it
[2] IMDEA Software Institute, Madrid, Spain
alessio.mansutti@imdea.org

Abstract. This paper focuses on succinctness results for fragments of Linear Temporal Logic with Past (LTL) devoid of binary temporal operators like *until*, and provides methods to establish them. We prove that there is a family of *cosafety* languages $(\mathcal{L}_n)_{n\geq 1}$ such that $\mathcal{L}_n$ can be expressed with a *pure future formula* of size $\mathcal{O}(n)$, but it requires formulae of size $2^{\Omega(n)}$ to be captured with *past formulae*. As a by-product, such a succinctness result shows the optimality of the *pastification algorithm* proposed in *[Artale et al., KR, 2023]*.

We show that, in the considered case, succinctness cannot be proven by relying on the classical automata-based method introduced in *[Markey, Bull. EATCS, 2003]*. In place of this method, we devise and apply a *combinatorial proof system* whose deduction trees represent LTL formulae. The system can be seen as a proof-centric (one-player) view on the games used by Adler and Immerman to study the succinctness of CTL.

Keywords: Temporal logics · LTL · Succinctness · Proof systems.

1 Introduction

Linear Temporal Logic with Past (LTL [17,23]) is the *de-facto* standard language for the specification, verification, and synthesis of reactive systems [19]. Concerning these reasoning tasks, two fundamental subsets of LTL-definable languages come into play, namely, *safety* and *cosafety* languages. Safety languages express properties stating that "something bad never happens"; cosafety languages, instead, express the fact that "something good will eventually happen". The crucial feature of cosafety (resp., safety) languages is that checking a *finite prefix* of an infinite trace suffices to establish whether the entire trace belongs (resp., does not belong) to the language. Such an ability of reducing reasoning over infinite words to the finite case plays a fundamental role in lowering the complexity of reasoning tasks [16]. Because of this, while LTL was commonly interpreted over infinite traces, recent work mainly considers its finite trace semantics [8,18,22].

In what follows, given a set of temporal operators S, we write $\mathsf{LTL}[S]$ for the set of all LTL formulae in *negation normal form* whose temporal operators are restricted to those in S. Similarly, we denote with $\mathsf{F}(\mathsf{LTL}[S])$ the set of formulae of the form $\mathsf{F}(\alpha)$, with $\alpha \in \mathsf{LTL}[S]$. Here, F is the *future* modality (a.k.a. *eventually*).

N. Kobayashi and J. Worrell (Eds.): FoSSaCS 2024, LNCS 14575, pp. 95–115, 2024.
https://doi.org/10.1007/978-3-031-57231-9_5

There are two notable syntactic characterizations of the cosafety languages of LTL. The first one is a *pure future* characterization given by the logic LTL[X, U] featuring modalities *next* X and *until* U. The second one is an *eventually pure past*[3] characterisation given by the logic F(pLTL), where pLTL is the *pure past* fragment of LTL, that is, the restriction of LTL to past modalities. Analogous characterizations have been provided for safety languages.

As for applications, F(pLTL) is considered to be much more convenient than LTL[X, U], because, starting from an (eventually) pure past formula of size n, it is possible to build an equivalent deterministic finite automaton of singly exponential size in n [7]. In the case of LTL[X, U], such an automaton may have size doubly exponential in n [16]. This computational advantage of pure past formulae originated a recent line of research that focuses on the *pastification problem*, i.e., the problem of translating an input *pure future* formula for a cosafety (or safety) language into an equivalent *pure past* (equivalently, *eventually pure past*) formula. While the best known algorithm for LTL[X, U] is triply exponential [7], a singly exponential pastification algorithm to transform LTL[X, F] formulae into $\mathsf{F}(\mathsf{LTL}[\mathsf{Y}, \widetilde{\mathsf{Y}}, \mathsf{O}])$ ones has been recently developed in [4]. Here, modalities *yesterday* Y and *once* O are the "temporal reverses" of modalities X and F, respectively, whereas the *weak yesterday* operator $\widetilde{\mathsf{Y}}$ is the dual of Y (we formally define the semantics of all these modalities in Section 2). No super-polynomial lower bounds for these pastification problems are known.

While the above two characterisations of cosafety languages have been thoroughly studied in the last decades in terms of expressiveness [6] and complexity [2], their *succinctness* is still poorly understood. To the best of our knowledge, the only known result is the one in [3] showing that F(pLTL) *can* be exponentially more succinct than LTL[X, U] — note that lower bounds to pastification problems require the opposite direction.[4]

In this paper, we study the succinctness of LTL[F] against $\mathsf{F}(\mathsf{LTL}[\mathsf{Y}, \widetilde{\mathsf{Y}}, \mathsf{O}, \mathsf{H}])$, where H is the dual of O, as well as the succinctness of their *reverse logics* [3], that is, the succinctness of F(LTL[O]) against $\mathsf{LTL}[\mathsf{X}, \widetilde{\mathsf{X}}, \mathsf{F}, \mathsf{G}]$. For these fragments of LTL, we establish the following two results.

Theorem 1. F(LTL[O]) *can be exponentially more succinct than* $\mathsf{LTL}[\mathsf{X}, \widetilde{\mathsf{X}}, \mathsf{F}, \mathsf{G}]$.

Theorem 2. LTL[F] *can be exponentially more succinct than* $\mathsf{F}(\mathsf{LTL}[\mathsf{Y}, \widetilde{\mathsf{Y}}, \mathsf{O}, \mathsf{H}])$.

The two theorems prove an *incomparability result* about the succinctness of the characterizations of cosafety languages in the pure future and eventually pure past fragments of LTL. Theorem 1 and Theorem 2 hold for both the finite and infinite trace semantics of LTL (however, due to lack of space, we report the proof of Theorem 1 only in the case of finite traces). As a corollary, Theorem 2 implies that the pastification algorithm proposed in [4] is optimal.

---

3 "Eventually pure past" refers to formulae of the form $\mathsf{F}(\alpha)$, with α pure past formula.

4 A logic $\mathbb{L}$ *can be exponentially more succinct* than a logic $\mathbb{L}'$ whenever there is a family of languages $(\mathcal{L}_n)_{n\geq 1}$ such that $\mathcal{L}_n$ can be expressed in $\mathbb{L}$ with a formula of size polynomial in n, whereas expressing $\mathcal{L}_n$ in $\mathbb{L}'$ requires formulae of size $2^{\Omega(n)}$.

Corollary 1. *The pastification of* $\mathsf{LTL}[\mathsf{X},\mathsf{F}]$ *into* $\mathsf{F}(\mathsf{LTL}[\mathsf{Y},\widetilde{\mathsf{Y}},\mathsf{O},\mathsf{H}])$ *is in* $2^{\Theta(n)}$.

To prove Theorem 1, we devise and apply a *combinatorial proof system.*[5] Given two sets of finite traces A and B, with the proof system one can establish whether there is a formula φ in $\mathsf{LTL}[\mathsf{X},\widetilde{\mathsf{X}},\mathsf{F},\mathsf{G}]$ that *separates* A from B, that is, φ is satisfied by all traces in A (written $A \models \varphi$) and violated by all traces in B (written $B \perp\!\!\!\perp \varphi$). A proof obtained by applying k rules of the proof system corresponds to the existence of one such separating formula φ of size k.

The proposed combinatorial proof system can be seen as a reformulation in terms of proofs of the games introduced by Adler and Immerman to show that CTL^+ is $\Theta(n)!$ more succinct than CTL [1]. They are two-player games that extend Ehrenfeucht–Fraïssé games for quantifier depth in a way that captures the notion of formula size instead. However, unlike Ehrenfeucht–Fraïssé ones, in Adler–Immerman games one of the two players (the duplicator) has always a trivial strategy. With our proof system, we show that removing the duplicator from the game yields a natural one-player game based on building proofs.

To prove Theorem 1 by applying the proposed proof system, we provide, for every $n \geq 1$, a formula Φ_n in $\mathsf{F}(\mathsf{LTL}[\mathsf{O}])$ of size linear in n and two sets of traces $\mathbf{A}_n$ and $\mathbf{B}_n$ such that $\mathbf{A}_n \models \Phi_n$ and $\mathbf{B}_n \perp\!\!\!\perp \Phi_n$, and then we show that the smallest deduction tree that separates $\mathbf{A}_n$ from $\mathbf{B}_n$ has size at least 2^n. This implies that all formulas of $\mathsf{LTL}[\mathsf{X},\widetilde{\mathsf{X}},\mathsf{F},\mathsf{G}]$ capturing Φ_n are of size at least 2^n.

Once Theorem 1 is established, one can prove Theorem 2 by "reversing" the direction of time, building correspondences between formulae of $\mathsf{LTL}[\mathsf{F}]$ and $\mathsf{FLTL}[\mathsf{O}]$, and between formulae of $\mathsf{F}(\mathsf{LTL}[\mathsf{Y},\widetilde{\mathsf{Y}},\mathsf{O},\mathsf{H}])$ and $\mathsf{LTL}[\mathsf{X},\widetilde{\mathsf{X}},\mathsf{F},\mathsf{G}]$.

In the context of LTL, the main technique to prove "future against past" succinctness discrepancies is arguably the automata method introduced by Markey in [20]. At its core, such a method exploits the fact that pure future formulae of LTL can be translated into nondeterministic Büchi automata of exponential size, and thus no property requiring a doubly exponential size automaton can be represented succinctly. The introduction of our proof system raises the question of whether Markey's method can be applied to establish our succinctness results. We prove that it cannot be used in our context. In order to obtain such a result, the key observation is that, given a cosafety formula $\mathsf{F}\psi$, a *deterministic* Büchi automaton (DBA) for $\mathsf{F}\psi$ of size ℓ, and a prefix Π consisting of k temporal operators among X, F, and G, the minimal DBA for the formula $\Pi\mathsf{F}\psi$ has size polynomial in k and ℓ.

Synopsis. Section 2 introduces the necessary background. Section 3 discusses the languages we use to prove Theorem 1. Section 4 introduces the combinatorial proof system. In Section 5 we prove Theorem 1. In Section 6 we prove Theorem 2 and Corollary 1. The limits of the automata-based method to prove succinctness lower bounds are discussed in Section 7. Related and future work are discussed in Section 8. An extended version of the paper, complete of all proofs, can be found in [13].

[5] We use the term "combinatorial" for our proof system to conform with the terminology from the Workshop "Combinatorial Games in Finite Model Theory", LICS'23.

2 Preliminaries

In this section, we introduce background knowledge on LTL focusing on finite traces. All definitions admit a natural extension to the setting of infinite traces.

Let Σ be a finite alphabet. We denote by Σ^* the set of all finite words over Σ and by Σ^+ the subset of finite non-empty words. We use the term *trace* as a synonym of word. A *language* $\mathcal{L}$ over Σ is a subset of Σ^*. Let $\sigma = \langle w_0, w_1, \dots, w_n \rangle$ be a word in Σ^*. We denote by $|\sigma|$ the *length* of σ, that is, $n+1$. A *position* in σ is an element in the set $\text{pos}(\sigma) := [0, n]$. For every $i \in \text{pos}(\sigma)$, we denote by $\sigma[i] \in \Sigma$ the letter w_i, and by $\sigma[i\rangle$ the word $\langle w_i, \dots, w_n \rangle$. We say that position j of σ has *type* $\tau \in \Sigma$ whenever $\sigma[j] = \tau$. Given two traces σ_1 and σ_2, we write $\sigma_1 \sqsubseteq \sigma_2$ whenever σ_1 is a *suffix* of σ_2, that is, there is $j \in \text{pos}(\sigma_2)$ such that $\sigma_1 = \sigma_2[j\rangle$. Given a word $\sigma' \in \Sigma^*$, we denote the *concatenation of* σ' *to* σ as $\sigma \cdot \sigma'$, or simply $\sigma\sigma'$. Given two languages $\mathcal{L}$ and $\mathcal{L}'$, we define $\mathcal{L} \cdot \mathcal{L}' := \{\sigma \cdot \sigma' \mid \sigma \in \mathcal{L}, \sigma' \in \mathcal{L}'\}$. We sometimes apply the concatenation to a word and a language; in these cases the word is implicitly converted into a singleton language, e.g., $\sigma \cdot \mathcal{L} := \{\sigma\} \cdot \mathcal{L}$. With $A \subseteq_{\mathit{fin}} B$ we denote the fact that A is a *finite* subset of the set B.

Linear Temporal Logic with Past. In the following, we introduce syntax and semantics of Linear Temporal Logic with Past (LTL) restricted to those operators that we are going to use throughout the paper. In particular, we omit the future operators *until* and *release*, and their past counterparts (*since* and *triggers*). Let $\mathcal{AP}$ be a finite set of atomic propositions. The syntax of the formulae over $\mathcal{AP}$ is generated by the following grammar:

$$\begin{aligned} \varphi := p \mid \neg p \mid \varphi \vee \varphi \mid \varphi \wedge \varphi & \qquad \text{Boolean connectives} \\ \mid \mathsf{X}\varphi \mid \widetilde{\mathsf{X}}\varphi \mid \mathsf{F}\varphi \mid \mathsf{G}\varphi & \qquad \text{future operators} \\ \mid \mathsf{Y}\varphi \mid \widetilde{\mathsf{Y}}\varphi \mid \mathsf{O}\varphi \mid \mathsf{H}\varphi & \qquad \text{past operators} \end{aligned}$$

where $p \in \mathcal{AP}$. The temporal operators are respectively called: X, *next*; $\widetilde{\mathsf{X}}$, *weak next*; F, *future*; G, *globally*; Y, *yesterday*; $\widetilde{\mathsf{Y}}$, *weak yesterday*; O, *once*; H, *historically*. For the rest of the paper, we let $\mathbb{OP} := \{\mathsf{X}, \widetilde{\mathsf{X}}, \mathsf{F}, \mathsf{G}, \mathsf{Y}, \widetilde{\mathsf{Y}}, \mathsf{O}, \mathsf{H}\}$.

For every formula φ, we define the *size of* φ, denoted by $\text{size}(\varphi)$, inductively defined as follows: (i) $\text{size}(p) := 1$ and $\text{size}(\neg p) := 1$, (ii) $\text{size}(\otimes\varphi) := \text{size}(\varphi) + 1$, for $\otimes \in \mathbb{OP}$, and (iii) $\text{size}(\varphi_1 \oplus \varphi_2) := \text{size}(\varphi_1) + \text{size}(\varphi_2) + 1$ for $\oplus \in \{\vee, \wedge\}$.

We focus on the interpretation of LTL formulae over *finite non-empty traces* over the alphabet $2^{\mathcal{AP}}$. From now on, we set the alphabet Σ to be $2^{\mathcal{AP}}$. Given a word $\sigma \in \Sigma^+$, the *satisfaction* of a formula φ by σ at time point / position $i \in \text{pos}(\sigma)$, denoted by $\sigma, i \models \varphi$, is defined as follows:

1. $\sigma, i \models p$ iff $p \in \sigma[i]$;
2. $\sigma, i \models \neg p$ iff $p \notin \sigma[i]$;
3. $\sigma, i \models \varphi_1 \vee \varphi_2$ iff $\sigma, i \models \varphi_1$ or $\sigma, i \models \varphi_2$;
4. $\sigma, i \models \varphi_1 \wedge \varphi_2$ iff $\sigma, i \models \varphi_1$ and $\sigma, i \models \varphi_2$;
5. $\sigma, i \models \mathsf{X}\varphi$ iff $i + 1 < |\sigma|$ and $\sigma, i + 1 \models \varphi$;
6. $\sigma, i \models \widetilde{\mathsf{X}}\varphi$ iff either $i + 1 = |\sigma|$ or $\sigma, i + 1 \models \varphi$;

7. $\sigma, i \models \mathsf{F}\varphi$ iff there exists $i \leq j < |\sigma|$ such that $\sigma, j \models \varphi$;
8. $\sigma, i \models \mathsf{G}\varphi$ iff for all $i \leq j < |\sigma|$, it holds $\sigma, j \models \varphi$;
9. $\sigma, i \models \mathsf{Y}\varphi$ iff $i > 0$ and $\sigma, i-1 \models \varphi$;
10. $\sigma, i \models \widetilde{\mathsf{Y}}\varphi$ iff either $i = 0$ or $\sigma, i-1 \models \varphi$;
11. $\sigma, i \models \mathsf{O}\varphi$ iff there exists $0 \leq j \leq i$ such that $\sigma, j \models \varphi$;
12. $\sigma, i \models \mathsf{H}\varphi$ iff for all $0 \leq j \leq i$, it holds $\sigma, j \models \varphi$.

For every formula φ, we say that a trace σ satisfies φ, written $\sigma \models \varphi$, if $\sigma, 0 \models \varphi$. The *language* of φ, denoted by $\mathcal{L}(\varphi)$, is the set of words $\sigma \in \Sigma^+$ such that $\sigma \models \varphi$. Given two formulae φ and ψ, we say that φ is *equivalent* to ψ, written $\varphi \equiv \psi$, whenever $\mathcal{L}(\varphi) = \mathcal{L}(\psi)$.

Fragments of LTL. Given a set of operators $S \subseteq \mathbb{OP}$, we denote by $\mathsf{LTL}[S]$ the set of formulae only using temporal operators from S. When dealing with a concrete S, we omit the curly brackets and write, e.g., $\mathsf{LTL}[\mathsf{X}, \mathsf{F}]$ instead of $\mathsf{LTL}[\{\mathsf{X}, \mathsf{F}\}]$. Whenever S contains only future operators (resp., past operators), the logic $\mathsf{LTL}[S]$ is called a *pure future* (resp., *pure past*) fragment of LTL. Finally, we denote by $\mathsf{F}(\mathsf{LTL}[S])$ (resp., $\mathsf{G}(\mathsf{LTL}[S])$) the set of formulae of the form $\mathsf{F}(\alpha)$ (resp., $\mathsf{G}(\alpha)$), where α is a formula of $\mathsf{LTL}[S]$. A language $\mathcal{L} \subseteq \Sigma^*$ is a *cosafety language* whenever $\mathcal{L} = K \cdot \Sigma^*$, for some $K \subseteq \Sigma^*$. A language $\mathcal{L}$ is a *safety language* whenever its complement $\overline{\mathcal{L}}$ is a cosafety language. For every formula φ in the fragments $\mathsf{LTL}[\mathsf{X}, \mathsf{F}]$ and $\mathsf{F}(\mathsf{LTL}[\mathsf{Y}, \widetilde{\mathsf{Y}}, \mathsf{O}, \mathsf{H}])$, it holds that $\mathcal{L}(\varphi)$ is a cosafety language. Similarly, for every formula φ in the fragments $\mathsf{LTL}[\widetilde{\mathsf{X}}, \mathsf{G}]$ and $\mathsf{G}(\mathsf{LTL}[\mathsf{Y}, \widetilde{\mathsf{Y}}, \mathsf{O}, \mathsf{H}])$, it holds that $\mathcal{L}(\varphi)$ is a safety language.

The pastification problem. Given two sets $S \subseteq \{\mathsf{X}, \widetilde{\mathsf{X}}, \mathsf{F}, \mathsf{G}\}$ and $S' \subseteq \{\mathsf{Y}, \widetilde{\mathsf{Y}}, \mathsf{O}, \mathsf{H}\}$, the *pastification problem for* $\mathsf{LTL}[S]$ *into* $\mathsf{F}(\mathsf{LTL}[S'])$ asks, given an input formula $\varphi \in \mathsf{LTL}[S]$, to return a formula ψ from $\mathsf{F}(\mathsf{LTL}[S'])$ such that $\varphi \equiv \psi$. An algorithm for the pastification problem is said to be of *k-exponential size* (for $k \in \mathbb{N}$ fixed) whenever the output formula ψ is such that $\text{size}(\psi) \in \exp_2^k(\text{poly}(\text{size}(\varphi)))$, where $\exp^k(.)$ is the k-th iteration of the base-2 tetration function given by $\exp^0(n) = n$ and $\exp^{i+1}(n) = 2^{\exp^i(n)}$. In [4], an exponential time, 1-exponential size, pastification algorithm for $\mathsf{LTL}[\mathsf{X}, \mathsf{F}]$ into $\mathsf{F}(\mathsf{LTL}[\mathsf{Y}, \widetilde{\mathsf{Y}}, \mathsf{O}])$ is presented.

Succinctness. Given two sets $S, S' \subseteq \mathbb{OP}$, we say that $\mathsf{LTL}[S]$ *can be exponentially more succinct than* $\mathsf{LTL}[S']$ if there is a family of languages $(\mathcal{L}_n)_{n \geq 1}$ such that, for every $n \geq 1$, $\mathcal{L}_n \subseteq \Sigma_n^+$, for some alphabet Σ_n, and:

- there is $\varphi \in \mathsf{LTL}[S]$ such that $\mathcal{L}(\varphi) = \mathcal{L}_n$ and $\text{size}(\varphi) \in \text{poly}(n)$, and
- for every $\psi \in \mathsf{LTL}[S']$, if $\mathcal{L}(\psi) = \mathcal{L}_n$ then $\text{size}(\psi) \in 2^{\Omega(n)}$.

It is worth noticing that the above-given syntax for LTL is already in *negation normal form*, that is, negation may only appear in front of atomic propositions. Allowing negations to occur freely in the formula neither increase expressiveness nor succinctness, as the grammar above is already closed under dual operators, e.g., $\mathsf{G}\varphi \equiv \neg\mathsf{F}\neg\varphi$, and the size of a formula does not depend on the number of negations occurring in literals. Because of this, all results given in the paper continue to hold when negation is added to the language.

3 A problematic cosafety language for $\mathsf{LTL}[\mathsf{X}, \widetilde{\mathsf{X}}, \mathsf{F}, \mathsf{G}]$

We now describe the property that we will exploit to prove that $\mathsf{F}(\mathsf{LTL}[\mathsf{O}])$ can be exponentially more succinct than $\mathsf{LTL}[\mathsf{X}, \widetilde{\mathsf{X}}, \mathsf{F}, \mathsf{G}]$ (Theorem 1). More precisely, we define a family of $\mathsf{F}(\mathsf{LTL}[\mathsf{O}])$ formulae $(\Phi_n)_{n \geq 1}$ such that, for every $n \geq 1$, Φ_n has size in $\mathcal{O}(n)$ and captures a property requiring a formula of size at least 2^n to be expressed in $\mathsf{LTL}[\mathsf{X}, \widetilde{\mathsf{X}}, \mathsf{F}, \mathsf{G}]$ (as we will see in Section 5).

Let $n \geq 1$. We consider the alphabet of $2n + 2$ distinct atomic propositions $\mathcal{AP} := \{\widetilde{p}, \widetilde{q}\} \cup P \cup Q$, with $P := \{p_1, \ldots, p_n\}$ and $Q := \{q_1, \ldots, q_n\}$. For all $n \geq 1$, the formula Φ_n of $\mathsf{F}(\mathsf{LTL}[\mathsf{O}])$ is defined as follows:

$$\Phi_n := \mathsf{F}\left(\widetilde{q} \wedge \bigwedge_{i=1}^{n} \Big((q_i \wedge \mathsf{O}(\widetilde{p} \wedge p_i)) \vee (\neg q_i \wedge \mathsf{O}(\widetilde{p} \wedge \neg p_i)) \Big) \right).$$

Observe that, for every $n \geq 1$, $\mathrm{size}(\Phi_n)$ belongs to $\mathcal{O}(n)$. The formula Φ_n is satisfied by those traces $\sigma \in \Sigma^+$ where there is a position $j \in \mathrm{pos}(\sigma)$ such that (i) $\widetilde{q} \in \sigma[j]$ and (ii) for every $i \in [1, n]$ there is a position $k_i \in [0, j]$ such that $\widetilde{p} \in \sigma[k_i]$ and $q_i \in \sigma[j]$ *if and only if* $p_i \in \sigma[k_i]$. Notice that each $k_i \in [0, j]$ depends on an index $i \in [1, n]$. Therefore, for distinct $i, j \in [1, n]$ the positions k_i and k_j might differ. This feature is crucial to get a language which has a compact definition in $\mathsf{F}(\mathsf{LTL}[\mathsf{O}])$, but is hard to capture for $\mathsf{LTL}[\mathsf{X}, \widetilde{\mathsf{X}}, \mathsf{F}, \mathsf{G}]$.

As a matter of fact, requiring the various k_i to coincide yields a formula Ψ_n characterising the property: "the trace σ has two positions $j \geq k$ such that $\widetilde{p} \in \sigma[k]$, $\widetilde{q} \in \sigma[j]$ and, for every $i \in [1, n]$, $q_i \in \sigma[j]$ if and only if $p_i \in \sigma[k]$". This formula is known to require exponential size in LTL [20], and therefore in $\mathsf{F}(\mathsf{LTL}[\mathsf{O}])$ as well. In a sense, the asymmetry obtained by relaxing the uniqueness of the position k above is what makes Φ_n being easily expressible in $\mathsf{F}(\mathsf{LTL}[\mathsf{O}])$, but difficult to characterise in $\mathsf{LTL}[\mathsf{X}, \widetilde{\mathsf{X}}, \mathsf{F}, \mathsf{G}]$. The same trick, applied to position j instead of position k, can be used to obtain a family of formulae that can be represented in an exponentially more succinct way in $\mathsf{LTL}[\mathsf{F}]$ than in $\mathsf{F}(\mathsf{LTL}[\mathsf{Y}, \widetilde{\mathsf{Y}}, \mathsf{O}, \mathsf{H}])$. This form of "temporal duality" is what we will ultimately exploit in Section 6 to prove Theorem 2.

The following lemma shows that Φ_n can be expressed in $\mathsf{LTL}[\mathsf{F}]$ (and thus in $\mathsf{LTL}[\mathsf{X}, \widetilde{\mathsf{X}}, \mathsf{F}, \mathsf{G}]$ as well) with a formula of exponential size.

Lemma 1. *For every $n \geq 1$, there is a formula Φ'_n in $\mathsf{LTL}[\mathsf{F}]$ such that $\Phi'_n \equiv \Phi_n$ and $\mathrm{size}(\Phi'_n) < 2^{n+1}(n+2)^2$.*

Proof sketch. Given $\tau \in 2^P$, we write $\overline{\tau}$ for the element of 2^Q such that $p_i \in \tau$ if and only if $q_i \in \overline{\tau}$, for every $i \in [1, n]$. Then, the formula Φ'_n is defined as follows:

$$\Phi'_n := \bigvee_{\tau \in 2^P} \Big(\bigwedge_{p \in \tau} \mathsf{F}(\widetilde{p} \wedge p \wedge \mathsf{F}(\widetilde{q} \wedge \psi_{\overline{\tau}})) \wedge \bigwedge_{p \in P \setminus \tau} \mathsf{F}(\widetilde{p} \wedge \neg p \wedge \mathsf{F}(\widetilde{q} \wedge \psi_{\overline{\tau}})) \Big),$$

where $\psi_{\overline{\tau}} := (\bigwedge_{q \in \overline{\tau}} q \wedge \bigwedge_{q \in Q \setminus \overline{\tau}} \neg q)$. □

4 A combinatorial proof system for $\mathsf{LTL}[\mathsf{X}, \widetilde{\mathsf{X}}, \mathsf{F}, \mathsf{G}]$

In this section, we introduce the proof system that we will later employ to prove Theorem 1, and discuss its connection with Adler–Immerman games [1].

Further notation. Let $A \subseteq \Sigma^+$, with $\Sigma := 2^{\mathcal{AP}}$ for some set of propositions $\mathcal{AP}$. We define $A^{\mathsf{X}} := \{\sigma[1\rangle : \sigma \in A \text{ s.t. } |\sigma| \geq 2\}$, i.e., the set of non-empty traces obtained from A by stepping each trace one position to the right. We define $A^{\mathsf{G}} := \{\sigma[j\rangle : \sigma \in A \text{ and } j \in \text{pos}(\sigma)\}$, i.e., the set of all suffixes of the traces in A. We say that a map $f : A \to \mathbb{N}$ is a *future point* for A whenever $f(\sigma) \in \text{pos}(\sigma)$ for every $\sigma \in A$. We write F_A for the set of all maps that are future points for A. Given a future point f for A and $\sigma \in A$ with $f(\sigma) = i$, we define $\sigma^f := \sigma[i\rangle$ and $A^f := \{\sigma^f : \sigma \in A\}$. Note that, by definition, $A^{\mathsf{G}} = \bigcup_{f \in \mathrm{F}_A} A^f$.

For a formula φ of LTL, we write $A \models \varphi$ whenever $(\sigma, 0) \models \varphi$ for every $\sigma \in A$, and $A \perp\!\!\!\perp \varphi$ whenever $(\sigma, 0) \not\models \varphi$ for every $\sigma \in A$. Given two sets of traces $A, B \subseteq \Sigma^+$ we say that φ *separates* A from B whenever $A \models \varphi$ and $B \perp\!\!\!\perp \varphi$. We write $\langle \cdot, \cdot \rangle_S \subseteq \Sigma^+ \times \Sigma^+$ for the *separable relation on* $S \subseteq \mathbb{OP}$, i.e., the binary relation holding on pairs (A, B) whenever there is some formula from $\mathsf{LTL}[S]$ that separates A from B. Note that, when A and B are finite sets and $\mathsf{X} \in S$, deciding whether $\langle A, B \rangle_S$ holds is trivial.

Lemma 2. *Let $A, B \subseteq \Sigma^+$ and $S \subseteq \mathbb{OP}$. Then, $\langle A, B \rangle_S$ implies $A \cap B = \varnothing$. Moreover, if A and B are finite sets and $\mathsf{X} \in S$, $A \cap B = \varnothing$ implies $\langle A, B \rangle_S$.*

Proof sketch. For the first statement, clearly if $A \cap B \neq \varnothing$ then it is not possible to separate A from B. To prove the second statement, one defines a disjunction φ of formulae, each characterising an element in A. For instance, for $\mathcal{AP} = \{p, q\}$, the trace $\{p\}\{q\}$ can be characterised with the formula $(p \wedge \neg q) \wedge \mathsf{X}(q \wedge \neg p \wedge \widetilde{\mathsf{X}} \bot)$, where $\bot := p \wedge \neg p$. Then, φ separates A from B. □

We mainly consider the relation $\langle \cdot, \cdot \rangle_S$ with S being the set $\{\mathsf{X}, \widetilde{\mathsf{X}}, \mathsf{F}, \mathsf{G}\}$, and thus from now on simply write $\langle \cdot, \cdot \rangle$ when considering this concrete choice of S.

4.1 The proof system

The combinatorial proof system that we define is a natural-deduction-style proof system. It is made of several inference rules of the form $\frac{H_1 \; H_2 \; \ldots \; H_n}{C}$, to be read as "if the hypotheses $H_1, \ldots, H_n$ hold, then the consequence C holds". As usual, proofs within the proof system have a tree-like presentation. An example of such a *deduction tree* is given in Figure 2, where $a := \{p\}$ and $b := \varnothing$, with $p \in \mathcal{AP}$. This is a deduction tree for the *term* $\langle \{abaa, aaaa\}, \{aaab\} \rangle$, which we call the *root* of the deduction tree. In Figure 2, to the root it is *applied* the rule OR, with hypotheses $\langle \{abaa\}, \{aaab\} \rangle$ and $\langle \{aaaa\}, \{aaab\} \rangle$. In turn, these two hypotheses are derived in the deduction tree by eventually reaching applications to the rule ATOMIC. A deduction tree is always *closed*: all maximal paths from the root ends with an application of the rule ATOMIC. This means that a rule of the proof system must be applied to each term $\langle A, B \rangle$ appearing in the tree. We call a tree a *partial deduction tree* if this property is not enforced, namely when there might be unproven terms $\langle A, B \rangle$. The *size* of a deduction tree is the number of rules in it. For instance, the tree in Figure 2 has size 5.

We define the inference rules of the proof system in Figure 1. Let us briefly describe these rules. The ATOMIC rule allows deriving $\langle A, B \rangle$ if every trace in A

$$\textsc{Atomic}\ \frac{A \models \alpha \quad B \perp\!\!\!\perp \alpha}{\langle A, B\rangle}\ \alpha \text{ literal} \qquad \textsc{Or}\ \frac{\langle A_1, B\rangle \quad \langle A_2, B\rangle}{\langle A_1 \uplus A_2, B\rangle} \qquad \textsc{And}\ \frac{\langle A, B_1\rangle \quad \langle A, B_2\rangle}{\langle A, B_1 \uplus B_2\rangle}$$

$$\textsc{Next}\ \frac{\langle A^{\mathsf{X}}, B^{\mathsf{X}}\rangle \quad A \subseteq \Sigma \cdot \Sigma^+}{\langle A, B\rangle} \qquad \textsc{WeakNext}\ \frac{\langle A^{\mathsf{X}}, B^{\mathsf{X}}\rangle \quad B \subseteq \Sigma \cdot \Sigma^+}{\langle A, B\rangle}$$

$$\textsc{Future}\ \frac{\langle A^f, B^{\mathsf{G}}\rangle}{\langle A, B\rangle}\ f \in \mathrm{F}_A \qquad \textsc{Globally}\ \frac{\langle A^{\mathsf{G}}, B^f\rangle}{\langle A, B\rangle}\ f \in \mathrm{F}_B$$

Fig. 1. The combinatorial proof system. Here, $A, B \subseteq \Sigma^+$.

$$\textsc{Or}\ \dfrac{\textsc{Next}\ \dfrac{\textsc{Atomic}\ \dfrac{\{baa\} \models \neg p \qquad \{aab\} \perp\!\!\!\perp \neg p}{\langle\{baa\}, \{aab\}\rangle}}{\langle\{abaa\}, \{aaab\}\rangle} \qquad \dfrac{\dfrac{\{aaaa, aaa, aa, a\} \models p \qquad \{b\} \perp\!\!\!\perp p}{\langle\{aaaa, aaa, aa, a\}, \{b\}\rangle}\ \textsc{Atomic}}{\langle\{aaaa\}, \{aaab\}\rangle}\ \textsc{Globally}}{\langle\{abaa, aaaa\}, \{aaab\}\rangle}$$

Fig. 2. A deduction tree proving $\langle\{abaa, aaaa\}, \{aaab\}\rangle$. Here, $a := \{p\}$ and $b := \varnothing$.

satisfies some literal α and every trace in B violates α. The Or rule corresponds the case of A being separable from B via a formula of the form $\varphi_1 \vee \varphi_2$. In this and the rule And, $\uplus$ stands for the union of disjoint sets. Intuitively, Or can be applied by proving that φ_1 separates a set $A_1 \subseteq A$ from B *and* that φ_2 separates the set $A \setminus A_1$ from B. The Next rule allows separating A from B with a formula of the form $\mathsf{X}\varphi$, by checking whether the sets obtained by stepping all traces in A and B to next time point are separable by φ. The condition $A \subseteq \Sigma \cdot \Sigma^+$ is necessary to ensure that all traces in A have a next time point. The Future rule separates A from B by following this principle: if the set obtained by choosing one suffix for every trace in A is separable from the set of all suffixes of the traces in B, then there is a formula of the form $\mathsf{F}\varphi$ separating A from B. The rules And, WeakNext and Globally are designed to be duals of the rules Or, Next and Future, respectively.

By using the proof system one can derive whether a pair of (finite or infinite) sets of traces (A, B) is in the separable relation $\langle\cdot,\cdot\rangle$. Because of Lemma 2, this is not, however, a particularly useful application. Instead, the proof system is to be used to derive non-trivial lower (or upper) bounds on the size of the minimal formula that separates A from B. This is done by studying the sizes of the possible deduction trees of $\langle A, B\rangle$ in the proof system.

For instance, the deduction tree of Figure 2 shows that there is a formula φ having $\text{size}(\varphi) = 5$ and separating $\{abaa, aaaa\}$ from $\{aaab\}$. This formula is found by simply reading bottom-up, starting from the root, the rules in the deduction tree, associating to each rule the homonymous operator of LTL. In the case of the tree in Figure 2 we have $\varphi := (\mathsf{X}\neg p) \vee \mathsf{G}p$. Note that the formula φ is not the smallest separating formula, because the formula $\mathsf{XXG}p$ also separates $\{abaa, aaaa\}$ from $\{aaab\}$ and corresponds to a tree of size 4.

The correspondence between deduction trees and formulae is formalised in the next theorem (we remark that A and B below do not need to be finite sets).

Theorem 3. *Consider $A, B \subseteq \Sigma^+$. Then, the term $\langle A, B\rangle$ has a deduction tree of size k if and only if there is a formula φ of* $\mathsf{LTL}[\mathsf{X}, \widetilde{\mathsf{X}}, \mathsf{F}, \mathsf{G}]$ *separating A from B and such that* $\mathrm{size}(\varphi) = k$.

Proof sketch. We leave to the reader the proof of the left to right direction of the theorem (shown by induction on k), as it is not required to establish lower bounds on the sizes of formulae, and focus instead on the right to left direction.

Consider a $\mathsf{LTL}[\mathsf{X}, \widetilde{\mathsf{X}}, \mathsf{F}, \mathsf{G}]$ formula φ that separates A and B. We construct a deduction tree of size $\mathrm{size}(\varphi)$. We proceed by structural induction on φ.

base case: φ literal. The deduction tree consists of a single rule Atomic.

induction step, case: $\varphi = \varphi_1 \vee \varphi_2$. Define $A_1 := \{a \in A : a \models \varphi_1\}$ and $A_2 := A \setminus A_1$. From $A \models \varphi$ and $B \perp\!\!\!\perp \varphi$ we get $A_i \models \varphi_i$ and $B \perp\!\!\!\perp \varphi_i$ for both $i \in \{1, 2\}$. By induction hypothesis $\langle A_i, B\rangle$ has a deduction tree of size $\mathrm{size}(\varphi_i)$. By applying the rule And, we obtain a deduction tree for $\langle A, B\rangle$ having size $\mathrm{size}(\varphi_1) + \mathrm{size}(\varphi_2) + 1 = \mathrm{size}(\varphi)$.

induction step, case: $\varphi = \mathsf{X}\psi$. Since $A \models \mathsf{X}\psi$, for every $\sigma \in A$ we have $|\sigma| \geq 2$ and $(\sigma, 1) \models \psi$. By definition of A^{X}, $A \subseteq \Sigma \cdot \Sigma^+$ and $A^{\mathsf{X}} \models \psi$. From $B \perp\!\!\!\perp \mathsf{X}\psi$, for every $\sigma' \in B$, if $|\sigma'| \geq 2$ then $(\sigma', 1) \not\models \psi$. By definition of B^{X}, we have $B^{\mathsf{X}} \perp\!\!\!\perp \psi$. By induction hypothesis, $\langle A^{\mathsf{X}}, B^{\mathsf{X}}\rangle$ has a deduction tree of size $\mathrm{size}(\psi)$. We apply the rule Next to obtain a deduction tree of $\langle A, B\rangle$ of size $\mathrm{size}(\psi) + 1 = \mathrm{size}(\varphi)$.

induction step: $\varphi = \mathsf{F}\psi$. Since $A \models \mathsf{F}\psi$, for every $\sigma \in A$ there is $j_\sigma \in \mathrm{pos}(\sigma)$ such that $(\sigma, j_\sigma) \models \psi$. Let $f \in \mathrm{F}_A$ be the map given by $f(\sigma) = j_\sigma$ for every $\sigma \in A$. We have $A^f \models \psi$. We show that $B^{\mathsf{G}} \perp\!\!\!\perp \psi$. *Ad absurdum*, suppose there is $\sigma_1 \in B^{\mathsf{G}}$ such that $\sigma_1 \models \psi$. By definition of B^{G} there is $\sigma_2 \in B$ such that $\sigma_1 \sqsubseteq \sigma_2$. Then, $(\sigma_2, 0) \models \mathsf{F}\psi$. However, this contradicts the fact that $B \perp\!\!\!\perp \mathsf{F}\psi$. Therefore, $B^{\mathsf{G}} \perp\!\!\!\perp \psi$. By induction hypothesis, $\langle A^f, B^{\mathsf{G}}\rangle$ has a deduction tree of size $\mathrm{size}(\psi)$. By applying the rule Future, we obtain a deduction tree for $\langle A, B\rangle$ of size $\mathrm{size}(\psi) + 1 = \mathrm{size}(\varphi)$.

induction step, cases $\varphi = \varphi_1 \wedge \varphi_2$, $\varphi = \widetilde{\mathsf{X}}\psi$ and $\varphi = \mathsf{G}\psi$. The cases for $\varphi = \varphi_1 \wedge \varphi_2$, $\varphi = \widetilde{\mathsf{X}}\psi$ and $\varphi = \mathsf{G}\psi$ are analogous to the cases $\varphi = \varphi_1 \vee \varphi_2$, $\varphi = \mathsf{X}\psi$ and $\varphi = \mathsf{F}\psi$, respectively. □

The right to left direction of Theorem 3 implies the following corollary that highlights how our proof system is used for formulae sizes lower bounds.

Corollary 2. *Consider a formula φ in* $\mathsf{LTL}[\mathsf{X}, \widetilde{\mathsf{X}}, \mathsf{F}, \mathsf{G}]$. *Suppose that (i) there are $A, B \subseteq \Sigma^+$ such that φ separates A from B, and (ii) every deduction tree of $\langle A, B\rangle$ has size at least k. Then,* $\mathrm{size}(\varphi) \geq k$.

4.2 Connections with the Adler–Immerman games

As outlined in Section 1, our proof system can be seen as an adaptation of the games for CTL introduced by Adler and Immerman in [1]. We now illustrate this connection. Readers that are mostly interested in seeing our proof system in action may want to skip to Section 5.

The *Adler–Immerman games* extend the classical Ehrenfeucht–Fraïssé games in order to bound the *sizes* of the formulae that separate two (sets of) structures, instead of their quantifier depths. As in the Ehrenfeucht–Fraïssé games, the Adler–Immerman games are two-player games between a *spoiler* and a *duplicator*. The game arena is a pair of sets of structures (A, B), and at each round of the game the spoiler choses a rule r to play (there is one rule for each Boolean connective and operator of the logic) and plays on one set of structures accordingly to what r dictates. The duplicator replies on the other set, again accordingly to r. The goal of the spoiler is to separate A from B (i.e., to show $\langle A, B\rangle$ in the context of CTL) in fewer rounds as possible, whereas the duplicator must prolong the game as much as she can. The length of the minimal game corresponds to the size of the minimal formula separating A from B. The main difference between an Adler–Immerman game and an Ehrenfeucht–Fraïssé game is that, in the former, in each round the duplicator is allowed to make copies of the structures in the set she is playing on, and to play differently in each of these copies. This extra power given to the duplicator is why the games end up capturing the notion of size of a formula.

In the setting of the Adler–Immerman games, the rule for the operator F in LTL would be spelled as follows: *"For each structure $\sigma \in A$, the spoiler moves to a future position of σ (i.e., $\sigma[j\rangle$ for some $j \in pos(\sigma)$). The duplicator answers by first making as many copies of elements in B as she wants, and then selects a future position for each of these copies"*. Because she can make copies, the duplicator has a trivial optimal strategy: at each round, copy the structures in B as much as possible, choosing a different position in each copy. The rule for F the simplifies to *"For each structure $\sigma \in A$, the spoiler moves to a future position of σ. The duplicator answers with B^{G}"*, which corresponds to our rule FUTURE.

While Adler and Immerman discuss the fact that the duplicator has a trivial optimal strategy, they do not restate the games with only one player (mainly to not lose the similarity with the Ehrenfeucht–Fraïssé games). Our work shows that removing the duplicator yields a natural one-player game based on building proofs within a proof system. We think that this proof-system view has a few merits over the games. When proving lower bounds, it reduces the clumsiness of discussing the various moves of the spoiler and the replies of the duplicator. The combinatorics is of course still there, but not the players, and this substantially simplifies the exposition. Second, the proof system resembles the way in which one reasons about the *algorithmic* problem of separating A from B. For instance, the algorithm presented in [21] uses decision trees for solving this problem. These decision trees, when they encode a formula from $\mathsf{LTL}[\mathsf{X}, \widetilde{\mathsf{X}}, \mathsf{F}, \mathsf{G}]$, can be easily translated into proofs in our proof system. We discuss more this line of work connected with LTL formulae learning and explainable planning in Section 8.

5 The exponential lower bound for Φ_n

In this section, we show that, for every $n \geq 1$, all formulae of $\mathsf{LTL}[\mathsf{X}, \widetilde{\mathsf{X}}, \mathsf{F}, \mathsf{G}]$ characterising the $\mathsf{F}(\mathsf{LTL}[\mathsf{O}])$ formula Φ_n defined in Section 3 have size at least 2^n.

According to the definition of Φ_n, we consider a set of $2n+2$ distinct atomic propositions $\mathcal{AP} := \{\widetilde{p}, \widetilde{q}\} \cup P \cup Q$, with $P := \{p_1, \dots, p_n\}$ and $Q := \{q_1, \dots, q_n\}$; and $\Sigma := 2^{\mathcal{AP}}$. Throughout the section, let $\alpha(n) := 2^{n+1}(n+2)^2$, i.e., the upper bound given in Lemma 1 for one of these formulae.

Following Corollary 2, to prove the exponential lower bound we

1. define $\mathbf{A}, \mathbf{B} \subseteq \Sigma^+$ such that Φ_n separates $\mathbf{A}$ from $\mathbf{B}$ (Section 5.1), and
2. prove that every deduction tree for $\langle \mathbf{A}, \mathbf{B} \rangle$ has size at least 2^n (Section 5.2).

5.1 Setting up the sets of traces A and B

We define the sets of *types* $T_P := \{\tau \in \Sigma : \widetilde{p} \in \tau \text{ and } \tau \subseteq P \cup \{\widetilde{p}\}\}$ and $T_Q := \{\tau \in \Sigma : \widetilde{q} \in \tau \text{ and } \tau \subseteq Q \cup \{\widetilde{q}\}\}$. Similarly to what done in the proof of Lemma 1, we write $\overline{(\cdot)}$ for the involution on $T_P \cup T_Q$ sending a type $\tau \in T_Q$ into the (only) type $\overline{\tau} \in T_P$ with $q_i \in \tau$ if and only if $p_i \in \overline{\tau}$, for every $i \in [1, n]$.

Throughout the section, we fix a (arbitrary) strict total order $\prec$ on the elements of T_Q. Then, we denote by $\mathcal{E} \in (\varnothing^{\alpha(n)} \cdot T_Q)^{2^n} \cdot \varnothing^{\alpha(n)}$ the (only) finite word enumerating all elements in T_Q, with respect to the order $\prec$. Note that, in $\mathcal{E}$, between any two subsequent elements of T_Q there are $\alpha(n)$ positions of type $\varnothing$. This "padding" added to the enumeration is required to handle the rules NEXT and WEAKNEXT. Given $\tau \in T_Q$, we write $\mathcal{E}|_{-\tau}$ for the trace obtained from $\mathcal{E}$ by eliminating the only position of type τ, together with the $\alpha(n)$ positions of type $\varnothing$ preceding it. So, $\mathcal{E}_{-\tau}$ belongs to $(\varnothing^{\alpha(n)} \cdot T_Q)^{2^n - 1} \cdot \varnothing^{\alpha(n)}$.

For instance, consider the case of $n = 2$, so $Q = \{q_1, q_2\}$ and $\alpha(n) = 128$. Suppose $\{\widetilde{q}\} \prec \{\widetilde{q}, q_1\} \prec \{\widetilde{q}, q_2\} \prec \{\widetilde{q}, q_1, q_2\}$ to be the strict order on T_Q. Then,

$$\mathcal{E} = \varnothing^{128} \cdot \{\widetilde{q}\} \cdot \varnothing^{128} \cdot \{\widetilde{q}, q_1\} \cdot \varnothing^{128} \cdot \{\widetilde{q}, q_2\} \cdot \varnothing^{128} \cdot \{\widetilde{q}, q_1, q_2\} \cdot \varnothing^{128},$$

$$\mathcal{E}|_{-\{\widetilde{q}, q_2\}} = \varnothing^{128} \cdot \{\widetilde{q}\} \cdot \varnothing^{128} \cdot \{\widetilde{q}, q_1\} \cdot \varnothing^{128} \cdot \{\widetilde{q}, q_1, q_2\} \cdot \varnothing^{128}.$$

For the rest of the paper, we denote with $\mathbf{A}$ and $\mathbf{B}$ the sets:

$$\mathbf{A} := \{\varnothing^j \cdot \overline{\tau} \cdot \mathcal{E} : j \in \mathbb{N}, \tau \in T_Q\}, \qquad \mathbf{B} := \{\varnothing^j \cdot \overline{\tau} \cdot (\mathcal{E}|_{-\tau}) : j \in \mathbb{N}, \tau \in T_Q\}.$$

Lemma 3. *The formula Φ_n separates* $\mathbf{A}$ *from* $\mathbf{B}$.

Proof. Let $j \in \mathbb{N}$ and $\tau \in T_Q$. In a nutshell, the fact that $\varnothing^j \cdot \overline{\tau} \cdot \mathcal{E} \models \Phi_n$ follows from the fact that τ occurs in $\mathcal{E}$, and from the position corresponding to τ one can refer back to $\overline{\tau}$ and find in this way a position satisfying p_i if and only if $q_i \in \tau$, for every $i \in [1, n]$. However, since τ is removed from $\mathcal{E}|_{-\tau}$, we see that $b := \varnothing^j \cdot \overline{\tau} \cdot (\mathcal{E}|_{-\tau}) \not\models \Phi_n$: indeed, $b[j] = \overline{\tau}$ corresponds to the only position in b satisfying $\widetilde{p}$, but τ does not appear in b (since it does not appear in $\mathcal{E}|_{-\tau}$). Therefore, $\mathbf{A} \models \Phi_n$ and $\mathbf{B} \perp\!\!\!\perp \Phi_n$. □

5.2 Separating A from B requires an exponential proof

We now show that every deduction tree for $\langle \mathbf{A}, \mathbf{B} \rangle$ has size at least 2^n. To do so, we use a relation $\approx$ that, roughly speaking, states what elements $(a, b) \in \mathbf{A}^{\mathsf{G}} \times \mathbf{B}^{\mathsf{G}}$ are similar enough to require a non-trivial proof in order to be separated using the proof system. Formally, for $a, b \in \Sigma^+$, we write $a \approx b$ whenever:

a and b are in the language $\varnothing^u \cdot \rho \cdot \varnothing^{\alpha(n)} \cdot \Sigma^*$, for some $u \in \mathbb{N}$ and $\rho \in T_Q \cup T_P$.

The central issue in the proof of the lower bound is counting how many of these pairs $a \approx b$ are preserved when applying the rules of the proof system. This count is done inductively on the size of the deduction tree, and allows us to derive the following lemma.

Lemma 4. *Let $r_1, t_1, \dots, r_m, t_m \in \mathbb{N}$ and let $\tau_1, \dots, \tau_m \in T_Q$ be pairwise distinct sets. Consider $A \subseteq \mathbf{A}^{\mathsf{G}}$, $B := \{(\varnothing^{t_i} \cdot \overline{\tau_i} \cdot \mathcal{E}|_{-\tau_i})[r_i\rangle : i \in [1, m]\}$, and $C := \{(a, b) \in A \times B : a \approx b\}$. Every deduction tree for $\langle A, B\rangle$ has size at least $|C| + 1$.*

Proof. Below, suppose that $\langle A, B\rangle$ has a deduction tree (else the statement is trivially true). In particular, let $\mathcal{T}$ be a minimal deduction tree for $\langle A, B\rangle$, and assume it has size s. Note that the hypothesis that $\tau_1, \dots, \tau_m$ are distinct implies $|B| \leq 2^n$, which in turn implies $|C| < 2^n$ (by definition of $\approx$, for every $b \in B$ there is at most one $a \in \mathbf{A}^{\mathsf{G}}$ such that $a \approx b$). Then, w.l.o.g. we can assume $s \leq \alpha(n)$; otherwise the lemma follows trivially.

During the proof, we write $\prec$ for the strict total order on elements of T_Q used to construct the trace $\mathcal{E}$ enumerating T_Q. Before continuing the proof of the lemma, we highlight a useful property of the elements of C.

Claim 1. Let $(a, b) \in C$ and $i \in [1, m]$ with $b = (\varnothing^{t_i} \cdot \overline{\tau_i} \cdot \mathcal{E}|_{-\tau_i})[r_i\rangle$. Then, $b = \varnothing^u \cdot \rho \cdot \varnothing^{\alpha(n)} \cdot \sigma$, for some $u \in \mathbb{N}$, $\rho \in \{\overline{\tau_i}\} \cup \{\tau \in T_Q : \tau \prec \tau_i\}$ and $\sigma \in \Sigma^*$.

In a nutshell, this claim tells us that for every $(a, b) \in C$ we have $b \not\sqsubseteq \mathcal{E}$.

Let us go back to the proof of Lemma 4. If $A = \varnothing$ or $m = 0$ then $C = \varnothing$ and the lemma follows trivially. Below, let us assume $A \neq \varnothing$ and $m \geq 1$. We prove the statement by induction on the size s of $\mathcal{T}$.

In the base case $s = 1$, $\mathcal{T}$ is a simple application of the rule Atomic. This means that for every $a \in A$ and $b \in B$ we have $a[0] \neq b[0]$. By definition of $\approx$, this implies $C = \varnothing$, and therefore $s \geq |C| + 1$.

Let us then consider the induction step $s \geq 2$. Note that if $|C| \leq 1$ then the statement follows trivially. Hence, below, we assume $|C| \geq 2$. We split the proof depending on the rule applied to the root $\langle A, B\rangle$ of $\mathcal{T}$. Since $s \geq 2$, this rule cannot be Atomic. We omit the cases for Or and And, as they simply follow the induction hypothesis, and focus on the rules related to temporal operators.

• **case: rules** Next **and** WeakNext**.** We consider Next and WeakNext together, as both require $\langle A^{\mathsf{X}}, B^{\mathsf{X}}\rangle$. Perhaps surprisingly, this case is non-trivial. The main difficulty stems from the fact that $C' := \{(a, b) \in A^{\mathsf{X}} \times B^{\mathsf{X}} : a \approx b\}$ might in principle even be empty, and thus applying the induction hypothesis on $\langle A^{\mathsf{X}}, B^{\mathsf{X}}\rangle$ is unhelpful for concluding that $s \geq |C| + 1$. We now show how to circumvent this issue. The minimal deduction tree for $\langle A^{\mathsf{X}}, B^{\mathsf{X}}\rangle$ has size $s - 1$. Within this deduction tree, consider the maximal partial deduction tree $\mathcal{T}'$ rooted at $\langle A^{\mathsf{X}}, B^{\mathsf{X}}\rangle$ and made solely of applications of the rules And, Or, Next, and WeakNext. Let $\langle A_1, B_1\rangle, \dots, \langle A_q, B_q\rangle$ be the leafs of such a tree. Let $j \in [1, q]$. In the tree $\mathcal{T}$, to $\langle A_j, B_j\rangle$ it is applied a rule among Atomic, Future and Globally. Let $\xi_j \geq 1$ be the number of Next and WeakNext rules used in the path of $\mathcal{T}$ from $\langle A, B\rangle$ to $\langle A_j, B_j\rangle$. Note that, from $s \leq \alpha(n)$, we have $\xi_j \leq \alpha(n)$. We define the following two sets C_j and N_j, whose role is essentially to "track" the evolution of pairs in C with respect to $A_j \times B_j$:

$$C_j := \{(a[\xi_j\rangle, b[\xi_j\rangle) \in A_j \times B_j : (a, b) \in C,\, a[\xi_j\rangle \approx b[\xi_j\rangle\},$$
$$N_j := \{(a[\xi_j\rangle, b[\xi_j\rangle) \in A_j \times B_j : (a, b) \in C,\, a[\xi_j\rangle \not\approx b[\xi_j\rangle\}.$$

The minimal deduction tree for $\langle A_j, B_j\rangle$ has size $s_j \geq |C_j| + 1$; by induction hypothesis. Claims 2 to 4 below highlight a series of properties on the sets C_j and N_j from which we derive $s \geq |C| + 1$.

Claim 2. For every $j \in [1, q]$, if $C_j \cup N_j \neq \varnothing$ then the rule applied to $\langle A_j, B_j\rangle$ in $\mathcal{T}$ is either FUTURE or GLOBALLY.

As already said, the rule applied to $\langle A_j, B_j\rangle$ is among the rules ATOMIC, FUTURE and GLOBALLY. Then, showing that $a[0] = b[0]$ for every $(a, b) \in C_j \cup N_j$ excludes the rule ATOMIC.

Claim 3. For every $j \in [1, q]$, $|N_j| \leq 1$.

The proof of this claim is by contradiction, assuming the existence of distinct $(a_1, b_1), (a_2, b_2) \in N_j$. In the proof, we analyse structural properties of the traces a_1, a_2, b_1 and b_2, and consider several cases depending on such properties (for instance, one case split depends on whether $a_1 \sqsubseteq a_2$). In all these cases, we reach a contradiction with either $(a_1, b_1) \neq (a_2, b_2)$ or Claim 2.

Claim 4. $|C| \leq \sum_{j=1}^{q} |C_j \cup N_j|$.

The claim follows as soon as one proves the following two statements:

1. for every $(a, b) \in C$ there is $j \in [1, q]$ such that $(a[\xi_j\rangle, b[\xi_j\rangle) \in C_j \cup N_j$,
2. for all distinct $(a_1, b_1), (a_2, b_2) \in C$, we have $(a_1[\ell\rangle, b_1[\ell\rangle) \neq (a_2[\ell\rangle, b_2[\ell\rangle)$ for every $\ell \leq \alpha(n)$ (recall that $\xi_j \leq \alpha(n)$, for every $j \in [1, q]$).

Item 1 is by induction on the size of $\mathcal{T}'$. Similarly to Claim 3, the proof of Item 2 again requires to consider many cases, and uses properties of $\approx$, $\mathcal{E}$ and $\mathcal{E}|_{-\mathcal{T}_i}$.

Thanks to Claims 3 and 4, one can then prove $s \geq |C| + 1$, concluding the proof for the rules NEXT and WEAKNEXT:

$$\begin{aligned} s &\geq 1 + \textstyle\sum_{j=1}^{q} s_j && \text{by definition of } \mathcal{T} \text{ and } \mathcal{T}' \\ &\geq 1 + \textstyle\sum_{j=1}^{q} (|C_j| + 1) && \text{by } s_j \geq |C_j| + 1 \text{ (induction hypothesis)} \\ &\geq 1 + \textstyle\sum_{j=1}^{q} (|C_j \cup N_j|) && \text{by } |N_j| \leq 1 \text{ (Claim 3)} \\ &\geq |C| + 1 && \text{by } |C| \leq \textstyle\sum_{j=1}^{q} |C_j \cup N_j| \text{ (Claim 4).} \end{aligned}$$

• **case: rule** FUTURE. Let $f \in \mathrm{F}_A$ be the future point used when applying this rule. Define $C' := \{(a', b') \in A^f \times B^{\mathsf{G}} : a' \approx b'\}$. The minimal deduction tree for $\langle A^f, B^{\mathsf{G}}\rangle$ has size $s - 1$. By induction hypothesis, $s - 1 \geq |C'| + 1$, i.e., $s \geq |C'| + 2$. We divide the proof into two cases.

Case 1: for every $a' \in A^f$, $a' \not\sqsubseteq \mathcal{E}$. By definition of $\approx$, every $(a,b) \in C$ is such that a and b belong to the language $\varnothing^u \cdot \overline{\tau_i} \cdot \varnothing^{\alpha(n)} \cdot \Sigma^*$ for some $u \in \mathbb{N}$, and $i \in [1,m]$. Since $a^f \not\sqsubseteq \mathcal{E}$, we must have $f(a) \leq u+1$. Then, $a^f \approx b[f(a)\rangle$. Note that distinct $(a,b) \in C$ concern distinct $\overline{\tau_i}$ with $i \in [1,m]$, and therefore, together with $b[f(a)\rangle \in B^G$, one concludes that $|C'| \geq |C|$; and so $s \geq |C|+2$.

Case 2: there is $a' \in A^f$ such that $a' \sqsubseteq \mathcal{E}$. Let us denote with $\widetilde{a}$ the element in A^f such that $\widetilde{a} \sqsubseteq a$ for every $a \in A^f$. The existence of such an element follows directly from the fact that $a' \sqsubseteq \mathcal{E}$ for some $a' \in A^f$.

Let $I \subseteq [1,m]$ be the subset of those indices $i \in [1,m]$ for which there is a pair $(a',b') \in C$ such that $b' = (\varnothing^{t_i} \cdot \overline{\tau_i} \cdot \mathcal{E}|_{\tau_i})[r_i\rangle$. Without loss of generality, suppose $I = \{1,\ldots,q\}$ for some $q \leq m$, and that $\tau_1 \prec \tau_2 \prec \cdots \prec \tau_q$; recall that all these types are pairwise distinct. By definition of $\approx$, for every $b' \in B$ there is at most one $a' \in \mathbf{A}^{\mathsf{G}}$ such that $a' \approx b'$, and therefore $q = |C|$. To conclude the proof it suffices to show $|C'| \geq q-1$. We do so by establishing a series of claims. Recall that we are assuming $|C| \geq 2$, so in particular C and I are non-empty.

Claim 5. There are $u \in \mathbb{N}$, $\rho \in T_Q$ and $\sigma \in (2^Q)^*$ s.t. $\widetilde{a} = \varnothing^u \cdot \rho \cdot \varnothing^{\alpha(n)} \cdot \sigma$. Moreover, $\rho \preceq \tau_i$ for every $i \in I$.

The first statement of this claim is established from the definition of $\widetilde{a}$. The second statement is proven by contradiction. In particular, assuming that there is $i \in I$ such that $\tau_i \prec \rho$ yields a contradiction with Claim 1.

Below, we write u, ρ and σ for the objects appearing in Claim 5. Note that, from $\tau_1 \prec \cdots \prec \tau_q$, the second statement of Claim 5 implies $\rho \prec \tau_2 \cdots \prec \tau_q$. For $i \in [2,q]$, let (a_i', b_i') denote the pair in C such that $b_i' = (\varnothing^{t_i} \cdot \overline{\tau_i} \cdot \mathcal{E}|_{\rho_i})[r_i\rangle$.

Claim 6. For each $i \in [2,q]$ there is $\ell \in \mathbb{N}$ such that $\widetilde{a} \approx b_i''$ with $b_i'' := b_i'[\ell\rangle$. Moreover, every type in $\{\tau_2,\ldots,\tau_q\} \setminus \{\tau_i\}$ appears in some position of b_i''.

This claim is proven using Claims 1 and 5 and properties of $\mathcal{E}|_{-\tau_i}$.

Since all types $\tau_2,\ldots,\tau_q$ are pairwise distinct, from the second statement in Claim 6 we conclude that $b_i'' \neq b_j''$ for every two distinct $i,j \in I \setminus \{i_1\}$. Then, the first statement in Claim 6 entails $|C'| \geq q-1$.

• **case: rule** GLOBALLY. Let $f \in \mathrm{F}_A$ be the future point used when applying this rule. The minimal deduction tree for $\langle A^{\mathsf{G}}, B^f\rangle$ has size $s-1$. We define $C' := \{(a',b') \in A^{\mathsf{G}} \times B^f : a' \approx b'\}$. By induction hypothesis, $s-1 \geq |C'|+1$, i.e., $s \geq |C'|+2$. To conclude the proof it suffices to show that $|C'| \geq |C|-1$ (in fact, we prove $|C'| \geq |C|$). Let $\{(a_1,b_1),\ldots,(a_{|C|},b_{|C|})\} = C$.

Claim 7. For every $j \in [1,|C|]$, b_j^f is not a suffix of $\mathcal{E}$. More precisely, given $i \in [1,m]$ such that $b_j = (\varnothing^{t_i} \cdot \overline{\tau_i} \cdot \mathcal{E}|_{-\tau_i})[r_i\rangle$, we have $b_j^f = \varnothing^u \cdot \rho \cdot \varnothing^{\alpha(n)} \cdot \sigma$, for some $u \in \mathbb{N}$, $\rho \in \{\overline{\tau_i}\} \cup \{\tau \in T_P : \tau \prec \tau_i\}$ and $\sigma \in \Sigma^*$.

See the similarities between this claim and Claim 1. The first statement is proven by contradiction, deriving an absurdum with the existence of $\mathcal{T}$. The second statement follows from the definition of $\mathcal{E}|_{-\tau_i}$.

Starting from Claim 7, we conclude that (i) for every $j \neq k \in [1, |C|]$, $b_j^f \neq b_k^f$, and (ii) for every $j \in [1, |C|]$ there is $\ell \in \mathbb{N}$ such that $a_j[\ell\rangle \approx b_j^f$. This directly implies $|C'| \geq |C|$. This concludes both the proof of the case GLOBALLY and the proof of the lemma. □

Together, Lemmas 3 and 4 yield an exponential lower bound for all formulae of $\mathsf{LTL}[\mathsf{X}, \widetilde{\mathsf{X}}, \mathsf{F}, \mathsf{G}]$ characterising Φ_n (Lemma 5), which in turn implies Theorem 1.

Lemma 5. *Let $\Psi_n \in \mathsf{LTL}[\mathsf{X}, \widetilde{\mathsf{X}}, \mathsf{F}, \mathsf{G}]$. If $\Psi_n \equiv \Phi_n$ then $\text{size}(\Psi_n) \geq 2^n$.*

Proof. We define the sets $A = \{\overline{\tau} \cdot \mathcal{E} : \tau \in T_Q\}$ and $B = \{\overline{\tau} \cdot \mathcal{E}_{-\tau} : \tau \in T_Q\}$. Observe that $A \subseteq \mathbf{A} \subseteq \mathbf{A}^{\mathsf{G}}$ and $B \subseteq \mathbf{B}$. Let $C = \{(a, b) \in A \times B : a \approx b\}$. From the definition of $\approx$, $|C| = 2^n$. We apply Lemma 4, and conclude that the minimal deduction tree for $\langle A, B\rangle$ has size at least 2^n (in fact, $2^n + 1$). Since $A \subseteq \mathbf{A}$ and $B \subseteq \mathbf{B}$, the same holds for the minimal deduction tree for $\langle \mathbf{A}, \mathbf{B}\rangle$. Then, the theorem follows from Corollary 2 and Lemma 3. □

While we do not prove it formally, we claim that Theorem 1 also holds for infinite traces. It is in fact quite simple to see this: all traces in $\mathbf{A}$ and $\mathbf{B}$, have a suffix of the form $\varnothing^{\alpha(n)}$. Roughly speaking, these suitably long suffixes are added to make the far-end of the traces in $\mathbf{A}$ and $\mathbf{B}$ indistinguishable at the level of formulae, so that they cannot be used in deduction trees to separate $\mathbf{A}$ from $\mathbf{B}$. Then, to prove Theorem 1 for infinite traces, it suffices to update the proof system to handle these structures and consider an infinite suffix $\varnothing^\omega$ instead. The proof of Lemma 4 goes through with no significant change.

A second observation: traces in $\mathbf{A}$ and $\mathbf{B}$ are closed under taking arbitrary long prefixes of the form $\varnothing^j$. This feature is not used to prove Lemma 5 (see the definition of A and B in the proof). However, these prefixes play a role in the next section, when studying the succinctness of $\mathsf{F}(\mathsf{LTL}[\mathsf{Y}, \widetilde{\mathsf{Y}}, \mathsf{O}, \mathsf{H}])$ on infinite traces.

6 Theorem 2: a 2^n lower bound for LTL[F] pastification

In this section, we rely on Lemma 5 to prove Theorem 2 and Corollary 1.

Theorem 2 is proven by relying on a "future–past duality" between future and past fragments of LTL. Given a trace $\sigma \in \Sigma^+$ we define the *reverse of* σ, written σ^-, as the trace satisfying $\sigma^-[i] = \sigma[|\sigma| - i]$ for every $i \in \text{pos}(\sigma)$. The *reverse of a language* $\mathcal{L} \subseteq \Sigma^+$ is defined as the language $\mathcal{L}^- := \{\sigma^- : \sigma \in \mathcal{L}\}$. Clearly, $(\mathcal{L}^-)^- = \mathcal{L}$. Given a set of temporal operators $S \subseteq \{\mathsf{X}, \widetilde{\mathsf{X}}, \mathsf{F}, \mathsf{G}\}$, we write S^- for the set of temporal operators among $\{\mathsf{Y}, \widetilde{\mathsf{Y}}, \mathsf{O}, \mathsf{H}\}$ such that S^- contains Y (resp. $\widetilde{\mathsf{Y}}$; O; H) if and only if S contains X (resp. $\widetilde{\mathsf{X}}$; F; G). For finite traces, the following lemma, proves that if there is a family of languages $(\mathcal{L}_n)_{n\geq 1}$ that can be compactly defined in $\mathsf{F}(\mathsf{LTL}[\mathsf{O}])$ but explodes in $\mathsf{LTL}[\mathsf{X}, \widetilde{\mathsf{X}}, \mathsf{F}, \mathsf{G}]$, then the family $(\mathcal{L}_n^-)_{n\geq 1}$ can be compactly defined in $\mathsf{LTL}[\mathsf{F}]$ but explodes in $\mathsf{F}(\mathsf{LTL}[\mathsf{Y}, \widetilde{\mathsf{Y}}, \mathsf{O}, \mathsf{H}])$.

Lemma 6. *Let $\mathcal{L} \subseteq \Sigma^+$, $S \subseteq \{\mathsf{X}, \widetilde{\mathsf{X}}, \mathsf{F}, \mathsf{G}\}$, and φ be a formula in $\mathsf{F}(\mathsf{LTL}[S^-])$. There is a formula ψ in $\mathsf{F}(\mathsf{LTL}[S])$ such that $\mathcal{L}(\psi) = \mathcal{L}(\varphi)^-$ and $\text{size}(\psi) = \text{size}(\varphi)$.*

Theorem 2 follows by applying Lemma 6 on the family of formulae $(\Phi_n)_{n\geq 1}$ defined in Section 3, and by relying on the exponential lower bounds of Lemma 5. The sequence of languages showing that $\mathsf{LTL}[\mathsf{F}]$ can be exponentially more succinct than $\mathsf{F}(\mathsf{LTL}[\mathsf{Y}, \widetilde{\mathsf{Y}}, \mathsf{O}, \mathsf{H}])$ is given by $(\mathcal{L}(\Phi_n)^{-})_{n\geq 1}$.

Next, we extend Theorem 2 to the case of infinite traces. As usual, let Σ^ω be the set of all infinite traces over the finite alphabet Σ. We denote with $\mathcal{L}^\omega(\varphi)$ the language of φ, when φ is interpreted over infinite traces (we refer the reader to, e.g., [2] for the semantics of LTL on infinite traces).

Lemma 7. *The family of languages of infinite traces* $(\mathcal{L}(\Phi_n)^{-} \cdot \Sigma^\omega)_{n\geq 1}$ *is such that, for every* $n \geq 1$, *(i) there is a formula* φ *of* $\mathsf{LTL}[\mathsf{F}]$ *such that* $\mathrm{size}(\varphi) \in \mathcal{O}(n)$ *and* $\mathcal{L}^\omega(\varphi) = \mathcal{L}(\Phi_n)^{-} \cdot \Sigma^\omega$, *and (ii) for every formula* ψ *in* $\mathsf{F}(\mathsf{LTL}[\mathsf{Y}, \widetilde{\mathsf{Y}}, \mathsf{O}, \mathsf{H}])$, *if* $\mathcal{L}^\omega(\psi) = \mathcal{L}(\Phi_n)^{-} \cdot \Sigma^\omega$ *then* $\mathrm{size}(\psi) \geq 2^n$.

Item (i) in the lemma above follows by applying Lemma 6 and exploiting the fact that formulae φ in $\mathsf{LTL}[\mathsf{F}]$ satisfy $\mathcal{L}^\omega(\varphi) = \mathcal{L}(\varphi)\cdot\Sigma^\omega$ and $\mathcal{L}(\varphi) = \mathcal{L}(\varphi) \cdot \Sigma^*$ (cf. [2, Definition 5 and Lemma 5]). The proof of Item (ii) is instead quite subtle. One would like to use the hypothesis $\mathcal{L}^\omega(\psi) = \mathcal{L}(\Phi_n)^{-}\cdot\Sigma^\omega$ and that $\mathcal{L}(\psi)$ is a cosafety language to derive $\mathcal{L}(\psi) = \mathcal{L}(\Phi_n)^{-}$. However, note that nothing prevents $\mathcal{L}(\psi)$ to be equal to $\mathcal{L}(\Phi_n)^{-}\cdot\Sigma$, and as it stands we do not have bounds for characterising this language. We apply instead the following strategy. We consider the family of structures $A' := \{a^{-} \cdot \varnothing^\omega : a \in \mathbf{A}\}$ and $B' := \{b^{-} \cdot \varnothing^\omega : b \in \mathbf{B}\}$. Note that $A' \subseteq \mathcal{L}^\omega(\psi)$ and $B' \cap \mathcal{L}^\omega(\psi) = \varnothing$. Since ψ is of the form $\mathsf{F}(\alpha)$ with $\alpha \in \mathsf{LTL}[\mathsf{Y}, \widetilde{\mathsf{Y}}, \mathsf{O}, \mathsf{H}]$, we can, roughly speaking, study the effects of applying to A' and B' a variant of the rule FUTURE for infinite words and that does not "forget the past", and then reverse all traces in the resulting sets $(A')^f$ and $(B')^{\mathsf{G}}$. In this way, we obtain two sets of finite traces $\widetilde{A} \subseteq \mathbf{A}$ and $\widetilde{B} \subseteq \mathbf{B}$ (this is where the prefixes $\varnothing^j$ discussed at the end of Section 5 play a role). We show that the hypotheses of Lemma 4 apply to $\widetilde{A}$ and a set $\widehat{B} \subseteq \widetilde{B}$ for which the set $\{(a, b) \in \widetilde{A} \times \widehat{B} : a \approx b\}$ has size at least $2^n - 1$. By Corollary 2, we get that α, and thus ψ, is of size at least 2^n.

Lemma 7 shows that Theorem 2 holds over infinite traces as well. Together with the $2^{\mathcal{O}(n)}$ upper bound for the pastification problem for $\mathsf{LTL}[\mathsf{X}, \mathsf{F}]$ into $\mathsf{F}(\mathsf{LTL}[\mathsf{Y}, \widetilde{\mathsf{Y}}, \mathsf{O}])$ established[6] in [4], this entails Corollary 1.

7 The automata method does not work for $\mathsf{F}(\mathsf{LTL}[\mathsf{O}])$

In this section we show that the classical method introduced by Markey in [20] to prove "future against past" succinctness discrepancies in fragments of LTL cannot be used to prove the results in Section 5, namely that $\mathsf{F}(\mathsf{LTL}[\mathsf{O}])$ can be exponentially more succinct than $\mathsf{LTL}[\mathsf{X}, \widetilde{\mathsf{X}}, \mathsf{F}, \mathsf{G}]$. Due to space constraints, we assume a basic familiarity with *non-deterministic Büchi automata* (NBAs) (and *deterministic Büchi automata*, DBAs), which are central tools in [20]. Moreover,

[6] To be more precise, in [4] the authors only provide a $2^{\mathcal{O}(n^2)}$ upper bound for their algorithm. Their analysis can however be easily improved to $2^{\mathcal{O}(n)}$.

we work on LTL on infinite traces, as done in [20], and write $\varphi \equiv_\omega \psi$ whenever $\mathcal{L}^\omega(\varphi) = \mathcal{L}^\omega(\psi)$. We write $\mathcal{L}^\omega(A)$ for the language of an NBA A.

Proposition 1 below summarises the method in [20], which is parametric on a fixed prefix Π of operators among X, F and G. Given two fragments F, F' of LTL, with F' pure future, to apply the method one has to provide the two families of formulae $(\Phi_n)_{n\geq 1} \in F$ and $(\Phi'_n)_{n\geq 1} \in F'$, as well as the family of minimal NBAs $(A_n)_{n\geq 1}$, satisfying the hypotheses of Proposition 1. In [20], this is done for $F = \mathsf{LTL}$ and F' set as the pure future fragment of LTL, using the prefix $\Pi = \mathsf{G}$.

Proposition 1 (Markey's method [20]). *Let F, F' be fragments of* LTL, *with F' pure future. Consider two families of formulae $(\Phi_n)_{n\geq 1} \in F$, $(\Phi'_n)_{n\geq 1} \in F'$, and a family of minimal* NBAs *$(A_n)_{n\geq 1}$, such that*

$$\mathrm{size}(A_n) \in 2^{2^{\Omega(n)}}, \quad \mathrm{size}(\Phi_n) \in \mathit{poly}(n), \quad \Phi_n \equiv_\omega \Phi'_n, \quad \mathcal{L}^\omega(\Pi(\Phi'_n)) = \mathcal{L}^\omega(A_n).$$

Then, $\mathrm{size}(\Phi'_n) \in 2^{\mathrm{size}(\Phi_n)^{\Omega(1)}}$ *and F can be exponentially more succinct than F'.*

The consequence $\mathrm{size}(\Phi'_n) \in 2^{\mathrm{size}(\Phi_n)^{\Omega(1)}}$ obtained in Proposition 1 follows directly from the fact that, from every pure future LTL formula φ, one can build an NBA A of size $2^{\mathcal{O}(\mathrm{size}(\varphi))}$ such that $\mathcal{L}^\omega(A) = \mathcal{L}^\omega(\varphi)$ [26]. Then, the hypotheses $\mathrm{size}(A_n) \in 2^{2^{\Omega(n)}}$ and $\mathcal{L}^\omega(\Pi(\Phi'_n)) = \mathcal{L}^\omega(A_n)$ imply $\mathrm{size}(\Phi'_n) \in 2^{\Omega(n)}$.

To show that Proposition 1 cannot be used to derive that $F := \mathsf{F}(\mathsf{LTL}[\mathsf{O}])$ can be exponentially more succinct than $F' := \mathsf{LTL}[\mathsf{X}, \widetilde{\mathsf{X}}, \mathsf{F}, \mathsf{G}]$, it suffices to show that no families $(\Phi_n)_{n\geq 1} \in F$, $(\Phi'_n)_{n\geq 1} \in F'$ and $(A_n)_{n\geq 1}$ achieve the hypotheses required by Proposition 1, no matter the temporal prefix Π. We do so by establishing that whenever $\mathrm{size}(\Phi_n) \in \mathrm{poly}(n)$ and $\Phi_n \equiv_\omega \Phi'_n$, the minimal *deterministic* Büchi automaton for $\mathcal{L}^\omega(\Pi(\Phi'_n))$ has size in $2^{\mathcal{O}(\mathrm{poly}(n))}$. Therefore, no family of minimal NBAs $(A_n)_{n\geq 1}$ such that $\mathrm{size}(A_n) \in 2^{2^{\Omega(n)}}$ can also satisfy the hypothesis $\mathcal{L}^\omega(\Pi(\Phi'_n)) = \mathcal{L}^\omega(A_n)$. Here is the formal statement:

Theorem 4. *Let Π be a prefix of k temporal operators among* X, F *and* G. *Let φ be a formula of* $\mathsf{F}(\mathsf{LTL}[\mathsf{O}])$, *and ψ be a formula of* $\mathsf{LTL}[\mathsf{X}, \widetilde{\mathsf{X}}, \mathsf{F}, \mathsf{G}]$, *with $\varphi \equiv_\omega \psi$. The minimal* DBA *for $\mathcal{L}^\omega(\Pi(\psi))$ is of size in $(k+1) \cdot 2^{\mathcal{O}(\mathrm{size}(\varphi))}$.*

The proof of this theorem is divided into three steps.

As a first step, one shows that $\psi \equiv_\omega \mathsf{F}\psi$; which essentially follows from the fact that $\psi \equiv_\omega \varphi$ with $\varphi \in \mathsf{F}(\mathsf{LTL}[\mathsf{O}])$. Together with tautologies of LTL such as $\mathsf{FGF}\psi' \equiv_\omega \mathsf{GF}\psi'$, $\mathsf{FX}\psi' \equiv_\omega \mathsf{XF}\psi'$ and $\mathsf{GX}\psi' \equiv_\omega \mathsf{XG}\psi'$, the equivalence $\psi \equiv_\omega \mathsf{F}\psi$ let us rearrange Π into a prefix of the form either $\mathsf{X}^j\mathsf{GF}$ or $\mathsf{X}^j\mathsf{F}$, for some $j \leq k$. Let us focus on the former (more challenging) case of $\Pi = \mathsf{X}^j\mathsf{GF}$.

The second step required for the proof is to bound the size of the minimal DBA A recognising $\mathcal{L}^\omega(\mathsf{F}\psi)$. Thanks to the equivalences $\varphi \equiv_\omega \psi \equiv_\omega \mathsf{F}\psi$, such a DBA has size exponential in $\mathrm{size}(\varphi)$ by the following lemma.

Lemma 8. *Let φ in* $\mathsf{F}(\mathsf{LTL}[\mathsf{O}])$. *There is a* DBA *for $\mathcal{L}^\omega(\varphi)$ of size $2^{\mathcal{O}(\mathrm{size}(\varphi))}$.*

Starting from A, the third and last step of the proof requires constructing a DBA for $\mathcal{L}^{\omega}(\mathsf{X}^{j}\mathsf{GF}\psi)$ of size in $(j+1) \cdot 2^{\mathcal{O}(\text{size}(\varphi))}$. The treatment for the prefix X^{j} is simple, so this step is mostly dedicated to constructing a DBA for $\mathcal{L}^{\omega}(\mathsf{GF}\psi)$. In the case of LTL, it is known that closing an NBA under the globally operator G might lead to a further exponential blow-up (in fact, this is one of the reasons Markey's method is possible in the first place). However, because φ is in F(LTL[O]), and it is thus a cosafety language (and so ψ is a cosafety language too), it turns out that the size of the minimal DBA for $\mathcal{L}^{\omega}(\mathsf{GF}\psi)$ is in $\mathcal{O}(\text{size}(A))$.

Lemma 9. *Let ψ be in* LTL, *such that $\mathcal{L}^{\omega}(\psi)$ is a cosafety language. Let A be a* DBA *for $\mathcal{L}^{\omega}(\mathsf{F}\psi)$. The minimal* DBA *for $\mathcal{L}^{\omega}(\mathsf{GF}\psi)$ has size in $\mathcal{O}(\text{size}(A))$.*

Thanks to Lemma 9, the proof of Theorem 4 can be easily completed. To prove this lemma, one modifies A by redirecting all transitions exiting a final state so that they mimic the transitions exiting the initial state of the automaton. The reason why the obtained automaton recognises $\mathcal{L}^{\omega}(\mathsf{GF}\psi)$ uses in a crucial way the fact that ψ and $\mathsf{F}\psi$ are cosafety languages.

8 Related and Future Work

The proof systems we use in this work to establish Theorem 2 and Theorem 1 are strongly related to recent work in two seemingly disconnected areas of computer science: (i) combinatorial games for formulae lower bounds of first-order logics and (ii) learning of LTL formulae in explainable planning and program synthesis.

Combinatorial games. We have already discussed the connections between our proof system and the CTL^{+} games by Adler and Immerman [1]. Recently, Fagin and coauthors [9,10] have looked at combinatorial games that allow to count the number of quantifiers required to express a property in first-order logic. While these games simplify Adler–Immerman games, they come with a drawback: by design, they implicitly look at how to express properties with first-order formulae in *prenex normal form*, and they are not able to give any bound on the quantifier-free part of such formulae. It seems then not possible to use these types of games in the context of LTL. One could in principle consider translations of LTL formulas into a prenex fragment of S1S. However, since S1S is both more expressive and more succinct than LTL [25], concluding that LTL[F] can be exponentially more succinct than F(LTL[Y, $\widetilde{\mathsf{Y}}$, O, H]) will ultimately require analysing structural properties of the quantifier-free part of the S1S formulae.

Closer to the case of LTL are the games on linear orders (implicitly) used by Grohe and Schweikartdt in [14]. These are formula-size games for a fixed number of variables of first-order logic. Using our notation, the method used to derive lower bounds in [14] relies on defining a function ω from terms of the form $\langle A, B\rangle$ to $\mathbb{N}$ that acts as a *sub-additive measure* with respect to the rules of the proof system. For instance, according to the rule OR, ω should satisfy $\omega(\langle A, B\rangle) \leq \omega(\langle A_1, B\rangle) + \omega(\langle A_2, B\rangle)$, whenever $A = A_1 \uplus A_2$. One can use a sub-additive measure ω to conclude that the minimal deduction tree for $\langle A, B\rangle$, if it exists, has size at least $\omega(\langle A, B\rangle)$. When restricted to the objects in Lemma 4, one can show that the function $\omega(\langle A, B\rangle) := |\{(a,b) \in A \times B : a \approx b\}| + 1$ is a sub-additive measure with respect to the rules Atomic, Or, And, Future and Globally (this is implicit in the proof of Lemma 4). However, it is not a sub-additive measure for the rules Next and WeakNext: as stressed in the proof, we might have $\omega(\langle A^{\mathsf{X}}, B^{\mathsf{X}}\rangle) = 1$ even for $\omega(\langle A, B\rangle)$ arbitrary large. This partially explains why the proof of Lemma 4 turned out to be quite involved.

In view of the optimality of the algorithm in [4] (Corollary 1), the main open problem regarding pastification is the optimality of the triply-exponential time procedure given in [7] for the pastification of $\mathsf{LTL}[\mathsf{X}, \mathsf{U}]$ into $\mathsf{F}(\mathsf{pLTL})$. As far as we are aware, no super-polynomial lower bound for this problem is known. Our proof system, properly extended with rules for the until and release operators, might be able to tackle this issue. Obtaining a matching triply-exponential lower bound might however be impossible: when restricted to propositional logic, our proof system is equivalent to the communication games introduced by Karchmer and Wigderson [15]. It is well-known that these games cannot prove super-quadratic lower bounds for formulae sizes, and one should expect similar limitations for temporal logics, albeit with respect to some function that is at least exponential.

LTL *formulae learning.* Motivated by the practical issue of understanding a complex system starting from its execution traces, several recent works study the algorithmic problem of finding an LTL formula separating two finite sets of traces, see e.g. [21,5,24,11,12]. In light of Theorem 3, this problem is equivalent to finding a proof in (possibly variations of) our combinatorial proof system. We believe that this simple observation will turn out to be quite fruitful for both the "combinatorial games" and the "formula learning" communities. From our experience, tools such as the one developed in [21,5,24] are quite useful when studying combinatorial lower bounds, as they can be used to empirically test whether families of structures are difficult to separate, before attempting a formal proof. In our case, we have used the tool in [21] while searching for the structures and formulae we ended up using in Section 5, and discarded several other candidates thanks to the evidences the tool gave us. On the other side of the coin, combinatorial proof systems can be seen as a common foundational framework for all these formulae-learning procedures. With this in mind, we believe that the techniques developed for proving lower bounds in works such as [14] might be of help for improving these procedures, for example using the minimization of a sub-additive measure as a heuristic.

Acknowledgements Luca Geatti and Angelo Montanari acknowledge the support from the 2023 Italian INdAM-GNCS project *"Analisi simbolica e numerica di sistemi ciberfisici"*, ref. no. CUP_E53C22001930001. Angelo Montanari acknowledges the support of the MUR PNRR project FAIR - Future AI Research (PE00000013) funded by the NextGenerationEU. Alessio Mansutti is supported by the César Nombela grant 2023-T1/COM-29001, funded by the Madrid Regional Government, and by the grant PID2022-138072OB-I00, funded by MCIN/AEI/10.13039/501100011033 (FEDER, EU).

References

1. M. Adler and N. Immerman. An n! lower bound on formula size. *ACM Transactions on Computational Logic*, 4(3):296–314, 2003.
2. A. Artale, L. Geatti, N. Gigante, A. Mazzullo, and A. Montanari. Complexity of safety and cosafety fragments of linear temporal logic. In *AAAI'23*, pages 6236–6244, 2023.
3. A. Artale, L. Geatti, N. Gigante, A. Mazzullo, and A. Montanari. LTL over finite words can be exponentially more succinct than pure-past LTL, and vice versa. In *TIME'23*, volume 278, pages 2:1–2:14, 2023.
4. A. Artale, L. Geatti, N. Gigante, A. Mazzullo, and A. Montanari. A singly exponential transformation of LTL[X, F] into pure past LTL. In *KR'23*, pages 65–74, 2023.
5. A. Camacho and S. A. McIlraith. Learning interpretable models expressed in linear temporal logic. In *ICAPS'19*, pages 621–630, 2019.
6. E. Y. Chang, Z. Manna, and A. Pnueli. Characterization of temporal property classes. In *ICALP'92*, pages 474–486, 1992.
7. G. De Giacomo, A. Di Stasio, F. Fuggitti, and S. Rubin. Pure-past linear temporal and dynamic logic on finite traces. In *IJCAI'21*, pages 4959–4965, 2021.
8. G. De Giacomo and M. Y. Vardi. Linear temporal logic and linear dynamic logic on finite traces. In *IJCAI'13*, pages 854–860, 2013.
9. R. Fagin, J. Lenchner, K. W. Regan, and N. Vyas. Multi-structural games and number of quantifiers. In *LICS'21*, pages 1–13, 2021.
10. R. Fagin, J. Lenchner, N. Vyas, and R. R. Williams. On the number of quantifiers as a complexity measure. In *MFCS'22*, pages 48:1–48:14, 2022.
11. M. Fortin, B. Konev, V. Ryzhikov, Y. Savateev, F. Wolter, and M. Zakharyaschev. Unique characterisability and learnability of temporal instance queries. In *KR'22*, 2022.
12. M. Fortin, B. Konev, V. Ryzhikov, Y. Savateev, F. Wolter, and M. Zakharyaschev. Reverse engineering of temporal queries mediated by LTL ontologies. In *IJCAI'23*, pages 3230–3238, 2023.
13. L. Geatti, A. Mansutti, and A. Montanari. Succinctness of Cosafety Fragments of LTL via Combinatorial Proof Systems (extended version). *arXiv, cs.LO/2401.09860*, 2024.
14. M. Grohe and N. Schweikardt. The succinctness of first-order logic on linear orders. *Log. Methods Comput. Sci.*, 1(1:6), 2005.
15. M. Karchmer. *Communication complexity - a new approach to circuit depth.* MIT Press, 1989.
16. O. Kupferman and M. Y. Vardi. Model checking of safety properties. *Formal Methods in System Design*, 19(3):291–314, 2001.
17. O. Lichtenstein, A. Pnueli, and L. Zuck. The glory of the past. In *Workshop on Logic of Programs*, pages 196–218, 1985.
18. F. M. Maggi, M. Montali, and R. Peñaloza. Temporal logics over finite traces with uncertainty. In *AAAI'20*, pages 10218–10225, 2020.
19. Z. Manna and A. Pnueli. *Temporal verification of reactive systems - safety.* Springer, 1995.
20. N. Markey. Temporal logic with past is exponentially more succinct, concurrency column. *Bull. EATCS*, 79:122–128, 2003.
21. D. Neider and I. Gavran. Learning linear temporal properties. In *FMCAD'18*, pages 1–10, 2018.

22. M. Pesic and W. M. P. van der Aalst. A declarative approach for flexible business processes management. In J. Eder and S. Dustdar, editors, *BPM'06*, pages 169–180, 2006.
23. A. Pnueli. The temporal logic of programs. In *FOCS (SFCS'77)*, pages 46–57, 1977.
24. R. Raha, R. Roy, N. Fijalkow, and D. Neider. Scalable anytime algorithms for learning fragments of linear temporal logic. In *TACAS'22*, pages 263–280, 2022.
25. L. J. Stockmeyer and A. R. Meyer. Cosmological lower bound on the circuit complexity of a small problem in logic. *J. ACM*, 49(6):753–784, 2002.
26. M. Y. Vardi and P. Wolper. An automata-theoretic approach to automatic program verification. In *LICS'86*, pages 322–331, 1986.

A Resolution-Based Interactive Proof System for UNSAT

Philipp Czerner, Javier Esparza, and Valentin Krasotin(✉)

Technical University of Munich, Munich, Germany
{czerner,esparza,krasotin}@in.tum.de

Abstract. Modern SAT or QBF solvers are expected to produce correctness certificates. However, certificates have worst-case exponential size (unless NP = coNP), and at recent SAT competitions the largest certificates of unsatisfiability are starting to reach terabyte size.
Recently, Couillard, Czerner, Esparza, and Majumdar have suggested to replace certificates with interactive proof systems based on the IP = PSPACE theorem. They have presented an interactive protocol between a prover and a verifier for an extension of QBF. The overall running time of the protocol is linear in the time needed by a standard BDD-based algorithm, and the time invested by the verifier is polynomial in the size of the formula. (So, in particular, the verifier never has to read or process exponentially long certificates). We call such an interactive protocol *competitive* with the BDD algorithm for solving QBF.
While BDD-algorithms are state-of-the-art for certain classes of QBF instances, no modern (UN)SAT solver is based on BDDs. For this reason, we initiate the study of interactive certification for more practical SAT algorithms. In particular, we address the question whether interactive protocols can be competitive with some variant of resolution. We present two contributions. First, we prove a theorem that reduces the problem of finding competitive interactive protocols to finding an *arithmetisation* of formulas satisfying certain commutativity properties. (Arithmetisation is the fundamental technique underlying the IP = PSPACE theorem.) Then, we apply the theorem to give the first interactive protocol for the Davis-Putnam resolution procedure.

1 Introduction

Automated reasoning tools should provide evidence of their correct behaviour. A substantial amount of research has gone into proof-producing automated reasoning tools [12,17,16,10,3]. These works define a notion of "correctness certificate" and adapt the reasoning engine to produce independently checkable certificates. For example, SAT solvers produce either a satisfying assignment or a proof of unsatisfiability in some proof system, e.g. resolution (see [12] for a survey).

Current tools may produce certificates for UNSAT with hundreds of GiB or even, in extreme cases, hundreds of TiB [13]. This makes checking the certificate, or even sending it to a verifier, computationally expensive. Despite much progress on reducing the size of proofs and improving the efficiency of checking proofs (see

N. Kobayashi and J. Worrell (Eds.): FoSSaCS 2024, LNCS 14575, pp. 116–136, 2024.
https://doi.org/10.1007/978-3-031-57231-9_6

e.g. [11,12]), this problem is of fundamental nature: unless $\mathsf{NP} = \mathsf{coNP}$, which is considered very unlikely, certificates for UNSAT have worst-case exponential size in the size of the formula.

The $\mathsf{IP} = \mathsf{PSPACE}$ theorem, proved in 1992 by Shamir [18], presents a possible fundamental solution to this problem: *interactive proofs*[1]. A language is in IP if there exists a sound and complete *interactive proof protocol* between two agents, Prover and Verifier, that Verifier can execute in randomised polynomial time [7,2,15,1]. Completeness means that, for any input in the language, an *honest prover* that truthfully follows the protocol will convince Verifier to accept the input. Soundness means that, for any input not in the language, Verifier will reject it with high probability, no matter how Prover behaves. "Conventional" certification is the special case of interactive proof in which Prover sends Verifier only one message, the certificate, and Verifier is convinced with probability 1. The $\mathsf{IP} = \mathsf{PSPACE}$ theorem implies the existence of interactive proof protocols for UNSAT in which Verifier only invests polynomial time *in the size of the formula*. In particular, Verifier must never check exponentially long certificates from Prover, as is the case for conventional certification protocols in which Prover generates a proof in the resolution, DRAT, or any other of the proof systems found in the literature, and Verifier checks each step of the proof.

Despite its theoretical promise, the automated reasoning community has not yet developed tools for UNSAT or QBF with interactive proof protocols. In a recent paper [5], Couillard, Czerner, Esparza, and Majumdar venture a possible explanation. They argue that all interactive certification protocols described in the literature have been designed to prove asymptotic complexity results, for which it suffices to use honest provers that construct the full truth table of the formula. Such provers are incompatible with automated reasoning tools, which use more sophisticated data structures and heuristics. To remedy this, Couillard *et al.* propose to search for interactive proof protocols based on algorithms closer to those used in automatic reasoning tools. In [5], they consider the standard BDD-based algorithm for QBF and design an interactive proof protocol based on it.

While BDDs are still considered interesting for QBF, the consensus is that they are not state-of-the-art for UNSAT due to their high memory consumption. In this paper we initiate the study of interactive certification for SAT-solving algorithms closer to the ones used in tools. For this, given an algorithm Alg and an interactive protocol P, both for UNSAT, we say that P is *competitive* for Alg if the ratio between the runtime of Prover in P and the runtime of Alg on inputs of length n is bounded by a polynomial in n. So, loosely speaking, if P is competitive with Alg, then one can add interactive verification to Alg with only polynomial overhead. The general question we address is: which algorithms for UNSAT have competitive interactive proof protocols?

Our first contribution is a generic technique that, given an algorithm for UNSAT satisfying certain conditions, constructs a competitive interactive protocol.

[1] In our context it would be more adequate to speak of interactive certification, but we use the standard terminology.

Let us be more precise. We consider algorithms for UNSAT that, given a formula φ_0, construct a sequence $\varphi_0, \varphi_1, ..., \varphi_k$ of formulas such that φ_i is equisatisfiable to φ_{i+1}, and there is a polynomial algorithm that decides if φ_k is unsatisfiable. Our interactive protocols are based on the idea of encoding the formulas in this sequence as polynomials over a finite field in such a way that the truth value of the formula for a given assignment is determined by the value of the polynomial on that assignment. The encoding procedure is called *arithmetisation.* We introduce the notion of an arithmetisation *compatible* with a given algorithm. Loosely speaking, compatibility means that for each step $\varphi_i \mapsto \varphi_{i+1}$, there is an operation on polynomials mimicking the operation on formulas that transforms φ_i into φ_{i+1}. We show that the problem of finding a competitive interactive protocol for a given algorithm *Alg* for UNSAT reduces to finding an arithmetisation compatible with *Alg*.

In our second contribution, we apply our technique to construct the first interactive protocol competitive with a simplified version of the well-known Davis-Putnam procedure (see e.g. section 2.9 of [9]). Our version fixes a total order on variables, resolves exhaustively with respect to the next variable, say x, and then "locks" all clauses containing x or $\neg x$, ensuring that they are never resolved again w.r.t. any variable. We show that, while standard arithmetisations are not compatible with Davis-Putnam, a non-standard arithmetisation is. In our opinion, this is the main insight of our paper: in order to find interactive protocols for sophisticated algorithms for UNSAT, one can very much profit from non-standard arithmetisations.

The paper is structured as follows. Section 2 contains preliminaries. Section 3 presents interactive proof systems and defines interactive proof systems competitive with a given algorithm. Section 4 defines our version of the Davis-Putnam procedure. Section 5 introduces arithmetisations, and defines arithmetisations compatible with a given algorithm. Section 6 presents an interactive proof system for Davis-Putnam. Section 7 contains conclusions.

2 Preliminaries

Multisets. A multiset over a set S is a mapping $m\colon S \to \mathbb{N}$. We also write multisets using set notation, for example we write $\{x, x, y\}$ or $\{2 \cdot x, y\}$. Given two multisets m_1 and m_2, we define $m_1 \oplus m_2$ as the multiset given by $(m_1 \oplus m_2)(s) = m_1(s) + m_2(s)$ for every $s \in S$, and $m_1 \ominus m_2$ as the multiset given by $(m_1 \ominus m_2)(s) = \max\{0, m_1(s) - m_2(s)\}$ for every $s \in S$.

Formulas, CNF, and resolution. A Boolean *variable* has the form x_i where $i = 1, 2, 3, ...$ Boolean *formulas* are defined inductively: *true*, *false* and variables are formulas; if φ and ψ are formulas, then so are $\neg\varphi$, $\varphi \vee \psi$, and $\varphi \wedge \psi$. A *literal* is a variable or the negation of a variable. A formula φ is in *conjunctive normal form (CNF)* if it is a conjunction of disjunctions of literals. We represent a formula in CNF as a *multiset* of *clauses* where a clause is a *multiset* of literals. For example, the formula $(x \vee x \vee x \vee \neg y) \wedge z \wedge z$ is represented by the multiset $\{\{3x, \neg y\}, 2\{z\}\}$.

Remark 1. Usually CNF formulas are represented as *sets* of clauses, which are defined as *sets* of literals. Algorithms that manipulate CNF formulas using the set representation are assumed to silently remove duplicate formulas or duplicate literals. In this paper, due to the requirements of interactive protocols, we need to make these steps explicit. In particular, we use multiset notation for clauses. For example, $C(x)$ denotes the number of occurrences of x in C.

We assume in the paper that formulas are in CNF. Abusing language, we use φ to denote both a (CNF) formula and its multiset representation.

Resolution. Resolution is a proof system for CNF formulas. Given a variable x, a clause C containing exactly one occurrence of x and a clause C' containing exactly one occurrence of $\neg x$, the *resolvent* of C and C' with respect to x is the clause $Res_x(C, C') := (C \ominus \{x\}) \oplus (C' \ominus \{\neg x\})$.

For example, $Res_x(\{x, \neg y, z\}, \{\neg x, \neg w\}) = \{\neg y, z, \neg w\}$. It is easy to see that $C \wedge C'$ and $Res_x(C, C')$ are equisatisfiable. A *resolution refutation* for a formula φ is a sequence of clauses ending in the empty clause whose elements are either clauses of φ or resolvents of two previous clauses in the sequence. It is well known that φ is unsatisfiable iff there exists a resolution refutation for it. There exist families of formulas, like the pigeonhole formulas, for which the length of the shortest resolution refutation grows exponentially in the size of the formula, see e.g. [8,4].

Polynomials. Interactive protocols make extensive use of polynomials over a finite field $\mathbb{F}$. Let X be a finite set of variables. We use $x, y, z, \ldots$ for variables and $p, p_1, p_2, \ldots$ for polynomials. We use the following operations on polynomials:

- *Sum, difference, and product*, denoted $p_1 + p_2$, $p_1 - p_2$, $p_1 \cdot p_2$, and defined as usual. For example, $(3xy - z^2) + (z^2 + yz) = 3xy + yz$ and $(x + y) \cdot (x - y) = x^2 - y^2$.
- *Partial evaluation.* Denoted $\pi_{[x:=a]}\, p$, it returns the result of setting the variable x to the field element a in the polynomial p, e.g. $\pi_{[x:=5]}(3xy - z^2) = 15y - z^2$.

A *(partial) assignment* is a (partial) mapping $\sigma : X \to \mathbb{F}$. We write $\Pi_\sigma\, p$ for $\pi_{[x_1:=\sigma(x_1)]} \cdots \pi_{[x_k:=\sigma(x_k)]}\, p$, where $x_1, \ldots, x_k$ are the variables for which σ is defined. Additionally, we call a (partial) assignment σ *binary* if $\sigma(x) \in \{0, 1\}$ for each $x \in X$.

The following lemma is at the heart of all interactive proof protocols. Intuitively, it states that if two polynomials are different, then they are different for almost every input. Therefore, by picking an input at random, one can check polynomial equality with high probability.

Lemma 1 (Schwartz-Zippel Lemma). *Let p_1, p_2 be distinct univariate polynomials over $\mathbb{F}$ of degree at most $d \geq 0$. Let r be selected uniformly at random from $\mathbb{F}$. The probability that $p_1(r) = p_2(r)$ holds is at most $d/|\mathbb{F}|$.*

Proof. Since $p_1 \neq p_2$, the polynomial $p := p_1 - p_2$ is not the zero polynomial and has degree at most d. Therefore p has at most d zeros, and so the probability of $p(r) = 0$ is at most $d/|\mathbb{F}|$. □

3 Interactive Proof Systems

An *interactive protocol* is a sequence of interactions between two parties: *Prover* and *Verifier*. Prover has unbounded computational power, whereas Verifier is a randomised, polynomial-time algorithm. Initially, the parties share an input x that Prover claims belongs to a given language L (e.g. UNSAT). The parties alternate in sending messages to each other according to a protocol. Intuitively, Verifier repeatedly asks Prover to send informations. At the end of the protocol, Verifier accepts or rejects the input.

Formally, let V, P denote (randomised) online algorithms, i.e. given a sequence of inputs $m_1, m_2, ... \in \{0,1\}^*$ they compute a sequence of outputs, e.g. $V(m_1), V(m_1, m_2), ...$ We say that $(m_1, ..., m_{2k})$ is a *k-round interaction*, with $m_1, ..., m_{2k} \in \{0,1\}^*$, if $m_{i+1} = V(m_1, ..., m_i)$ for odd i and $m_{i+1} = P(m_1, ..., m_i)$ for even i.

The *output* $\mathsf{out}_{V,P,k}(x)$ is m_{2k}, where $(m_1, ..., m_{2k})$ is a k-round interaction with $m_1 = x$. We also define the *Verifier-time* $\mathsf{vtime}_{V,P,k}(x)$ as the expected time it takes V to compute $m_2, m_4, ..., m_{2k}$ for any k-round interaction $(m_1, ..., m_{2k})$ with $m_1 = x$. We define the *Prover-time* $\mathsf{ptime}_{V,P,k}(x)$ analogously.

Let L be a language and $p : \mathbb{N} \to \mathbb{N}$ a polynomial. A tuple (V, P_H, p) is an *interactive protocol for* L if for each $x \in \{0,1\}^*$ of length n we have $\mathsf{vtime}_{V,P_H,p(n)}(x) \in \mathcal{O}(\text{poly}\, n)$ and:

1. (*Completeness*) $x \in L$ implies $\mathsf{out}_{V,P_H,p(n)}(x) = 1$ with probability 1, and
2. (*Soundness*) $x \notin L$ implies that for all P we have $\mathsf{out}_{V,P,p(n)}(x) = 1$ with probability at most 2^{-n}.

The completeness property ensures that if the input belongs to the language L, then there is an "honest" Prover P_H who can always convince Verifier that indeed $x \in L$. If the input does not belong to the language, then the soundness property ensures that Verifier rejects the input with high probability no matter how a (dishonest) Prover tries to convince it.

IP is the class of languages for which there exists an interactive protocol. It is known that IP = PSPACE [15,18], that is, every language in PSPACE has a polynomial-round interactive protocol. The proof exhibits an interactive protocol for the language QBF of true quantified boolean formulas; in particular, the honest Prover is a polynomial-space, exponential-time algorithm.

3.1 Competitive Interactive Protocols

In an interactive protocol there are no restrictions on the running time of Prover. The existence of an interactive protocol for some coNP-complete problem in which Prover runs in polynomial time would imply e.g. NP $\subseteq$ BPP. Since this is widely believed to be false, Provers are allowed to run in exponential time, as in the proofs of [15,18]. However, while all known approaches for UNSAT use exponential time in the worst case, some perform much better in practice than others. For example, the Provers of [15,18] run in exponential time *in the best case*. This motivates our next definition: instead of stating that Prover must always be efficient, we say that it must have a bounded overhead *compared* to some given algorithm *Alg*.

Formally, let $L \subseteq \{0,1\}^*$ be a language, let *Alg* be an algorithm for L, and let (V, P_H, p) be an interactive protocol for L. We say that (V, P_H, p) is *competitive with Alg* if for every input $x \in \{0,1\}^*$ of length n we have $\mathsf{ptime}_{V,P_H,p(n)}(x) \in \mathcal{O}(\text{poly}(n)T(x))$, where $T(x)$ is the time it takes *Alg* to run on input x.

Recently, Couillard, Czerner, Esparza and Majumdar [5] have constructed an interactive protocol for QBF that is competitive with BDDSOLVER, the straightforward BDD-based algorithm that constructs a BDD for the satisfying assignments of each subformula, starting at the leaves of the syntax tree and progressively moving up. In this paper, we will investigate UNSAT and give an interactive protocol that is competitive with DAVISPUTNAM, a decision procedure for UNSAT based on a restricted version of resolution.

4 The Davis-Putnam Resolution Procedure

We introduce the variant of resolution for which we later construct a competitive interactive protocol. It is a version of the Davis-Putnam procedure [6,9][2]. Recall that in our setting, clauses are multisets, and given a clause C and a literal l, $C(l)$ denotes the number of occurrences of l in C.

Definition 1. *Let x be a variable.* Full x-resolution *is the procedure that takes as input a formula φ satisfying $C(x)+C(\neg x) \leq 1$ for every clause C, and returns the formula $R_x(\varphi)$ computed as follows:*

1. *For every pair C_1, C_2 of clauses of φ such that $x \in C_1$ and $\neg x \in C_2$, add to φ the resolvent w.r.t. x of C_1 and C_2 (i.e. set $\varphi := \varphi \oplus Res_x(C_1, C_2)$).*
2. *Remove all clauses containing x or $\neg x$.*

Full x-cleanup *is the procedure that takes as input a formula φ satisfying $C(x)+C(\neg x) \leq 2$ for every clause C, and returns the formula $C_x(\varphi)$ computed as follows:*

1. *Remove from φ all clauses containing both x and $\neg x$.*
2. *Remove from each remaining clause all duplicates of x or $\neg x$.*

The Davis-Putnam resolution procedure *is the algorithm for UNSAT that, given a total order $x_1 \prec x_2 \prec \cdots \prec x_n$ on the variables of an input formula φ, executes Algorithm 1. The algorithm assumes that φ is a set of sets of literals, that is, clauses contain no duplicate literals, and φ contains no duplicate clauses. We let $\square$ denote the empty clause.*

[2] In Harrison's book [9], the Davis-Putnam procedure consists of three rules. The version in Definition 1 uses only Rule III, which is sometimes called the Davis-Putnam resolution procedure. Unfortunately, at the time of writing this paper, the Wikipedia article for the Davis-Putnam algorithm uses a different terminology (even though it cites [9]): it calls the three-rule procedure the Davis-Putnam *algorithm*, and the algorithm consisting only of Rule III the Davis-Putnam *procedure*.

Algorithm 1 DAVISPUTNAM(φ)

```
for i = 1, ..., n do
    φ := R_{x_i}(φ)
    for j = i + 1, ..., n do
        φ := C_{x_j}(φ)
if □ ∈ φ then
    return "unsatisfiable"
else
    return "satisfiable"
```

Step	Formula	Arithmetisation
Inp.	$\varphi = \{\{x,y\},\{x,\neg y,\neg z\},\{\neg x,\neg z\},$ $\{\neg x,\neg y,\neg z\},\{y,z\},\{\neg y,z\}\}$	$\mathcal{B}(\varphi) = (1-x)(1-y) + (1-x)y^3z^3 + x^3z^3$ $+x^3y^3z^3 + (1-y)(1-z) + y^3(1-z)$
R_x	$\varphi_1 = \{\{y,\neg z\},\{y,\neg y,\neg z\},\{\neg y,\neg z,\neg z\}$ $\{\neg y,\neg z,\neg y,\neg z\},\{y,z\},\{\neg y,z\}\}$	$\mathcal{B}(\varphi_1) = (1-y)z^3 + (1-y)y^3z^3 + y^3z^6$ $+y^6z^6 + (1-y)(1-z) + y^3(1-z)$
C_y	$\varphi_2 = \{\{y,\neg z\}, 2\cdot\{\neg y,\neg z,\neg z\},$ $\{y,z\},\{\neg y,z\}\}$	$\mathcal{B}(\varphi_2) = (1-y)z^3 + 2y^3z^6$ $+(1-y)(1-z) + y^3(1-z)$
C_z	$\varphi_3 = \{\{y,\neg z\}, 2\cdot\{\neg y,\neg z\}$ $\{y,z\},\{\neg y,z\}\}$	$\mathcal{B}(\varphi_3) = (1-y)z^3 + 2y^3z^3$ $+(1-y)(1-z) + y^3(1-z)$
R_y	$\varphi_4 = \{2\cdot\{\neg z,\neg z\}, 3\cdot\{\neg z,z\},\{z,z\}\}$	$\mathcal{B}(\varphi_4) = 2z^6 + 3z^3(1-z) + (1-z)^2$
C_z	$\varphi_5 = \{2\cdot\{\neg z\},\{z\}\}$	$\mathcal{B}(\varphi_5) = 2z^3 + (1-z)$
R_z	$\varphi_6 = \{2\cdot\Box\}$	$\mathcal{B}(\varphi_6) = 2$

Table 1. Run of DAVISPUTNAM on an input φ, and arithmetisation of the intermediate formulas.

Observe that while the initial formula contains no duplicate clauses, the algorithm may create them, and they are not removed.

Example 1. Table 1 shows on the left a run of DAVISPUTNAM on a formula φ with three variables and six clauses. The right column is explained in Section 6.1.

It is well-known that the Davis-Putnam resolution procedure is complete, but we give a proof suitable for our purposes. Let $\varphi[x := \mathit{true}]$ denote the result of replacing all occurrences of x in φ by *true* and all occurrences of $\neg x$ by *false*. Define $\varphi[x := \mathit{false}]$ reversely. Further, let $\exists x\varphi$ be an abbreviation of $\varphi[x := \mathit{true}] \vee \varphi[x := \mathit{false}]$. We have:

Lemma 2. *Let x be a variable and φ a formula in CNF such that $C(x) + C(\neg x) \leq 1$ for every clause C. Then $R_x(\varphi) \equiv \exists x\varphi$.*

Proof. Let $C_1, ..., C_k$ be the clauses of φ. We have

$$\exists x\varphi \equiv \varphi[x := \mathit{true}] \vee \varphi[x := \mathit{false}]$$

$$\equiv \Big(\bigwedge_{i \in [k]} C_i[x := \mathit{true}] \Big) \vee \Big(\bigwedge_{j \in [k]} C_j[x := \mathit{false}] \Big)$$

$$\equiv \bigwedge_{i,j \in [k]} (C_i[x := \mathit{true}] \vee C_j[x := \mathit{false}])$$

$$\equiv \bigwedge_{i \in [k],\ x, \neg x \notin C_i} C_i \wedge \bigwedge_{i,j \in [k], \neg x \in C_i, x \in C_j} (C_i[x := \mathit{true}] \vee C_j[x := \mathit{false}])$$

$$\equiv R_x(\varphi).$$

For the second-to-last equivalence, consider a clause C_i containing neither x nor $\neg x$. Then $C_i \vee C_i$ is a clause of $\bigwedge_{i,j \in [k]} \big(C_i[x := \mathit{true}] \vee C_j[x := \mathit{false}]\big)$, and it subsumes any other clause of the form $C_i \vee C_j$. The first conjunct of the penultimate line contains these clauses. Furthermore, if C_i contains x or if C_j contains $\neg x$, then the disjunction $C_i[x := \mathit{true}] \vee C_j[x := \mathit{false}]$ is a tautology and can thus be ignored. It remains to consider the pairs (C_i, C_j) of clauses such that $\neg x \in C_i$ and $x \in C_j$. This is the second conjunct. □

Lemma 3. *Let x be a variable and φ a formula in CNF such that $C(x) + C(\neg x) \leq 2$ for every clause C. Then $C_x(\varphi) \equiv \varphi$.*

Proof. Since $x \vee \neg x \equiv \mathit{true}$, a clause containing both x and $\neg x$ is valid and thus can be removed. Furthermore, duplicates of x in a clause can be removed because $x \vee x \equiv x$. □

Theorem 1. DAVISPUTNAM *is sound and complete.*

Proof. Let φ be a formula over the variables $x_1, ..., x_n$. By Lemmas 2 and 3, after termination the algorithm arrives at a formula without variables equivalent to $\exists x_n \cdots \exists x_1 \varphi$. This final formula is equivalent to the truth value of whether φ is satisfiable; that is, φ is unsatisfiable iff the final formula contains the empty clause. □

5 Constructing Competitive Interactive Protocols for UNSAT

We consider algorithms for UNSAT that, given a formula, execute a sequence of *macrosteps*. Throughout this section, we use DAVISPUTNAM as running example.

Definition 2. *A* macrostep *is a partial mapping M that transforms a formula φ into a formula $M(\varphi)$ equisatisfiable to φ.*

The first macrostep is applied to the input formula. The algorithm accepts if the formula returned by the last macrostep is equivalent to *false*. Clearly, all these algorithms are sound.

Example 2. The macrosteps of DAVISPUTNAM are R_x and C_x for each variable x. On a formula with n variables, DAVISPUTNAM executes exactly $\frac{n(n+1)}{2}$ macrosteps.

We present an abstract design framework to obtain competitive interactive protocols for these macrostep-based algorithms. As in [15,18,5], the framework is based on *arithmetisation* of formulas. Arithmetisations are mappings that assign to a formula a polynomial with integer coefficients. In protocols, Verifier asks Prover to return the result of evaluating polynomials obtained by arithmetising formulas not over the integers, but over a prime field $\mathbb{F}_q$, where q is a sufficiently large prime. An arithmetisation is useful for the design of protocols if the value of the polynomial on a *binary input*, that is, an assignment that assigns 0 or 1 to every variable, determines the truth value of the formula under the assignment. We are interested in the following class of arithmetisations, just called *arithmetisations* for brevity:

Definition 3. *Let $\mathcal{F}$ and $\mathcal{P}$ denote the sets of formulas and polynomials over a set of variables. An* arithmetisation *is a mapping $\mathcal{A}\colon \mathcal{F} \to \mathcal{P}$ such that for every formula φ and every assignment σ to its variables:*

(a) σ satisfies φ iff $\Pi_\sigma \mathcal{A}(\varphi) = 0$,[3] and
(b) $\Pi_\sigma \mathcal{A}(\varphi) \pmod q$ can be computed in time $\mathcal{O}(|\varphi| \operatorname{polylog} q)$ for any prime q.

In particular, two formulas φ, ψ over the same set of variables are equivalent if and only if for every binary assignment σ, $\Pi_\sigma \mathcal{A}(\varphi)$ and $\Pi_\sigma \mathcal{A}(\psi)$ are either both zero or both nonzero.

Example 3. Let $\mathcal{A}$ be the mapping inductively defined as follows:

$$\begin{array}{lll} \mathcal{A}(\mathit{true}) := 0 & \mathcal{A}(\neg x) := x & \mathcal{A}(\varphi_1 \wedge \varphi_2) := \mathcal{A}(\varphi_1) + \mathcal{A}(\varphi_2) \\ \mathcal{A}(\mathit{false}) := 1 & \mathcal{A}(x) := 1 - x & \mathcal{A}(\varphi_1 \vee \varphi_2) := \mathcal{A}(\varphi_1) \cdot \mathcal{A}(\varphi_2). \end{array}$$

For example, $\mathcal{A}((x \vee \mathit{false}) \wedge \neg x)) = ((1-x) \cdot 1) + x = 1$. It is easy to see that $\mathcal{A}$ is an arithmetisation in the sense of Definition 3. Notice that $\mathcal{A}$ can map equivalent formulas to different polynomials. For example, $\mathcal{A}(\neg x) = x$ and $\mathcal{A}(\neg x \wedge \neg x) = 2x$.

We define when an arithmetisation $\mathcal{A}$ is compatible with a macrostep M.

Definition 4. *Let $\mathcal{A}\colon \mathcal{F} \to \mathcal{P}$ be an arithmetisation and let $M\colon \mathcal{F} \to \mathcal{F}$ be a macrostep. $\mathcal{A}$ is* compatible with M *if there exists a partial mapping $P_M \colon \mathcal{P} \to \mathcal{P}$ and a* pivot variable $x \in X$ *satisfying the following conditions:*

(a) P_M simulates M*: For every formula φ where $M(\varphi)$ is defined, we have $\mathcal{A}(M(\varphi)) = P_M(\mathcal{A}(\varphi))$.*

[3] In most papers one requires that σ satisfies φ iff $\Pi_\sigma \mathcal{A}(\varphi) = 1$. Because of our later choice of arithmetisations, we prefer $\Pi_\sigma \mathcal{A}(\varphi) = 0$.

(b) P_M commutes with partial evaluations*: For every polynomial* p *and every assignment* $\sigma\colon X \setminus \{x\} \to \mathbb{Z}$*:* $\Pi_\sigma(P_M(p)) = P_M(\Pi_\sigma(p))$*.*
(c) $P_M(p \pmod q) = P_M(p) \pmod q$ *for any prime* q. [4]
(d) P_M *can be computed in polynomial time.*

An arithmetisation $\mathcal{A}$ *is* compatible *with Alg if it is compatible with every macrostep executed by Alg.*

Graphically, an arithmetisation $\mathcal{A}$ is compatible with M if there exists a mapping P_M such that the following diagram commutes:

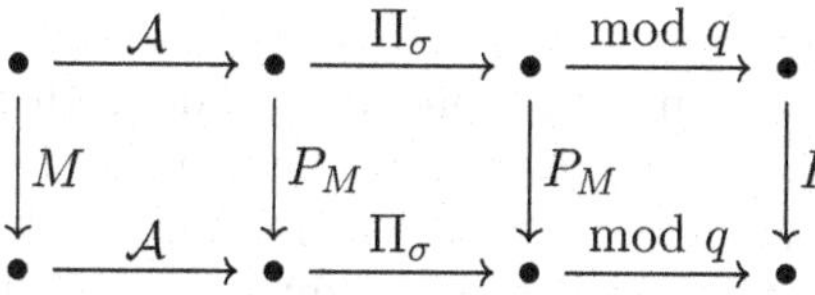

We can now state and prove the main theorem of this section.

Theorem 2. *Let Alg be an algorithm for UNSAT and let* $\mathcal{A}$ *be an arithmetisation compatible with Alg such that for every input* φ

(a) Alg executes a sequence of $k \in \mathcal{O}(\text{poly}|\varphi|)$ *macrosteps, which compute a sequence* $\varphi_0, \varphi_1, ..., \varphi_k$ *of formulas with* $\varphi_0 = \varphi$*,*
(b) $\mathcal{A}(\varphi_i)$ *has maximum degree at most* $d \in \mathcal{O}(\text{poly}|\varphi|)$*, for any* i*, and*
(c) $\mathcal{A}(\varphi_k)$ *is a constant polynomial such that* $|\mathcal{A}(\varphi_k)| \leq 2^{2^{\mathcal{O}(|\varphi|)}}$.

Then there is an interactive protocol for UNSAT that is competitive with Alg.

To prove Theorem 2, we first define a generic interactive protocol for UNSAT depending only on *Alg* and $\mathcal{A}$, and then prove that it satisfies the properties of an interactive proof system: if φ is unsatisfiable and Prover is honest, Verifier always accepts; and if φ is satisfiable, then Verifier accepts with probability at most $2^{-|\varphi|}$, regardless of Prover.

5.1 Interactive Protocol

The interactive protocol for a given algorithm *Alg* operates on polynomials over a prime finite field, instead of the integers. Given a prime q, we write $\mathcal{A}_q(p) := \mathcal{A}(p) \pmod q$ for the polynomial over $\mathbb{F}_q$ (the finite field with q elements) that one obtains by taking the coefficients of $\mathcal{A}(p)$ modulo q.

At the start of the protocol, Prover sends Verifier a prime q, and then exchanges messages with Verifier about the values of polynomials over $\mathbb{F}_q$, with the goal of convincing Verifier that $\mathcal{A}(\varphi_k) \neq 0$ by showing $\mathcal{A}_q(\varphi_k) \neq 0$ instead. The

[4] We implicitly extend P_M to polynomials over $\mathbb{F}_q$ in the obvious way: we consider the input p as a polynomial over $\mathbb{Z}$ by selecting the smallest representative in $\mathbb{N}$ for each coefficient, apply P_M, and then take the coefficients of the output polynomial modulo q.

following lemma demonstrates that this is both sound and complete; (a) shows that a dishonest Prover cannot cheat in this way, and (b) shows that an honest Prover can always convince Verifier.

Lemma 4. *Let φ_k be the last formula computed by Alg.*

(a) For every prime q, we have that $\mathcal{A}_q(\varphi_k) \neq 0$ implies that φ is unsatisfiable.
(b) If φ is unsatisfiable, then there exists a prime q s.t. $\mathcal{A}_q(\varphi_k) \neq 0$.

Proof. For every prime q, if $\mathcal{A}_q(\varphi_k) \neq 0$ then $\mathcal{A}(\varphi_k) \neq 0$. For the converse, pick any prime q larger than $\mathcal{A}(\varphi_k)$. □

We let $\varphi = \varphi_0, \varphi_1, ..., \varphi_k$ denote the sequence of formulas computed by *Alg*, and d the bound on the polynomials $\mathcal{A}(\varphi_i)$ of Theorem 2. Observe that the formulas in the sequence can be exponentially larger than φ, and so Verifier cannot even read them. For this reason, during the protocol Verifier repeatedly sends Prover partial assignments σ chosen at random, and Prover sends back to Verifier *claims* about the formulas of the sequence of the form $\Pi_\sigma \mathcal{A}_q(\varphi_i) = w$. The first claim is about φ_k, the second about φ_{k-1}, and so on. Verifier stores the current claim by maintaining variables i, w, and σ. The protocol guarantees that the claim about φ_i *reduces* to the claim about φ_{i-1}, in the following sense: if a dishonest Prover makes a false claim about φ_i but a true claim about φ_{i+1}, Verifier detects with high probability that the claim about φ_i is false and rejects. Therefore, in order to make Verifier accept a satisfiable formula φ, a dishonest Prover must keep making false claims, and in particular make a false last claim about $\varphi_0 = \varphi$. The protocol also guarantees that a false claim about φ_0 is always detected by Verifier.

The protocol is described in Table 2. It presents the steps of Verifier and an *honest* Prover.

Example 4. In the next section we use the generic protocol of Table 2 to give an interactive protocol for *Alg* := DavisPutnam, using an arithmetisation called $\mathcal{B}$. Table 3 shows a possible run of this protocol on the formula φ of Table 1. We can already explain the shape of the run, even if $\mathcal{B}$ is not defined yet.

Recall that on input φ, DavisPutnam executes six steps, shown on the left column of Table 1, that compute the formulas $\varphi_1, ..., \varphi_6$. Each row of Table 3 corresponds to a round of the protocol. In round i, Prover sends Verifier the polynomial p_i corresponding to the claim $\Pi_\sigma \mathcal{A}_q(\varphi_i)$ (column Honest Prover). Verifier performs a check on the claim (line with $\stackrel{?}{=}$). If the check passes, Verifier updates σ and sends it to Prover as the assignment to be used for the next claim.

5.2 The interactive protocol is correct and competitive with *Alg*

We need to show that the interactive protocol of Table 2 is correct and competitive with *Alg*. We do so by means of a sequence of lemmas. Lemmas 6-8 bound the error probability of Verifier and the running time of both Prover and Verifier as a function of the prime q. Lemma 9 shows that Prover can efficiently compute a suitable prime. The last part of the section combines the lemmas to prove Theorem 2.

1. Prover picks an appropriate prime q; i.e. a prime s.t. $\mathcal{A}_q(\varphi_k) \neq 0$, where φ_k is the last formula computed by *Alg*. (The algorithm to compute q is given later.)
2. Prover sends both q and $\mathcal{A}_q(\varphi_k)$ to Verifier. If Prover sends $\mathcal{A}_q(\varphi_k) = 0$, Verifier rejects.
3. Verifier sets $i := k$, $w := \mathcal{A}_q(\varphi_k)$ (sent by Prover in the previous step), and σ to an arbitrary assignment. (Since initially $\mathcal{A}_q(\varphi_k)$ is a constant, σ is irrelevant.)
4. For each $i = k, ..., 1$, the claim about φ_i is reduced to a claim about φ_{i-1}:
   - 4.1 Let x denote the pivot variable of M_i and set σ' to the partial assignment that is undefined on x and otherwise matches σ. Prover sends the polynomial $p := \Pi_{\sigma'} \mathcal{A}_q(\varphi_{i-1})$, which is a univariate polynomial in x.
   - 4.2 If the degree of p exceeds d or $\pi_{[x:=\sigma(x)]} P_{M_i}(p) \neq w$, Verifier rejects. Otherwise, Verifier chooses an $r \in \mathbb{F}_q$ uniformly at random and updates $w := \pi_{[x:=r]} p$ and $\sigma(x) := r$.
5. Finally, Verifier checks the claim $\Pi_\sigma \mathcal{A}_q(\varphi_0) = w$ by itself and rejects if it does not hold. Otherwise, Verifier accepts.

Table 2. An interactive protocol for an algorithm for UNSAT describing the behaviour of Verifier and the honest Prover.

Completeness. We start by establishing that an honest Prover can always convince Verifier.

Lemma 5. *If φ is unsatisfiable and Prover is honest (i.e. acts as described in Table 2), then Verifier accepts with probability 1.*

Proof. We show that Verifier accepts. First we show that Verifier does not reject in step 2, i.e. that $\mathcal{A}_q(\varphi_k) \neq 0$. Since φ is unsatisfiable by assumption, by Definition 2 we have that φ_k is unsatisfiable. Then, Definition 3(a) implies $\mathcal{A}_q(\varphi_k) \neq 0$ (note that $\mathcal{A}_q(\varphi_k)$ is constant, by Theorem 2(c)).

Let us now argue that the claim Verifier tracks (i.e., the claim given by the current values of the variables) is always true. In step 3, it is initialised with $w := \mathcal{A}_q(\varphi_k)$, so the claim is true at that point.

In each step 4.2, Verifier checks $\pi_{[x:=\sigma(x)]} P_{M_i}(p) \stackrel{?}{=} w$. As Prover is honest, it sent $p := \Pi_{\sigma'} \mathcal{A}_q(\varphi_{i-1})$ in the previous step; so the check is equivalent to

$$\begin{aligned} w &\stackrel{?}{=} \pi_{[x:=\sigma(x)]} P_{M_i}(\Pi_{\sigma'} \mathcal{A}_q(\varphi_{i-1})) && \text{(Definition 4(b))} \\ &= \Pi_\sigma P_{M_i}(\mathcal{A}_q(\varphi_{i-1})) && \text{(Definition 4(a,c))} \\ &= \Pi_\sigma \mathcal{A}_q(M_i(\varphi_{i-1})) = \Pi_\sigma \mathcal{A}_q(\varphi_i) \end{aligned}$$

By induction hypothesis $w = \Pi_\sigma \mathcal{A}_q(\varphi_i)$ holds, and thus Verifier does not reject.

When Verifier updates the claim, it selects a random number r. Due to $p = \Pi_{\sigma'} \mathcal{A}_q(\varphi_{i-1})$, the new claim will hold for all possible values of r.

In step 5, we still have the invariant that the claim is true, so Verifier will accept. □

Probability of error. Establishing soundness is more involved. The key idea of the proof (which is the same idea as for other interactive protocols) is that for Verifier to accept erroneously, the claim it tracks must at some point be true. However, initially the claim is false. It thus suffices to show that each step of the algorithm is unlikely to turn a false claim into a true one.

Lemma 6. *Let $\mathcal{A}, d, k$ as in Theorem 2. If φ is satisfiable, then for any Prover, honest or not, Verifier accepts with probability at most $dk/q \in \mathcal{O}(\text{poly}(|\varphi|)/q)$.*

Proof. Let $i \in \{k, ..., 1\}$, let σ, w denote the values of these variables at the beginning of step 4.1 in iteration i, and let σ', w' denote the corresponding (updated) values at the end of step 4.2.

We say that Prover *tricks* Verifier at iteration i if the claim tracked by Verifier was false at the beginning of step 4 and is true at the end, i.e. $\Pi_\sigma \mathcal{A}_q(\varphi_i) \neq w$ and $\Pi_{\sigma'} \mathcal{A}_q(\varphi_{i-1}) = w'$.

The remainder of the proof is split into three parts.

(a) If Verifier accepts, it was tricked.
(b) For any i, Verifier is tricked at iteration i with probability at most d/q.
(c) Verifier is tricked with probability at most $dk/q \in \mathcal{O}(\text{poly}(|\varphi|)/q)$.

Part (a). If φ is satisfiable, then so is φ_k (Definition 2), and thus $\Pi_\sigma \mathcal{A}_q(\varphi_k) = 0$ (Definition 3(a); also note that $\Pi_\sigma \mathcal{A}_q(\varphi_k)$ is constant). Therefore, in step 2 Prover either claims $\Pi_\sigma \mathcal{A}_q(\varphi_k) = 0$ and Verifier rejects, or the initial claim in step 3 is false.

If Verifier is never tricked, the claim remains false until step 5 is executed, at which point Verifier will reject. So to accept, Verifier must be tricked.

Part (b). Let $i \in \{k, ..., 1\}$ and assume that the claim is false at the beginning of iteration i of step 4. Now there are two cases. If Prover sends the polynomial $p = \Pi_{\sigma'} \mathcal{A}_q(\varphi_{i-1})$, then, as argued in the proof of Lemma 5, Verifier's check is equivalent to $w \stackrel{?}{=} \Pi_\sigma \mathcal{A}_q(\varphi_i)$, which is the current claim. However, we have assumed that the claim is false, so Verifier would reject. Hence, Prover must send a polynomial $p \neq \Pi_{\sigma'} \mathcal{A}_q(\varphi_{i-1})$ (of degree at most d) to trick Verifier.

By Lemma 1, the probability that Verifier selects an r with $\pi_{[x:=r]} p = \pi_{[x:=r]} \Pi_{\sigma'} \mathcal{A}_q(\varphi_{i-1})$ is at most d/q. Conversely, with probability at least $1 - d/q$, the new claim is false as well and Verifier is not tricked in this iteration.

Part (c). We have shown that the probability that Verifier is tricked in one iteration is at most d/q. By union bound, Verifier is thus tricked with probability at most dk/q, as there are k iterations. By conditions (a) and (b) of Theorem 2, we get $dk/q \in \mathcal{O}(\text{poly}(|\varphi|)/q)$. □

Running time of Verifier. The next lemma estimates Verifier's running time in terms of $|\varphi|$ and q.

Lemma 7. *Verifier runs in time $\mathcal{O}(\text{poly}(|\varphi| \log q))$.*

Proof. Verifier performs operations on polynomials of degree at most d with coefficients in $\mathbb{F}_q$. So a polynomial can be represented using $d \log q$ bits, and arithmetic operations are polynomial in that representation. Additionally, Verifier needs to execute P_{M_i} for each i, which can also be done in polynomial time (Definition 4(c)). There are $k \in \mathcal{O}(\text{poly}|\varphi|)$ iterations.

Finally, Verifier checks the claim $\Pi_\sigma \mathcal{A}_q(\varphi) = w$ for some assignment σ and $w \in \mathbb{F}_q$. Definition 3 ensures that this takes $\mathcal{O}(|\varphi| \operatorname{polylog} q)$ time. The overall running time is therefore in $\mathcal{O}(\text{poly}(|\varphi| d \log q))$. The final result follows from condition (b) of Theorem 2. □

Running time of Prover. We give a bound on the running time of Prover, excluding the time needed to compute the prime q.

Lemma 8. *Assume that $\mathcal{A}$ is an arithmetisation satisfying the conditions of Theorem 2. Let T denote the time taken by Alg on φ. The running time of Prover, excluding the time needed to compute the prime q, is $\mathcal{O}(T \operatorname{poly}|\varphi| \log q))$.*

Proof. After picking the prime q, Prover has to compute $\Pi_\sigma \mathcal{A}_q(\varphi_i)$ for different $i \in [k]$ and assignments σ. The conditions of Theorem 2 guarantee that this can be done in time $\mathcal{O}(|\varphi_i| \operatorname{polylog} q) \subseteq \mathcal{O}(|\varphi_i| \operatorname{poly}(|\varphi| \log q))$. We have $\sum_i |\varphi_i| \leq T$, as *Alg* needs to write each φ_i during its execution. The total running time follows by summing over i. □

Computing the prime q. The previous lemmas show the dependence of Verifier's probability of error and the running times of Prover and Verifier as a function of $|\varphi|$ and q. Our final lemma gives a procedure for Prover to compute a suitable prime q. Together with the previous lemmas, this will easily yield Theorem 2.

Lemma 9. *For every $c > 0$ there exists a procedure for Prover to find a prime $q \in 2^{\mathcal{O}(|\varphi|)}$ such that $q \geq 2^{c|\varphi|}$ and $\mathcal{A}_q(\varphi_k) \neq 0$ in expected time $\mathcal{O}(T|\varphi|)$, where T is the running time of Alg.*

Proof. Assume wlog. that $c > 1$. Prover first runs *Alg* to compute φ_k and then chooses a prime q with $2^{c|\varphi|} \leq q < 2^{c|\varphi|+1}$ uniformly at random; thus $q \in 2^{\mathcal{O}(|\varphi|)}$ is guaranteed. If Prover arrives at $\mathcal{A}_q(\varphi_k) = 0$, Prover chooses another prime q in the same way, until one is chosen s.t. $\mathcal{A}_q(\varphi_k) \neq 0$.

Since $|\mathcal{A}(\varphi_k)| \leq 2^{2^{|\varphi|}}$, $\mathcal{A}(\varphi_k)$ is divisible by at most $2^{|\varphi|}$ different primes. Using the prime number theorem, there are $\Omega(2^{c|\varphi|}/c|\varphi|)$ primes $2^{c|\varphi|} \leq q < 2^{c|\varphi|+1}$, so the probability that the picked q divides $\mathcal{A}(\varphi_k)$ is $\mathcal{O}(c|\varphi|/2^{(c-1)|\varphi|})$.

Therefore, for any $c > 1$ this probability is at most, say, 1/2 for sufficiently large $|\varphi|$. In expectation, Prover thus needs to test 2 primes q, and each test takes time $\mathcal{O}(|\varphi_k| \operatorname{polylog} q)$ (see Definition 3(b)), which is in $\mathcal{O}(T|\varphi|)$. □

Proof of Theorem 2. We can now conclude the proof of the theorem.

Completeness was already proved in Lemma 5.

Soundness. We need to ensure that the error probability is at most $2^{-|\varphi|}$. By Lemma 6, the probability p of error satisfies $p \leq dk/q$, where $dk \in \mathcal{O}(\text{poly}(|\varphi|))$. So there is a $\xi > 0$ with $dk \leq 2^{\xi|\varphi|}$. Using $c := 1 + \xi$ as constant for Lemma 9, we are done.

Verifier's running time. By Lemma 7, Verifier runs in time $\mathcal{O}(\text{poly}(|\varphi| \log q))$. Using the prime $q \in 2^{\mathcal{O}(|\varphi|)}$ of Lemma 9, the running time is $\mathcal{O}(\text{poly}(|\varphi|)$.

Competitivity. By Lemma 8, Prover runs in time $\mathcal{O}(T \,\text{poly}(|\varphi| \log q))$ plus the time need to compute the prime, which, by Lemma 9, is in $\mathcal{O}(T \,\text{poly}(|\varphi|))$. Again using $q \in \mathcal{O}(2^{|\varphi|})$, we find that the protocol is competitive with *Alg*. □

6 An Interactive Proof System Competitive with the Davis-Putnam Resolution Procedure

In order to give an interactive proof system for the Davis-Putnam resolution procedure, it suffices to find an arithmetisation which is compatible with the full x-resolution step R_x and the full x-cleanup step C_x such that all properties of Theorem 2 are satisfied. In this section, we present such an arithmetisation.

6.1 An arithmetisation compatible with R_x and C_x

We find an arithmetisation compatible with both R_x and C_x. Let us first see that the arithmetisation of Example 3 does not work.

Example 5. The arithmetisation $\mathcal{A}$ of Example 3 is not compatible with R_x. To see this, let $\varphi = (\neg x \vee \neg y) \wedge (x \vee \neg z) \wedge \neg w$. We have $R_x(\varphi) = (\neg y \vee \neg z) \wedge \neg w$, $\mathcal{A}(R_x(\varphi)) = yz + w$, and $\mathcal{A}(\varphi) = xy + (1-x)z + w = x(y-z) + z + w$. If $\mathcal{A}$ were compatible with R_x, then there would exist an operation P_{R_x} on polynomials such that $P_{R_x}(x(y-z) + z + w) = yz + w$ by Definition 4(a), and from Definition 4(b), we get $P_{R_x}(\Pi_\sigma(x(y-z) + z + w)) = \Pi_\sigma(yz + w)$ for all partial assignments $\sigma : \{y, z, w\} \to \mathbb{Z}$. For $\sigma := \{y \mapsto 1, z \mapsto 0, w \mapsto 1\}$, it follows that $P_{R_x}(x+1) = 1$, but for $\sigma := \{y \mapsto 2, z \mapsto 1, w \mapsto 0\}$, it follows that $P_{R_x}(x+1) = 2$, a contradiction.

We thus present a non-standard arithmetisation.

Definition 5. *The arithmetisation $\mathcal{B}$ of a CNF formula φ is the recursively defined polynomial*

$$\mathcal{B}(true) := 0 \quad \mathcal{B}(x) := 1 - x \quad \mathcal{B}(\varphi_1 \wedge \varphi_2) := \mathcal{B}(\varphi_1) + \mathcal{B}(\varphi_2)$$
$$\mathcal{B}(false) := 1 \quad \mathcal{B}(\neg x) := x^3 \quad \mathcal{B}(\varphi_1 \vee \varphi_2) := \mathcal{B}(\varphi_1) \cdot \mathcal{B}(\varphi_2).$$

Example 6. The right column of Table 1 shows the polynomials obtained by applying $\mathcal{B}$ to the formulas on the left. For example, we have $\mathcal{B}(\varphi_5) = \mathcal{B}(\neg z \wedge \neg z \wedge z) = 2\mathcal{B}(\neg z) + \mathcal{B}(z) = 2z^3 + 1 - z$.

We first prove that $\mathcal{B}$ is indeed an arithmetisation.

Proposition 1. *For every formula φ and every assignment $\sigma : X \to \{0,1\}$ to the variables X of φ, we have that σ satisfies φ iff $\Pi_\sigma \mathcal{B}(\varphi) = 0$.*

Proof. We prove the statement by induction on the structure of φ. The statement is trivially true for $\varphi \in \{\mathit{true}, \mathit{false}, x, \neg x\}$. For $\varphi = \varphi_1 \vee \varphi_2$, we have

$$\begin{aligned}
&\sigma \text{ satisfies } \varphi \Leftrightarrow \sigma \text{ satisfies } \varphi_1 \vee \varphi_2 \Leftrightarrow \sigma \text{ satisfies } \varphi_1 \text{ or } \sigma \text{ satisfies } \varphi_2\\
&\overset{IH}{\Leftrightarrow} \Pi_\sigma \mathcal{B}(\varphi_1) = 0 \vee \Pi_\sigma \mathcal{B}(\varphi_2) = 0 \Leftrightarrow \Pi_\sigma \mathcal{B}(\varphi_1) \cdot \Pi_\sigma \mathcal{B}(\varphi_2) = 0\\
&\Leftrightarrow \Pi_\sigma \mathcal{B}(\varphi_1 \vee \varphi_2) = 0 \Leftrightarrow \Pi_\sigma \mathcal{B}(\varphi) = 0,
\end{aligned}$$

and for $\varphi = \varphi_1 \wedge \varphi_2$, we have

$$\begin{aligned}
&\sigma \text{ satisfies } \varphi \Leftrightarrow \sigma \text{ satisfies } \varphi_1 \wedge \varphi_2 \Leftrightarrow \sigma \text{ satisfies } \varphi_1 \text{ and } \sigma \text{ satisfies } \varphi_2\\
&\overset{IH}{\Leftrightarrow} \Pi_\sigma \mathcal{B}(\varphi_1) = 0 \wedge \Pi_\sigma \mathcal{B}(\varphi_2) = 0 \Leftrightarrow \Pi_\sigma \mathcal{B}(\varphi_1) + \Pi_\sigma \mathcal{B}(\varphi_2) = 0\\
&\Leftrightarrow \Pi_\sigma \mathcal{B}(\varphi_1 \wedge \varphi_2) = 0 \Leftrightarrow \Pi_\sigma \mathcal{B}(\varphi) = 0.
\end{aligned}$$

The equivalence $\Pi_\sigma \mathcal{B}(\varphi_1) = 0 \wedge \Pi_\sigma \mathcal{B}(\varphi_2) = 0 \Leftrightarrow \Pi_\sigma \mathcal{B}(\varphi_1) + \Pi_\sigma \mathcal{B}(\varphi_2) = 0$ holds because $\Pi_\sigma \mathcal{B}(\varphi)$ cannot be negative for binary assignments σ. □

$\mathcal{B}$ is compatible with R_x. We exhibit a mapping $\gamma_x \colon \mathcal{P} \to \mathcal{P}$ satisfying the conditions of Definition 4 for the macrostep R_x. Recall that R_x is only defined for formulas φ in CNF such that $C(x) + C(\neg x) \leq 1$ for every clause C. Since arithmetisations of such formulas only have an x^3 term, an x term, and a constant term, it suffices to define γ_x for polynomials of the form $a_3x^3 + a_1x + a_0$.

Lemma 10. *Let $\gamma_x \colon \mathcal{P} \to \mathcal{P}$ be the partial mapping defined by $\gamma_x(a_3x^3 + a_1x + a_0) := -a_3a_1 + a_1 + a_0$. The mapping γ_x witnesses that $\mathcal{B}$ is polynomially compatible with the full resolution macrostep R_x.*

Proof. We show that γ_x satisfies all properties of Definition 4. Let φ be a formula in CNF such that $C(x) + C(\neg x) \leq 1$ for every clause C (see Definition 1). Then φ is of the form

$$\varphi = \Big(\bigwedge_{i \in [k]} x \vee a_i\Big) \wedge \Big(\bigwedge_{j \in [l]} \neg x \vee b_j\Big) \wedge c$$

where a_i, b_j are disjunctions not depending on x and c is a conjunction of clauses not depending on x. We have $R_x(\varphi) = \bigwedge_{i \in [k],\, j \in [l]} (a_i \vee b_j) \wedge c$. Now

$$\mathcal{B}(\varphi) = \sum_{i \in [k]} (1 - x)a_i + \sum_{j \in [l]} x^3 b_j + c = \sum_{j \in [l]} b_j x^3 - \sum_{i \in [k]} a_i x + \sum_{i \in [k]} a_i + c$$

and thus

$$\gamma_x(\mathcal{B}(\varphi)) = \Big(\sum_{j\in[l]} b_j\Big)\Big(\sum_{i\in[k]} a_i\Big) - \sum_{i\in[k]} a_i + \sum_{i\in[k]} a_i + c$$
$$= \sum_{i\in[k],\, j\in[l]} a_i b_j + c = \mathcal{B}(R_x(\varphi)).$$

This proves (a). Since γ_x does not depend on variables other than x, (b) is also given. (c) and (d) are trivial. □

$\mathcal{B}$ is compatible with C_x. We exhibit a mapping $\delta_x\colon \mathcal{P} \to \mathcal{P}$ satisfying the conditions of Definition 4 for the cleanup macrostep C_x. Recall that C_x is only defined for formulas φ in CNF such that $C(x) + C(\neg x) \leq 2$ for every clause C. Arithmetisations of such formulas are polynomials of degree at most 6 in each variable, and so it suffices to define δ_x for these polynomials.

Lemma 11. *Let $\delta_x\colon \mathcal{P} \to \mathcal{P}$ be the partial mapping defined by*

$$\delta_x(a_6x^6 + a_5x^5 + \cdots + a_1x + a_0) := (a_6 + a_4 + a_3)x^3 + (a_2 + a_1)x + a_0.$$

The mapping δ_x witnesses that $\mathcal{B}$ is polynomially compatible with C_x.

Proof. We show that δ_x satisfies all properties of Definition 4. We start with (a). Since $\mathcal{B}(C \wedge C') = \mathcal{B}(C) + \mathcal{B}(C')$ for clauses C, C' and $\delta_x(p_1+p_2) = \delta_x(p_1)+\delta_x(p_2)$, it suffices to show that $\delta_x(\mathcal{B}(C)) = \mathcal{B}(C_x(C))$ for all clauses C of φ. Now let C be a clause of φ. We assume that $C(x) + C(\neg x) \leq 2$ (see Definition 1).

- If $C(x) + C(\neg x) \leq 1$, then $\delta_x(\mathcal{B}(C)) = \mathcal{B}(C) = \mathcal{B}(C_x(C))$.
- If $C = x \vee x \vee C'$, then $\mathcal{B}(C) = (1-x)^2\mathcal{B}(C') = (1 - 2x + x^2)\mathcal{B}(C')$, so $\delta_x\mathcal{B}(C) = (1 - 2x + x)\mathcal{B}(C') = (1 - x)\mathcal{B}(C') = \mathcal{B}(x \vee C') = \mathcal{B}(C_x(C))$.
- If $C = \neg x \vee \neg x \vee C'$, then $\mathcal{B}(C) = x^6\mathcal{B}(C')$, so $\delta_x\mathcal{B}(C) = x^3\mathcal{B}(C') = \mathcal{B}(\neg x \vee C') = \mathcal{B}(C_x(C))$.
- If $C = x \vee \neg x \vee C'$, then $\mathcal{B}(C) = (1 - x)x^3\mathcal{B}(C') = x^3\mathcal{B}(C') - x^4\mathcal{B}(C')$, so $\delta_x\mathcal{B}(C) = x^3\mathcal{B}(C') - x^3\mathcal{B}(C') = 0 = \mathcal{B}(C_x(C))$.

This proves (a). Since δ_x does not depend on variables other than x, (b) is also given. Parts (c) and (d) are trivial. □

As observed earlier, DavisPutnam does not remove duplicate clauses; that is, Prover maintains a multiset of clauses that may contain multiple copies of a clause. We show that the number of copies is at most double-exponential in $|\varphi|$.

Lemma 12. *Let φ be the input formula, and let φ_k be the last formula computed by* DavisPutnam. *Then $\mathcal{A}(\varphi_k) \in 2^{2^{\mathcal{O}(|\varphi|)}}$.*

Proof. Let $n_C(\psi)$ be the number of clauses in a formula ψ, let x be a variable. Then $n_C(C_x(\psi)) \leq n_C(\psi)$ because a cleanup step can only change or delete clauses. Moreover, $n_C(R_x(\psi)) = n_x n_{\neg x} - n_x - n_{\neg x} + n_C(\psi)$ where n_x and $n_{\neg x}$ are the numbers of clauses in ψ which contain x and $\neg x$, respectively. We get

Round	Honest Prover	Verifier
Initial	$q := 15871$ $p_6 := \mathcal{B}_q(\varphi_6) = 2$ send q, p_6	$w := p_6 = 2$ $\sigma := \{x \mapsto 3, y \mapsto 4, z \mapsto 3\}$ send σ
$k = 6$	$\sigma' := \{x \mapsto 3, y \mapsto 4\}$ $p_5 := \Pi_{\sigma'}(\mathcal{B}_q(\varphi_5))$ $= 2z^3 - z + 1$ send p_5	$\pi_{[z:=3]}\gamma_z(p_5) = \pi_{[z:=3]}2 \stackrel{?}{=} 2$ $\sigma(z) := 4$ $w := \pi_{[z:=4]}p_5 = 125$ send σ
$k = 5$	$\sigma' := \{x \mapsto 3, y \mapsto 4\}$ $p_4 := \Pi_{\sigma'}(\mathcal{B}_q(\varphi_4))$ $= 2z^6 - 3z^4 + 3z^3 + z^2 - 2z + 1$ send p_4	$\pi_{[z:=4]}\delta_z(p_4) = \pi_{[z:=4]}2z^3 - z + 1 \stackrel{?}{=} 125$ $\sigma(z) := 2$ $w := \pi_{[z:=2]}p_4 = 105$ send σ
$k = 4$	$\sigma' := \{x \mapsto 3, z \mapsto 2\}$ $p_3 := \Pi_{\sigma'}(\mathcal{B}_q(\varphi_3))$ $= 15y^3 - 7y + 7$ send p_3	$\pi_{[y:=4]}\gamma_y(p_3) = \pi_{[y:=4]}105 \stackrel{?}{=} 105$ $\sigma(y) := 2$ $w := \pi_{[y:=2]}p_3 = 113$ send σ
$k = 3$	$\sigma' := \{x \mapsto 3, y \mapsto 2\}$ $p_2 := \Pi_{\sigma'}(\mathcal{B}_q(\varphi_2))$ $= 16z^6 - z^3 - 7z + 7$ send p_2	$\pi_{[z:=2]}\delta_z(p_2) = \pi_{[z:=2]}15z^3 - 7z + 7 \stackrel{?}{=} 113$ $\sigma(z) := 3$ $w := \pi_{[z:=3]}p_2 = 11623$ send σ
$k = 2$	$\sigma' := \{x \mapsto 3, z \mapsto 2\}$ $p_1 := \Pi_{\sigma'}(\mathcal{B}_q(\varphi_1))$ $= 729y^6 - 27y^4 + 754y^3 - 25y + 25$ send p_1	$\pi_{[y:=2]}\delta_y(p_1) = \pi_{[y:=2]}1456y^3 - 25y + 25 \stackrel{?}{=} 11623$ $\sigma(y) := 1$ $w := \pi_{[y:=1]}p_1 = 1456$ send σ
$k = 1$	$\sigma' := \{y \mapsto 1, z \mapsto 2\}$ $p_0 := \Pi_{\sigma'}(\mathcal{B}_q(\varphi_0))$ $= 54x^3 - 27x + 25$ send p_0	$\pi_{[x:=3]}\gamma_x(p_0) = \pi_{[x:=3]}1456 \stackrel{?}{=} 1456$ $\sigma(x) := 2$ $w := \pi_{[x:=2]}p_0 = 493$ send σ
Final		$\Pi_\sigma \mathcal{B}_q(\varphi) \stackrel{?}{=} 493$

Table 3. Run of the instance of the interactive protocol of Table 2 for DavisPutnam, using the arithmetisation $\mathcal{B}$ of Definition 5.

$n_C(R_x(\psi)) \leq (n_x + n_{\neg x})^2 - (n_x + n_{\neg x}) + n_C(\psi)$. Since $n_x + n_{\neg x} \leq n_C(\psi)$, it follows that $n_C(R_x(\psi)) \leq (n_C(\psi))^2$. Now let n be the number of variables. Since φ_k is reached after n resolution steps, it follows that $\mathcal{B}(\varphi_k) = n_C(\varphi_k) \leq n_C(\varphi)^{2^n} \in 2^{2^{\mathcal{O}(|\varphi|)}}$. □

Proposition 2. *There exists an interactive protocol for UNSAT that is competitive with* DavisPutnam.

Proof. We show that the $\mathcal{B}$ satisfies all properties of Theorem 2. On an input formula φ over n variables, DavisPutnam executes n resolution steps R_x and $n(n-1)/2$ cleanup steps C_x, which gives $n(n+1)/2$ macrosteps in total and proves (a).

Since φ does not contain any variable more than once per clause and since cleanup steps w.r.t. all remaining variables are applied after every resolution step, resolution steps can only increase the maximum degree of $\mathcal{B}(\varphi_i)$ to at most

6 (from 3). Hence the maximum degree of $\mathcal{B}(\varphi_i)$ is at most 6 for any i, showing (b).

Furthermore, since $R_x(\varphi_i)$ does not contain any occurrence of x, and resolution steps are performed w.r.t. all variables, φ_k does not contain any variables, so $\varphi_k = \{a \cdot \Box\}$ for some $a \in \mathbb{N}$ where $\Box$ is the empty clause. Together with Lemma 12, (c) follows. $\Box$

Instantiating Theorem 2 with $\mathcal{B}$ yields an interactive protocol competitive with DavisPutnam. Table 3 shows a run of this protocol on the formula φ of Table 1. Initially, Prover runs DavisPutnam on φ, computing the formulas $\varphi_1, ..., \varphi_6$. Then, during the run of the protocol, it sends to Verifier polynomials of the form $\Pi_{\sigma'} \mathcal{B}_q(\varphi_{i-1})$ for the assignments σ' chosen by Verifier.

7 Conclusions

We have presented the first technique for the systematic derivation of interactive proof systems competitive with a given algorithm for UNSAT. More precisely, we have shown that such systems can be automatically derived from arithmetisations satisfying a few commutativity properties. In particular, this result indicates that non-standard arithmetisations can be key to obtaining competitive interactive proof systems for practical algorithms. We have applied our technique to derive the first interactive proof system for the Davis-Putnam resolution procedure, opening the door to interactive proof systems for less restrictive variants of resolution.

Lovasz et al. have shown that given a refutation by the Davis-Putnam resolution procedure, one can extract a multi-valued decision diagram, polynomial in the size of the refutation, in which the path for a given truth assignment leads to a clause false under that assignment (that is, to a clause witnessing that the assignment does not satisfy the formula) [14]. This suggests a possible connection between our work and the work of Couillard et al. in [5]. As mentioned in the introduction, [5] presents an interactive proof system competitive with the algorithm for UNSAT that iteratively constructs a BDD for the formula (starting at the leaves of its syntax tree, and moving up at each step), and returns "unsatisfiable" iff the BDD for the root of the tree only contains the node 0. We conjecture that a future version of our systematic derivation technique could subsume both [5] and this paper.

Acknowledgments. We thank the anonymous reviewers for their comments and Albert Atserias for helpful discussions.

References

1. Arora, S., Barak, B.: Computational Complexity: A Modern Approach. Cambridge University Press (2006), https://theory.cs.princeton.edu/complexity/book.pdf

2. Babai, L.: Trading group theory for randomness. In: Sedgewick, R. (ed.) Proceedings of the 17th Annual ACM Symposium on Theory of Computing, May 6-8, 1985, Providence, Rhode Island, USA. pp. 421–429. ACM (1985). https://doi.org/10.1145/22145.22192, https://doi.org/10.1145/22145.22192
3. Barbosa, H., Reynolds, A., Kremer, G., Lachnitt, H., Niemetz, A., Nötzli, A., Ozdemir, A., Preiner, M., Viswanathan, A., Viteri, S., Zohar, Y., Tinelli, C., Barrett, C.W.: Flexible proof production in an industrial-strength SMT solver. In: Blanchette, J., Kovács, L., Pattinson, D. (eds.) Automated Reasoning - 11th International Joint Conference, IJCAR 2022, Haifa, Israel, August 8-10, 2022, Proceedings. Lecture Notes in Computer Science, vol. 13385, pp. 15–35. Springer (2022). https://doi.org/10.1007/978-3-031-10769-6_3, https://doi.org/10.1007/978-3-031-10769-6_3
4. Buss, S.R., Turán, G.: Resolution proofs of generalized pigeonhole principles. Theor. Comput. Sci. **62**(3), 311–317 (1988)
5. Couillard, E., Czerner, P., Esparza, J., Majumdar, R.: Making IP = PSPACE practical: Efficient interactive protocols for BDD algorithms. In: Enea, C., Lal, A. (eds.) Computer Aided Verification - 35th International Conference, CAV 2023, Paris, France, July 17-22, 2023, Proceedings, Part III. Lecture Notes in Computer Science, vol. 13966, pp. 437–458. Springer (2023). https://doi.org/10.1007/978-3-031-37709-9_21, https://doi.org/10.1007/978-3-031-37709-9_21
6. Davis, M., Putnam, H.: A computing procedure for quantification theory. J. ACM **7**(3), 201–215 (1960). https://doi.org/10.1145/321033.321034, https://doi.org/10.1145/321033.321034
7. Goldwasser, S., Micali, S., Rackoff, C.: The knowledge complexity of interactive proof-systems (extended abstract). In: Sedgewick, R. (ed.) Proceedings of the 17th Annual ACM Symposium on Theory of Computing, May 6-8, 1985, Providence, Rhode Island, USA. pp. 291–304. ACM (1985). https://doi.org/10.1145/22145.22178, https://doi.org/10.1145/22145.22178
8. Haken, A.: The intractability of resolution. Theor. Comput. Sci. **39**, 297–308 (1985)
9. Harrison, J.: Handbook of Practical Logic and Automated Reasoning. Cambridge University Press (2009)
10. Henzinger, T., Jhala, R., Majumdar, R., Necula, G., Sutre, G., Weimer, W.: Temporal-safety proofs for systems code. In: CAV 02: Computer-Aided Verification, pp. 526–538. Lecture Notes in Computer Science 2404, Springer-Verlag (2002)
11. Heule, M., Jr., W.A.H., Kaufmann, M., Wetzler, N.: Efficient, verified checking of propositional proofs. In: ITP. Lecture Notes in Computer Science, vol. 10499, pp. 269–284. Springer (2017)
12. Heule, M.J.H.: Proofs of unsatisfiability. In: Biere, A., Heule, M., van Maaren, H., Walsh, T. (eds.) Handbook of Satisfiability - Second Edition, Frontiers in Artificial Intelligence and Applications, vol. 336, pp. 635–668. IOS Press (2021). https://doi.org/10.3233/FAIA200998, https://doi.org/10.3233/FAIA200998
13. Heule, M.J.H., Kullmann, O., Marek, V.W.: Solving and verifying the boolean pythagorean triples problem via cube-and-conquer. CoRR **abs/1605.00723** (2016)
14. Lovász, L., Naor, M., Newman, I., Wigderson, A.: Search problems in the decision tree model. SIAM J. Discret. Math. **8**(1), 119–132 (1995)
15. Lund, C., Fortnow, L., Karloff, H.J., Nisan, N.: Algebraic methods for interactive proof systems. J. ACM **39**(4), 859–868 (1992). https://doi.org/10.1145/146585.146605, https://doi.org/10.1145/146585.146605
16. Namjoshi, K.: Certifying model checkers. In: CAV 01: Computer Aided Verification, pp. 2–13. Lecture Notes in Computer Science 2102, Springer-Verlag (2001)

17. Necula, G.: Proof-carrying code. In: Principles of Programming Languages. pp. 106–119. ACM Press (1997)
18. Shamir, A.: IP = PSPACE. J. ACM **39**(4), 869–877 (1992). https://doi.org/10.1145/146585.146609, https://doi.org/10.1145/146585.146609

Craig Interpolation for Decidable First-Order Fragments

Balder ten Cate[1] and Jesse Comer[2(✉)]

[1] ILLC, University of Amsterdam, Amsterdam 1098 XH, The Netherlands
[2] University of Pennsylvania, Philadelphia, PA 19104, USA
jacomer@seas.upenn.edu

Abstract. We show that the guarded-negation fragment (GNFO) is, in a precise sense, the smallest extension of the guarded fragment (GFO) with Craig interpolation. In contrast, we show that the smallest extension of the two-variable fragment (FO^2), and of the forward fragment (FF) with Craig interpolation, is full first-order logic. Similarly, we also show that all extensions of FO^2 and of the fluted fragment (FL) with Craig interpolation are undecidable.

Keywords: Craig interpolation · Decidability · Abstract model theory.

1 Introduction

The study of decidable fragments of first-order logic (FO) is a topic with a long history, dating back to the early 1900s ([40,52], cf. also [16]), and more actively pursued since the 1990s. Inspired by Vardi [55], who asked "what makes modal logic so robustly decidable?" and Andreka et al. [1], who asked "what makes modal logic tick?" many decidable fragments have been introduced and studied over the last 25 years that take inspiration from modal logic (ML), which itself can be viewed as a fragment of FO that features a restricted form of quantification. These include the following fragments, each of which naturally generalizes modal logic in a different way: the *two-variable fragment* (FO^2) [42], the *guarded fragment* (GFO) [1], and the *unary negation fragment* (UNFO) [22]. Further decidable extensions of these fragments were subsequently identified, including the *two-variable fragment with counting quantifiers* (C^2) [29] and the *guarded negation fragment* (GNFO) [4]. The latter can be viewed as a common generalization of GFO and UNFO. Many decidable logics used in computer science and AI, including various description logics and rule-based languages, can be translated into GNFO and/or C^2. In this sense, GNFO and C^2 are convenient tools for explaining the decidability of other logics. Extensions of GNFO have been studied that push the decidability frontier even further (for instance with fixed-point operators and using clique-guards), but these fall outside the scope of this paper.

In an earlier line of investigation, Quine identified the decidable *fluted fragment* (FL) [51], the first of several *ordered logics* which have been the subject of recent interest [47,48,49,50,44]. The idea behind ordered logics is to restrict the order in which variables are allowed to occur in atomic formulas and quantifiers.

N. Kobayashi and J. Worrell (Eds.): FoSSaCS 2024, LNCS 14575, pp. 137–159, 2024.
https://doi.org/10.1007/978-3-031-57231-9_7

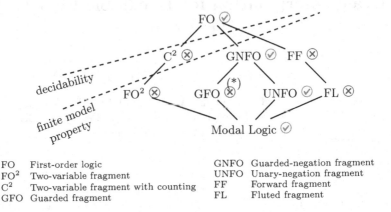

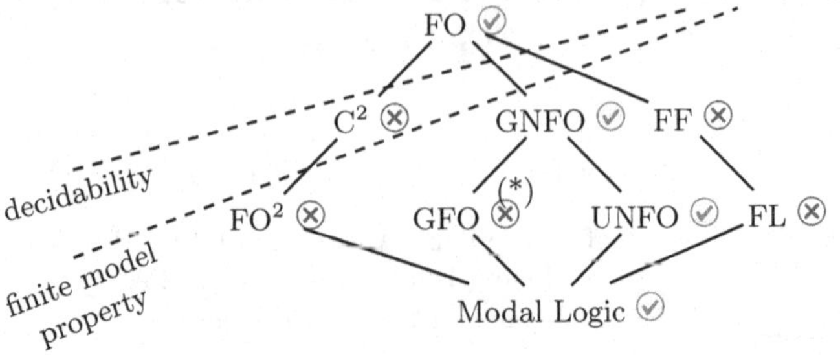

FO	First-order logic	GNFO	Guarded-negation fragment
FO^2	Two-variable fragment	UNFO	Unary-negation fragment
C^2	Two-variable fragment with counting	FF	Forward fragment
GFO	Guarded fragment	FL	Fluted fragment

Fig. 1. Landscape of decidable fragments of FO with (✓) and without (⊗) CIP. The inclusion marked (∗) holds only for sentences and self-guarded formulas.

Another recently introduced decidable fragment that falls in this family is the *forward fragment* (FF), whose syntax strictly generalizes that of FL. Both FL and FF have the finite model property (FMP) [44] and embed ML [34], but are incomparable in expressive power to GFO [45], FO^2, and UNFO.[3]

Ideally, an FO-fragment is not only decidable, but also model-theoretically well behaved. A particularly important model-theoretic property of logics is the *Craig Interpolation Property* (CIP). It states that, for all formulas φ, ψ, if $\varphi \models \psi$, then there exists a formula ϑ such that $\varphi \models \vartheta$ and $\vartheta \models \psi$, and such that all non-logical symbols occurring in ϑ occur both in φ and in ψ. Craig [24] proved in 1957 that FO itself has this property (hence the name). Several refinements of Craig's result have subsequently been obtained (e.g., [43,10]). These have found applications in various areas of computer science and AI, including formal verification, modular hard/software specification and automated deduction [41,18,31], and are emerging as a new prominent technology in databases [53,12] and knowledge representation [39,21,37]. While we have described CIP here as a model theoretic property, it also has a proof-theoretic interpretation. Indeed, it has been argued that CIP is an indicator for the existence of nice proof systems [32].

Turning our attention to the decidable fragments of FO we mentioned earlier, it turns out that, although GFO is in many ways model-theoretically well-behaved [1], it lacks CIP [33]. Likewise, FO^2 lacks CIP [23] and the same holds for C^2 ([35, Example 2] yields a counterexample). Both FF and FL lack CIP [7]. On the other hand, UNFO and GNFO have CIP [22,3]. Figure 1 summarizes these known results. Note that we restrict attention to relational signatures without constant symbols and function symbols. Some of the results depend on this restriction. Other known results not reflected in Figure 1 (to avoid clutter) are that

[3] Specifically, the FO-sentence $\exists xy(R(x,y) \wedge R(y,x))$ belongs to GFO, FO^2 and UNFO, but is not expressible in FF, since the structure consisting of two points with symmetric edges and the structure $(\mathbb{Z}, S)$ with S the successor relation, are "infix bisimilar," as described in [7].

the intersection of GFO and FO^2 (also known as GFO^2) has CIP [33]. Similarly, the intersection of FF with GFO and the intersection of FL with GFO (known as G_{FF} and G_{FL}, respectively) have CIP [7].

When a logic L lacks CIP, the question naturally arises as to whether there exists a more expressive logic L' that has CIP. If such an L' exists, then, in particular, interpolants for valid L-implications can be found in L'. This line of analysis is sometimes referred to as *Repairing Interpolation* [2]. If L' is an FO-fragment, and our aim is to repair interpolation by extension, then there is a trivial solution: FO itself is an extension of L satisfying CIP. We will instead consider the following refinement of the question: can a natural extension L' of L be identified which satisfies CIP while retaining decidability? We will answer this question for three of the fragments depicted in Figure 1 that lack CIP, by identifying the minimal natural extension L' of L satisfying CIP. Our main results can be stated informally as follows:

1. The smallest logic extending GFO that has CIP is GNFO.
2. The smallest logic extending FO^2 that has CIP is FO, and no decidable extension of FO^2 has CIP.
3. The smallest logic extending FF that has CIP is FO, and no decidable extension of FL has CIP.

The precise statements of these results will be given in the respective sections. They involve some natural closure assumptions on the logics in question, and, for the undecidability results, some assumptions regarding the effective computability of the translation between the extension and the logic that it extends.

These results give us a clear sense of where, in the larger landscape of decidable fragments of FO, we may find logics that enjoy CIP. What makes the above results remarkable is that, from the definition of the Craig interpolation property, it doesn't appear to follow that a logic without CIP would have a unique minimal extension with CIP. Note that a valid implication may have many possible interpolants, and the Craig interpolation property merely requires the existence of one such interpolant. Nevertheless, the above results show that, in the case FO^2, GFO, and FF, such a unique minimal extension indeed exists (assuming suitable closure properties, which will be spelled out in detail in the next sections).

Related Work. Several other approaches have been proposed for dealing with logics that lack CIP. One approach is to weaken CIP. For example, it was shown in [33] that GFO satisfies a weak, "modal" form of Craig interpolation, where, roughly speaking, only the relation symbols that occur in non-guard positions in the interpolant are required to occur both in the premise and the conclusion. As it turns out, this weakening of CIP is strong enough to entail the (non-projective) *Beth Definability Property*, which is one important use case of CIP. See also Section 7 for further discussion of weak forms of CIP.

Another recent approach [35] is to develop algorithms for testing whether an interpolant exists for a given entailment. That is, rather than viewing Craig interpolation as a property of logics, the existence of interpolants is studied as an algorithmic problem at the level of individual entailments. The interpolant existence problem turns out to be indeed decidable (although of higher complexity than the satisfiability problem) for both GFO and FO^2 [35].

Additional results are known for UNFO and GNFO beyond the fact that they have CIP. In particular, CIP holds for their fixed-point extensions [9,8], interpolants can be constructed effectively, and tight bounds are known on the size of interpolants and the computational complexity of computing them [11].

Our paper can be viewed as an instance of *abstract model theory* for fragments of FO. One large driving force behind the development of abstract model theory was the identification of *extensions* of FO which satisfy desirable model-theoretic properties, such as the compactness theorem, the Löwenheim-Skolem, and Craig interpolation. One takeaway from this line of research is that CIP is scarce among many "reasonable" FO-extensions. An early result of Lindström showed that FO-extensions with finitely many generalized quantifiers and satisfying the downward Löwenheim-Skolem property do not have the Beth property (and hence fail to satisfy CIP) [38]. Similarly, Caicedo [17], generalizing an early result by Friedman [26], established a strong negative CIP result that applies to arbitrary proper FO-extensions with monadic generalized quantifiers. For a survey of negative interpolation results among FO-extensions, see [54]. These negative results not only show that CIP is scarce among extensions of FO, they also provide clues as to where, within the space of all extensions, one may hope to find logics with CIP. Our results can be viewed similarly, except that they pertain to (extensions of) fragments of FO.

Our results can also be appreciated as characterizations of GNFO and of FO. While traditional Lindström-style characterizations are maximality theorems (e.g., FO is a maximal logic having the compactness and Löwenheim-Skolem properties), our results can be viewed as minimality theorems (e.g., GNFO is the minimal logic extending GFO and having CIP).

Some prior work exists that studies abstract model theory for (extensions of) fragments of FO. Most closely related is [19], which studies modal logics and hybrid logics. Among other things, it was shown in [19] that the smallest extension of modal logic with the difference operator (ML(D)) which satisfies CIP is full first-order logic. Additionally, in [28], the authors identified minimal extensions of various fragments of propositional linear temporal logic (PLTL) with CIP. Furthermore, it was shown in [19] that every abstract logic extending GFO with CIP can express all FO sentences and formulas with one free variable, and is thus undecidable. A crucial difference between this result and ours is that [19] assumes signatures with constant symbols and concerns a stronger version of CIP, interpolating not only over relation symbols but also over constant symbols. In contrast, we only consider purely relational signatures without constant symbols. Other prior work on abstract model theory for fragments of FO are [13,15,27]. Repairing interpolation has also been pursued in the context of quantified modal logics, which typically lack CIP; in [2], the authors showed that CIP can be repaired for such logics by adding nominals, @-operators and the $\downarrow$-binder.

Outline. Section 2 introduces the abstract model-theoretic framework. In Sections 3, 4, and 5, we repair interpolation for FO^2, GFO, and FF, respectively. In Section 6, we provide results showing that, even with weak expressive assumptions, extensions of FO^2 and FL with CIP are undecidable. In Section 7, we discuss the implications and limitations of our results, and future directions.

2 Preliminaries

We assume familiarity with the syntax and semantics of FO. Signatures are denoted by σ and τ, and are assumed to be relational and finite. If φ contains only relation symbols occurring in σ, then we write $M, g \models \varphi$ to denote that a σ-structure M satisfies φ under the variable assignment g. We write x_i, y_i, z_i, u_i to denote variables, and $\overline{x}, \overline{y}, \overline{z}, \overline{u}$ to denote tuples of variables. We write a_i, b_i, c_i to denote elements of structures and $\overline{a}, \overline{b}, \overline{c}$ to denote tuples of such elements. Given a tuple of elements $\overline{a} = a_1, \ldots, a_n$ in a structure M, a tuple of variables $\overline{x} = x_1, \ldots, x_n$, and a variable assignment g, we write $g[\overline{x}/\overline{a}]$ to denote the variable assignment which is the same as g except that $g(x_i) = a_i$ for each $i \leq n$. In order to state our main results precisely, we must formally define what we mean by *extensions L' of L* (where L is some fragment of FO that lacks CIP). One option is to let L' range over fragments of FO that syntactically include L. However, as it turns out, our main results apply even to extensions that are not themselves contained in FO. We therefore opt, instead, to work with an abstract definition of logics, as typically used in abstract model theory.

Abstract Logics. An *abstract logic* (or *logic*) is a pair $(L, \models_L)$, where L is a map from relational signatures σ to collections of *formulas*, and $\models_L$ is a ternary *satisfaction relation*. A *formula* of an abstract logic $(L, \models_L)$ is an element of $L(\sigma)$ for some finite relational signature σ. L must be monotone: if $\sigma \subseteq \tau$, then $L(\sigma) \subseteq L(\tau)$. Each formula φ has an associated finite set of free variables $\mathrm{free}(\varphi)$, and we write $\varphi(\overline{x})$ or $\varphi(x_1, \ldots, x_k)$ to denote that the free variables of φ are exactly those in the tuple $\overline{x} = x_1, \ldots, x_k$. As in the case of FO, a formula φ is a *sentence* if $\mathrm{free}(\varphi) = \emptyset$. We write $\mathrm{sig}(\varphi)$ to denote the least signature σ such that $\varphi \in L(\sigma)$. The ternary *satisfaction relation* $\models_L$ is defined over triples (M, g, φ), where φ is an L-formula, M is a τ-structure such that $\mathrm{sig}(\varphi) \subseteq \tau$, and g is a variable assignment with $\mathrm{free}(\varphi) \subseteq \mathrm{dom}(g)$; we write $M, g \models_L \varphi$ if this relation holds between these objects. The notions of logical consequence and logical equivalence for abstract logics are defined completely analogously to FO. In later sections, we will prefer to suppress the subscript L in the notation for the satisfaction relation and write L to denote an abstract logic $(L, \models_L)$. Furthermore, we often write $\varphi \in L$ rather than $\varphi \in L(\sigma)$, leaving the signature implicit.

All abstract logics L are assumed to satisfy the *reduct property* and the *renaming property*. The *reduct property* states that if $\sigma \subseteq \tau$, then for all $\varphi \in L(\sigma)$, all τ-structures M, and all assignments g, if $M, g \models_L \varphi$, then $M \upharpoonright \sigma, g \models_L \varphi$. In other words, the truth of a formula of an abstract logic L in a structure depends only on the interpretations of the symbols in the signature of that formula. The *renaming property* states that if $\rho : \sigma \to \tau$ is an injective map preserving the arity of relation symbols, then for each formula $\varphi \in L(\sigma)$, there is a formula $\psi \in L(\tau)$ such that for all τ-structures M, we have that $M, g \models_L \psi$ if and only if $\rho^{-1}[M], g \models_L \varphi$, where $\rho^{-1}[M]$ is the σ-structure with the same domain as M where, for each $R \in \sigma$, we have that $R^{\rho^{-1}[M]} = \rho(R)^M$. Intuitively, the renaming property states that if a formula over a signature σ can be expressed in a logic L, then the formula obtained by renaming all of its relation symbols can also be expressed in L.

For arbitrary abstract logics L, the Craig interpolation property states that if $\varphi \models_L \psi$ for L-formulas φ and ψ, then there exists a formula $\vartheta \in L(\text{sig}(\varphi) \cap \text{sig}(\psi))$ with $\text{free}(\vartheta) = \text{free}(\varphi) \cap \text{free}(\psi)$ such that $\varphi \models_L \vartheta$ and $\vartheta \models_L \psi$.

We say a formula φ of a logic L expresses a formula ψ of a logic L' if $\text{free}(\varphi) = \text{free}(\psi)$, $\text{sig}(\varphi) = \text{sig}(\psi)$, and for all structures M and assignments g, we have that $M, g \models_L \varphi$ if and only if $M, g \models_{L'} \psi$. We say that a logic L' is an *extension* of a logic L (notation: $L \preceq L'$) if L' can express all formulas of L. An FO-fragment can then be precisely defined, without reference to syntax, as a logic of which FO is an extension. We say that L' is a *sentential extension* of L (notation: $L \preceq_{sent} L'$) if L' can express all sentences of L.

Let L be a logic and $\psi(x_1, \ldots, x_n)$ be an L-formula. We write $[\![\psi]\!]^M$ for the collection of tuples $(a_1, \ldots, a_n) \in M^n$ such that there exists an assignment g where $M, g \models \psi$ and $g(x_i) = a_i$ for each $i \leq n$. Given formulas $\psi_1, \ldots, \psi_k \in L(\sigma)$, a σ-structure M, and relation symbols $R_1, \ldots, R_k \in \sigma$ with $|\text{free}(\psi_i)| = \text{arity}(R_i)$ for each $i \leq k$, we define $M[R_1/\psi_1, \ldots, R_k/\psi_k]$ to be the σ-structure with the same domain as M and such that $R_i^{M[R_1/\psi_1,\ldots,R_k/\psi_k]} = [\![\psi_i]\!]^M$ for each $i \leq k$. We now describe a syntax-free notion of uniform substitution for formulas of an abstract logic.

Definition 2.1. *Let L be a logic and $\varphi \in L(\sigma)$ with $R_1, \ldots, R_k \in sig(\varphi)$, where for each $i \leq k$, we have that R_i is a k_i-ary relation symbol. Furthermore, let $\psi_1, \ldots, \psi_k \in L(\sigma)$ be formulas with $|free(\psi_i)| = k_i$ for each $i \leq k$. We say that L* expresses the substitution of $\psi_1, \ldots, \psi_k$ for $R_1, \ldots, R_k$ in φ *if there exists a formula $\chi \in L(\sigma)$ such that, for every σ-structure M,*

$$M, g \models \chi \iff M[R_1/\psi_1, \ldots, R_k/\psi_k], g \models \varphi.$$

Most studies in abstract logic assume that the logics under study are *regular*, roughly meaning that they can express atomic formulas, Boolean connectives, and existential quantification. In other words, to study regular logics is to study extensions of FO. Since we are interested in a more fine-grained view of logics including FO-fragments, these assumptions are too strong. As a result, the first step of studying extensions of FO-fragments from the perspective of abstract logic is to identify natural expressive assumptions for those extensions which are strictly weaker than regularity. We do this in the respective sections.

Some of our proofs will use second-order quantification (for expository reasons only), and we recall the semantics of these quantifiers here. Given a formula $\varphi \in L(\sigma \cup \{P\})$ of some abstract logic L, we can form new formulas $\exists P \varphi$ and

$\forall P\varphi$ with signature σ and the same free variables as φ. Given a σ-structure M and an assignment g, the semantics of these formulas are defined as follows:

$$\begin{aligned} M, g \models \exists P\varphi \quad & \text{if there is a } \sigma \cup \{P\}\text{-expansion } M' \text{ of } M \\ & \text{such that } M', g \models \varphi, \text{ and} \\ M, g \models \forall P\varphi \quad & \text{if for all } \sigma \cup \{P\}\text{-expansions } M' \text{ of } M, \\ & \text{we have that } M', g \models \varphi. \end{aligned}$$

If L itself does not allow second-order quantification, we can view $\exists P\varphi$ and $\forall P\varphi$ as elements of $L'(\sigma)$ for a suitable extension L' of L. In particular, if φ is an FO-formula, then $\exists P\varphi$ and $\forall P\varphi$ are formulas of second-order logic (SO).

3 Repairing Interpolation for FO^2

The two-variable fragment (FO^2) consists of all FO-formulas containing only two variables, say, x and y, where we allow for nested quantifiers that reuse the same variable (as in $\exists xy(R(x, y) \wedge \exists x(R(y, x)))$, expressing the existence of a path of length 2). In this context, as is customary, we restrict attention to relations of arity at most 2. It is known that FO^2 is decidable [42] but does not have CIP [23].

3.1 Natural Extensions of FO^2

While FO^2 is restricted to only two variables and predicates of arity as most 2, it has no restriction on its connectives: it is fully closed under Boolean connectives and existential and universal quantification. Because of this fact, we will consider in this section those abstract logics which are *strong extensions* of FO^2.

Definition 3.1. *We say that a logic L'* strongly extends *a logic L if L' extends L and, for each formula $\varphi \in L'$ with $R_1, \dots, R_k \in sig(\varphi)$, where φ expresses some $\psi \in L$, and all formulas $\psi_1, \dots, \psi_k \in L'$, we have that L' expresses the substitution of $\psi_1, \dots, \psi_k$ for $R_1, \dots, R_k$ in φ (cf. Definition 2.1).*

Intuitively, Definition 3.1 means that L' can express uniform substitutions of its formulas into formulas of L. In other words, the notion of a strong extension is a syntax-free way to say that L' extends L and is closed under the connectives of L. In particular, if L strongly extends FO^2, then L can express all of the usual first-order connectives: for ψ_0 and ψ_1 expressible in L, it must also be the case that $\neg\psi_0$, $\psi_0 \wedge \psi_1$, and $\exists x\psi_0$ are expressible in L, under the usual semantics of these connectives. Clearly FO^2 is the smallest strong extension of itself.

3.2 Finding the Minimal Extension of FO^2 with CIP

Recall that we write $L \preceq_{sent} L'$ if every sentence of L is expressible in L'. Our main result in this section is the following.

Theorem 3.1. *If L is a strong extension of* FO^2 *with CIP, then* $\text{FO} \preceq_{sent} L$.

Proof. We will show by induction on the complexity of formulas that, for every FO-formula $\varphi(x_1 \ldots, x_n)$ there is a sentence $\psi \in L$ over an extended signature containing additional unary predicates $P_1, \ldots, P_n$, that is equivalent to

$$\exists x_1 \ldots x_n((\bigwedge_{i=1\ldots n} P_i(x_i) \land \forall y(P_i(y) \to y = x_i)) \land \varphi(x_1, \ldots, x_n)).$$

In other words, ψ is a sentence expressing that φ holds under an assignment of its free variables to some tuple of elements which uniquely satisfy the P_i predicates. In the case that $n = 0$ (i.e., the case that φ is a sentence), we then have that ψ is equivalent to φ, which shows that $FO \preceq_{sent} L$.

The base case of the induction is straightforward (recall that we restrict attention to relations of arity at most 2). The induction step for the Boolean connectives is straightforward as well (using the fact that L is a strong extension of FO^2, and thus can express all connectives of FO^2). In fact, the only non-trivial part of the argument is the induction step for the existential quantifier. Let $\varphi(x_1, \ldots, x_n)$ be of the form $\exists x_{n+1}\varphi'(x_1 \ldots, x_n, x_{n+1})$. By the inductive hypothesis, there is an L-sentence ψ with $\text{sig}(\psi) = \text{sig}(\varphi') \cup \{P_1, \ldots, P_{n+1}\}$, where $P_1, \ldots, P_{n+1}$ are unary predicates not in $\text{sig}(\varphi')$, which is equivalent to

$$\exists x_1 \ldots x_n \exists x_{n+1}((\bigwedge_{i \leq n+1} P_i(x_i) \land \forall y(P_i(y) \to y = x_i)) \land \varphi'(x_1, \ldots, x_n, x_{n+1})).$$

Now, let ψ' be obtained from ψ by replacing every occurrence of P_{n+1} by P' for some fresh unary predicate P'; this is expressible in L by the renaming property. Furthermore, let

$$\gamma(x) := \psi \land P_{n+1}(x), \text{ and}$$
$$\chi(x) := (P'(x) \land \forall y(P'(y) \to y = x)) \to \psi'.$$

(where x is either of the two variables we have at our disposal; it does not matter which). Since L strongly extends FO^2, both can be written as an L-formula. Then

$$\gamma(x) \models \chi(x).$$

Let $\vartheta(x) \in L$ be an interpolant. Observe that since P_{n+1} occurs only in $\gamma(x)$ and P' only in $\chi(x)$, the following second-order entailment is also valid:

$$\exists P_{n+1}\gamma(x) \models \vartheta(x) \models \forall P'\chi(x).$$

It is not hard to see that $\exists P_{n+1}\gamma(x)$ and $\forall P'\chi(x)$ are equivalent. Indeed, both are satisfied in a structure M under an assignment g precisely if $M', g \models \varphi$, where M' is the expansion of M in which P_{n+1} denotes the singleton set $\{g(x_{n+1})\}$. It then follows that $\vartheta(x)$, being sandwiched between the two, is also equivalent to $\exists P_{n+1}\gamma(x)$. This implies that $\vartheta(x)$ is the unique interpolant (up to logical equivalence) of the entailment $\gamma(x) \models \chi(x)$, and so it is expressible in L. Then since L strongly extends FO^2, it can express $\exists x\vartheta(x)$. We claim that this sentence satisfies the requirement of our claim. To see this, observe that $\exists x\vartheta(x)$ is equivalent to $\exists x\exists P_{n+1}\gamma(x)$, which is equivalent to $\exists P_{n+1}\psi$, which clearly satisfies the requirement of our claim. □

4 Repairing Interpolation for GFO

The guarded fragment (GFO) [1] allows formulas in which all quantifiers are "guarded." Formally, a *guard* for a formula φ is an atomic formula α whose free variables include all free variables of φ. Following [30], we allow α to be an equality. More generally, by an $\exists$-*guard* for φ, we will mean a possibly-existentially-quantified atomic formula $\exists\overline{x}\beta$ whose free variables include all free variables of φ. The formulas of GFO are generated by the following grammar:

$$\varphi := \top \mid R(\overline{x}) \mid x = y \mid \varphi \wedge \psi \mid \varphi \vee \psi \mid \neg\varphi \mid \exists\overline{x}(\alpha \wedge \varphi),$$

where, in the last clause, α is a guard for φ. Note again that we do not allow constants and function symbols.

In the guarded-negation fragment (GNFO) [4], arbitrary existential quantification is allowed, but every negation is required to be guarded. More precisely, the formulas of GNFO are generated by the following grammar:

$$\varphi := \top \mid R(\overline{x}) \mid x = y \mid \varphi \wedge \varphi \mid \varphi \vee \varphi \mid \exists x\varphi \mid \alpha \wedge \neg\varphi,$$

where, in the last clause, α is a guard for φ.

As is customary, the above definitions are phrased in terms of ordinary guards α. However, it is easy to see that if we allow for $\exists$-guards, this would not affect the expressive power (or computational complexity) of these logics in any way. This is because, when the variables in the tuple $\overline{x}$ do not occur free in φ, as is the case when $\exists\overline{x}\beta$ is an $\exists$-guard for φ, then we can write $\exists\overline{x}\beta \wedge \varphi$ equivalently as $\exists\overline{x}(\beta \wedge \varphi)$. In other words, an $\exists$-guard is as good as an ordinary guard. We call an FO-formula *self-guarded* if it is either a sentence or it is of the form $\alpha \wedge \varphi$ where α is an $\exists$-guard for φ.

In this section, we will require the notions of *conjunctive queries* (CQs) and *unions of conjunctive queries* (UCQs). A CQ is an FO-formula of the form

$$\varphi(x_1, \ldots, x_n) := \exists y_1 \ldots \exists y_m (\bigwedge_{i \in I} \alpha_i),$$

where each α_i is an atomic relation, possibly an equality, whose free variables are among $\{x_1, \ldots, x_n, y_1, \ldots, y_m\}$. The collection of all CQs is expressively equivalent to the fragment $FO_{\exists,\wedge}$ of first-order logic, which is generated by the following grammar:

$$\varphi := R(x_1, \ldots, x_k) \mid x = y \mid \varphi \wedge \varphi \mid \exists x\varphi.$$

A UCQ is a finite disjunction of CQs. Importantly, GNFO can be alternatively characterized as the smallest logic which can express every UCQ and is closed under guarded negation [4]. This is made explicit in the following expressively equivalent grammar for GNFO:

$$\varphi := \top \mid R(\overline{x}) \mid x = y \mid \alpha \wedge \neg\varphi \mid q[R_1/\varphi_1, \ldots, R_n/\varphi_n],$$

where q is a UCQ with relation symbols $R_1, \ldots, R_n$ and $\varphi_1, \ldots, \varphi_n$ are self-guarded formulas with the appropriate number of free variables and generated by the same recursive grammar. We refer to this as the UCQ syntax for GNFO.

4.1 Natural Extensions of GFO

Unlike FO^2, guarded fragments are peculiar in that they are not closed under substitution. For example, $\exists xy(R(x, y) \wedge \neg S(x, y))$ belongs to GFO, but if we substitute $x = x \wedge y = y$ for $R(x, y)$, we obtain $\exists xy(x = x \wedge y = y \wedge \neg S(x, y))$, which does not belong to GFO (and is not even expressible in GNFO). GFO and GNFO are, however, closed under *self-guarded substitution*: we can uniformly substitute self-guarded formulas for atomic relations. We generalize the notion of a self-guarded formula to abstract logics L as follows: a formula $\varphi(\overline{x}) \in L(\sigma)$ with $\text{free}(\varphi) = \{x_1, \ldots, x_k\}$ is *self-guarded* if there is a n-ary relation symbol $G \in \sigma$, where $n \geq k$, and a tuple of variables $\overline{y}$ containing exactly the variables $\text{free}(\varphi) \cup \{z_1, \ldots, z_m\}$, such that for all σ-structures M and assignments g,

$$M, g \models \varphi \implies M, g \models \exists z_1 \ldots \exists z_m G(\overline{y}).$$

Intuitively, we can think of a self-guarded L-formula as a conjunction of the form $\alpha \wedge \psi$, where α is an $\exists$-guard for ψ. We can then capture the notion of self-guarded substitution for abstract logics by the following definition.

Definition 4.1. *We say that an abstract logic L* expresses self-guarded substitutions *if, for each formula $\varphi \in L$ with $R_1, \ldots, R_k \in sig(\varphi)$, and all self-guarded formulas $\psi_1, \ldots, \psi_k \in L$, we have that L can express the substitution of $\psi_1, \ldots, \psi_k$ for $R_1, \ldots, R_k$ in φ (cf. Definition 2.1).*

It was shown in [4] that every self-guarded GFO-formula is expressible in GNFO. In particular, this applies to all GFO-sentences and GFO-formulas with at most one free variable (since all such formulas can be equivalently written as $x = x \wedge \varphi$). It is therefore common to treat GNFO as an extension of GFO. To make this precise, we say that L' is a *self-guarded extension* of L if L' can express all *self-guarded* formulas of L (notation: $L \preceq_{sg} L'$). In Figure 1, the line marked (*) indicates that GNFO extends GFO in this weaker sense. Furthermore, it is worth noting that GNFO is also not closed under implication, while GFO is. If it were, then GNFO would be able to express full negation (using formulas of the form $\varphi \to \bot$). However, GFO and GNFO both have disjunction and conjunction in common. We formalize all of these considerations into the following notion.

Definition 4.2. *A* guarded logic *is a logic L such that*

1. *GFO $\preceq_{sg} L$,*
2. *L expresses self-guarded substitutions, and*
3. *L expresses conjunction and disjunction.*

Clearly, GFO and GNFO are both guarded logics. Furthermore, observe that the *smallest* guarded logic consists of all conjunctions and disjunctions of self-guarded formulas of GFO.

4.2 Finding the Minimal Extension of GFO with CIP

Our main result in this section is the following.

Theorem 4.1. *Let L be a guarded logic with CIP. Then* GNFO $\preceq L$.

In other words, loosely speaking, GNFO is the smallest extension of GFO with CIP. It is based on similar ideas as the proof of Theorem 3.1, but the argument is more intricate. The main thrust of the argument will be to show that our abstract logic L can express all positive existential formulas, from which it will follow easily that L is able to express all formulas in the UCQ syntax for GNFO. Toward this end, the main technical result is the following proposition.

Proposition 4.1. *Let L be a logic with CIP that can express atomic formulas, guarded quantification, conjunction, and unary implication. Then* $\mathrm{FO}_{\exists,\wedge} \preceq L$.

Here, we say that a logic L can *express guarded quantification* if, whenever $\varphi \in L$ and α is a guard for φ, L can express $\exists\overline{x}(\alpha \wedge \varphi)$; we say that L can *express unary implications* if, whenever $\varphi \in L$ and α is an atomic formula with only one free variable, L can express $\alpha \to \varphi$.

The following definition is used in the proof of Proposition 4.1.

Definition 4.3. *Let φ be a formula in $\mathrm{FO}_{\exists,\wedge}$, let $\overline{y} = y_1, \ldots, y_n$ be a tuple of distinct variables, and let $\overline{P} = P_1, \ldots, P_n$ be a tuple of unary predicates of the same length. Then $\mathit{BIND}_{\overline{y}\mapsto\overline{P}}(\varphi)$ is defined recursively as follows:*

$$\begin{aligned}
\mathit{BIND}_{\overline{y}\mapsto\overline{P}}(\alpha) &= \exists\overline{y}(\alpha \wedge \textstyle\bigwedge_{1\le i\le n} P_i(y_i))\\
\mathit{BIND}_{\overline{y}\mapsto\overline{P}}(\phi\wedge\psi) &= \mathit{BIND}_{\overline{y}\mapsto\overline{P}}(\phi) \wedge \mathit{BIND}_{\overline{y}\mapsto\overline{P}}(\psi)\\
\mathit{BIND}_{\overline{y}\mapsto\overline{P}}(\exists z\psi) &= \exists z(\mathit{BIND}_{\overline{y}\mapsto\overline{P}}(\psi)),
\end{aligned}$$

where α is an atomic formula (possibly an equality).

The $\mathsf{BIND}_{\overline{y}\to\overline{P}}$ operation applied to a formula $\varphi \in \mathrm{FO}_{\exists,\wedge}$ wraps each atomic subformula of φ with quantifiers for the variables in $\overline{y}$, and adds additional unary predicates for these variables. Thus, the free variables of $\mathsf{BIND}_{\overline{y}\mapsto\overline{P}}(\varphi)$, for $\overline{y} = y_1, \ldots, y_n$, are exactly $\mathrm{free}(\varphi) \setminus \{y_1, \ldots, y_n\}$, which justifies our use of the word "BIND". The utility of this definition is due to the following fact: for any $\varphi \in \mathrm{FO}_{\exists,\wedge}$, whenever $M, g \models \mathsf{BIND}_{\overline{y}\to\overline{P}}(\varphi)$, and the interpretation in M of each unary predicate P_i in $\overline{P}$ is a singleton, then $M, g' \models \varphi$, where g' is the extension of g which maps each y_i to the unique element satisfying P_i (cf. Propositions 4.3, 4.4). The following proposition is a simple consequence of the definition of BIND.

Proposition 4.2. *For all $\mathrm{FO}_{\exists,\wedge}$-formulas φ and for all $\overline{x},\overline{y}$ and $\overline{P},\overline{Q}$, if $\overline{x}$ and $\overline{y}$ are disjoint, then $\mathit{BIND}_{\overline{xy}\mapsto\overline{PQ}}(\varphi) \equiv \mathit{BIND}_{\overline{x}\mapsto\overline{P}}(\mathit{BIND}_{\overline{y}\mapsto\overline{Q}}(\varphi))$.*

A formula φ is *clean* if no free variable of φ also occurs bound in φ, and φ does not contain two quantifiers for the same variable. Every FO-formula is equivalent to a clean FO-formula, and all subformulas of a clean formula are also clean. We now state two technical propositions, whose proofs can be found in the full version of this paper [20].

Proposition 4.3. *For every clean* $\mathrm{FO}_{\exists,\wedge}$*-formula* φ*, for every tuple of distinct variables* $\overline{y} = y_1, \ldots, y_n$ *(with each* $y_i \in$ *free*(φ)*), and for every tuple of unary predicates* $\overline{P} = P_1, \ldots, P_n$*, we have that*

$$(\bigwedge_{i=1,\ldots,n} P_i(y_i)) \models \varphi \rightarrow \mathit{BIND}_{\overline{y}\mapsto\overline{P}}(\varphi).$$

Proposition 4.4. *For every clean* $\mathrm{FO}_{\exists,\wedge}$*-formula* $\varphi(x, \overline{y})$ *with* $\overline{y} = y_1, \ldots, y_n$ *distinct from* x*, and for every* n*-tuple of unary predicates* $\overline{P} = P_1, \ldots, P_n$ *not occurring in* φ*, we have that*

$$\exists x \varphi(x, \overline{y}) \equiv \forall \overline{P}\Big((\bigwedge_{i=1\ldots n} P_i(y_i)) \rightarrow \exists x \mathit{BIND}_{\overline{y}\mapsto\overline{P}}(\varphi(x, \overline{y}))\Big).$$

The following lemma enables the proof of Proposition 4.1.

Lemma 4.1. *Let* L *be an* FO*-fragment which can express atomic formulas and is closed under guarded quantification, conjunction, and unary implication. If* L *can express a formula* $\varphi \in \mathrm{FO}_{\exists,\wedge}$ *and all of its subformulas, then for all tuples* $\overline{y}$ *of variables, we have that* L *can express* $\mathit{BIND}_{\overline{y}\mapsto\overline{P}}(\varphi)$*.*

Proof. We show by strong induction on the complexity of clean $\mathrm{FO}_{\exists,\wedge}$-formulas φ that this proposition holds.

Base Case
Suppose φ is an atomic formula. Fix an arbitrary tuple $\overline{y} = y_1 \ldots, y_n$. Then

$$\mathsf{BIND}_{\overline{y}\mapsto\overline{P}}(\varphi) \equiv \exists \overline{y}(\varphi \wedge \bigwedge_{1\leq i\leq n} P_i(y_i)),$$

which L can express by closure under conjunction and guarded quantification.

Inductive Step
Suppose inductively that, for all formulas ψ of lesser complexity than φ, and all tuples $\overline{z}$ of variables, we have that L can express $\mathsf{BIND}_{\overline{z}\mapsto\overline{P}}(\psi)$. Fix an arbitrary tuple $\overline{y}$ of variables.

Suppose that $\varphi = \psi_1 \wedge \psi_2$. Since L can express φ and all of its subformulas, it can also express ψ_1, ψ_2, and all of their subformulas. Then by the inductive hypothesis, L can express $\mathsf{BIND}_{\overline{y}\mapsto\overline{P}}(\psi_1)$ and $\mathsf{BIND}_{\overline{y}\mapsto\overline{P}}(\psi_2)$. Then by closure under conjunctions, L can express $\mathsf{BIND}_{\overline{y}\mapsto\overline{P}}(\psi_1) \wedge \mathsf{BIND}_{\overline{y}\mapsto\overline{P}}(\psi_2)$, which is the same as $\mathsf{BIND}_{\overline{y}\mapsto\overline{P}}(\varphi)$ (cf. Definition 4.3).

Now suppose that $\varphi(\overline{x}, \overline{y}) = \exists z \psi(\overline{x}, \overline{y}, z)$, where the (possibly empty) tuple $\overline{x}$ consists of all free variables of φ not in the tuple $\overline{y}$. We need to show that L can express $\mathsf{BIND}_{\overline{y}\mapsto\overline{P}}(\varphi(\overline{x}, \overline{y}))$, which is the same as $\exists z(\mathsf{BIND}_{\overline{y}\mapsto\overline{P}}(\psi(\overline{x}, \overline{y}, z)))$ (cf. Definition 4.3). Since L can express φ and all of its subformulas, it can also express ψ and all of its subformulas. Then, by the inductive hypothesis, L can express $\mathsf{BIND}_{\overline{y}\mapsto\overline{P}}(\psi)$, whose free variables are those in the tuple $\overline{x}z$, as well as $\mathsf{BIND}_{\overline{xy}\mapsto\overline{QP}}(\psi)$, whose only free variable is z. Since L is closed under conjunction and guarded quantification, it follows that L can express

$$\gamma(\overline{x}) := \exists z(G(\overline{x}, z) \wedge \mathsf{BIND}_{\overline{y}\mapsto\overline{P}}(\psi)) \quad \text{and} \quad \exists z(z = z \wedge \mathsf{BIND}_{\overline{xy}\mapsto\overline{QP}}(\psi)),$$

where G is a fresh relation symbol not occurring in ψ. Then by closure under unary implications, we have that L can also express

$$\chi(\overline{x}) := \big(\bigwedge_i Q_i(x_i)\big) \to \exists z(z = z \wedge \mathsf{BIND}_{\overline{xy}\mapsto\overline{QP}}(\psi)).$$

Claim: $\gamma(\overline{x}) \models \chi(\overline{x})$

Proof of claim: By Proposition 4.2,

$$\mathsf{BIND}_{\overline{xy}\mapsto\overline{QP}}(\psi) \equiv \mathsf{BIND}_{\overline{x}\mapsto\overline{Q}}(\mathsf{BIND}_{\overline{y}\mapsto\overline{P}}(\psi)). \tag{1}$$

Then by applying Proposition 4.3 and inverting the hypotheses, we have

$$\mathsf{BIND}_{\overline{y}\mapsto\overline{P}}(\psi) \models \big(\bigwedge_i Q_i(x_i)\big) \to \mathsf{BIND}_{\overline{xy}\mapsto\overline{QP}}(\psi).$$

From this, it follows (because z is distinct from x_i variables) that

$$\exists z(\mathsf{BIND}_{\overline{y}\mapsto\overline{P}}(\psi)) \models \big(\bigwedge_i Q_i(x_i)\big) \to \exists z\mathsf{BIND}_{\overline{xy}\mapsto\overline{QP}}(\psi),$$

and therefore $\gamma(\overline{x}) \models \chi(\overline{x})$. This concludes the proof of the claim.

Since L can express both $\gamma(\overline{x})$ and $\chi(\overline{x})$, we have by the Craig interpolation property that L can express some Craig interpolant $\vartheta(\overline{x})$. Since G and the Q_i predicates do not occur in φ, they do not occur in $\vartheta(\overline{x})$, and therefore, the following second-order implication is valid:

$$\exists G\gamma(\overline{x}) \models \vartheta(\overline{x}) \models \forall\overline{Q}\chi(\overline{x}).$$

It is easy to see that $\exists G\gamma(\overline{x}) \equiv \exists z\mathsf{BIND}_{\overline{y}\mapsto\overline{P}}(\psi)$. Similarly, it follows from Proposition 4.4 and equation (1) that $\forall\overline{Q}\chi(\overline{x}) \equiv \exists z\mathsf{BIND}_{\overline{y}\mapsto\overline{P}}(\psi)$. Therefore, $\vartheta(\overline{x}) \equiv \exists z\mathsf{BIND}_{\overline{y}\mapsto\overline{P}}(\psi)$. In particular, $\exists z\mathsf{BIND}_{\overline{y}\mapsto\overline{P}}(\psi)$ is expressible in L. □

We are now ready to prove Proposition 4.1, restated below.

Proposition 4.1. *Let L be a logic with CIP that can express atomic formulas, guarded quantification, conjunction, and unary implication. Then* $\mathrm{FO}_{\exists,\wedge} \preceq L$.

Proof. By strong induction on the complexity of $\mathrm{FO}_{\exists,\wedge}$-formulas. The base case is immediate, since L can express all atomic formulas. For the inductive step, if $\varphi := \psi_1 \wedge \psi_2$, then by the inductive hypothesis, L can express ψ_1 and ψ_2, and so by closure under conjunction, L can express φ. Now suppose $\varphi(\overline{y}) := \exists x(\psi(x, \overline{y}))$. By the inductive hypothesis, together with closure under guarded quantification, L can express

$$\gamma(\overline{y}) := \exists x(G(x, \overline{y}) \wedge \psi).$$

Furthermore, by Lemma 4.1, L can express $\mathsf{BIND}_{\overline{y}\mapsto\overline{P}}(\psi)$, and therefore, by closure under guarded quantification and unary implications, L can express

$$\chi(\overline{y}) := (\bigwedge_i P_i(y_i)) \to \exists x(x = x \wedge \mathsf{BIND}_{\overline{y}\mapsto\overline{P}}(\psi)).$$

Claim: $\gamma(\overline{y}) \models \chi(\overline{y})$.

Proof of claim: It is clear that $\gamma(\overline{y}) \models \exists x\psi$. Furthermore, by Proposition 4.3, $\psi \models (\bigwedge_i P_i(y_i)) \mapsto \mathsf{BIND}_{\overline{y}\mapsto\overline{P}}(\psi)$, from which it follows that $\exists x\psi \models \chi(\overline{y})$ (since the variable x is distinct from $y_1, \ldots, y_n$). Therefore, $\gamma(\overline{y}) \models \chi(\overline{y})$.

Let $\vartheta(\overline{y})$ be an interpolant for $\gamma(\overline{y}) \models \chi(\overline{y})$ in L. Since G and the predicates in $\overline{P}$ do not occur in ψ, the following second-order entailments are valid:

$$\exists G\exists x(G(x, \overline{y}) \wedge \psi) \models \vartheta(\overline{y}) \models \forall \overline{P}((\bigwedge_i P_i(y_i)) \to \exists x\mathsf{BIND}_{\overline{y}\mapsto\overline{P}}(\psi)).$$

It is easy to see that

$$\exists G\exists x(G(x, \overline{y}) \wedge \psi) \equiv \exists x\psi.$$

Furthermore, by Proposition 4.4,

$$\psi \equiv \forall \overline{P}((\bigwedge_i P_i(y_i)) \to \mathsf{BIND}_{\overline{y}\mapsto\overline{P}}(\psi)).$$

from which it follows (since x is distinct from $y_1, \ldots, y_n$) that

$$\exists x\psi \equiv \forall \overline{P}((\bigwedge_i P_i(y_i)) \to \exists x\mathsf{BIND}_{\overline{y}\mapsto\overline{P}}(\psi)).$$

Therefore, $\vartheta(\overline{y}) \equiv \varphi(\overline{y})$, and so we are done. □

Our main result follows easily from Proposition 4.1, the closure properties of guarded logics, and the UCQ characterization of GNFO.

Theorem 4.1. *Let L be a guarded logic with CIP. Then* GNFO $\preceq L$.

Proof. L can express self-guarded GFO-formulas, so it can express formulas of the form $\exists\overline{x}\beta$, where β is an atomic formula. Then since L can express self-guarded substitution, L can express guarded quantification. Furthermore, L can express all self-guarded formulas of the form $\alpha \wedge \neg\beta$, where α and β are atomic formulas such that free$(\alpha) =$ free(β). Furthermore, for every formula φ expressible in L with free$(\varphi) \subseteq$ free(α), $\alpha \wedge \varphi$ is a self-guarded formula. Thus by expressibility of self-guarded substitution, L can also express $\alpha \wedge \neg(\alpha \wedge \varphi)$, which is equivalent to $\alpha \wedge \neg\varphi$; hence L can express guarded negation. If L can express φ, then by expressibility of guarded negation and disjunction, it can also express the formula $(x = x \wedge \neg P(x)) \vee \varphi$, which is equivalent to $P(x) \to \varphi$. Hence L can express unary implications. Therefore, by Proposition 4.1, L can express all formulas in $\mathrm{FO}_{\exists,\wedge}$. Then by expressibility of disjunction, L can express all unions of conjunctive queries. The result then follows immediately from the UCQ-syntax for GNFO, by closure under self-guarded substitution. □

5 Repairing Interpolation for FF

The fluted fragment (FL) [51] is an ordered logic, in which all occurrences of variables in atomic formulas and quantifiers must follow a fixed order. In the context of ordered logics, we assume a fixed infinite sequence of variables $X = \langle x_i \rangle_{i \in \mathbb{Z}^+}$. A *suffix n-atom* is an atomic formula of the form $R(x_j, \ldots, x_n)$, where $x_j, \ldots, x_n$ is a finite contiguous subsequence of X. FL is defined by the following recursion.

Definition 5.1. *For each $n \in \mathbb{N}$, define collections of formulas FL^n as follows:*

1. *FL^n contains all suffix n-atoms,*
2. *FL^n is closed under Boolean combinations, and*
3. *If φ is in FL^{n+1}, then $\exists x_{n+1}\varphi$ and $\forall x_{n+1}\varphi$ are in FL^n.*

We set $\mathrm{FL} = \bigcup_{n \in \mathbb{N}} \mathrm{FL}^n$.

The forward fragment (FF), introduced in [6], is a syntactic generalization of FL. We say that $R(x_j, \ldots, x_k)$ is an *infix n-atom* if $x_j, \ldots, x_n$ is a finite contiguous subsequence of X and $k \leq n$. FF is defined by the following recursion.

Definition 5.2. *For each $n \in \mathbb{N}$, define collections of formulas FF^n as follows:*

1. *FF^n contains all infix n-atoms,*
2. *FF^n is closed under Boolean combinations, and*
3. *If φ is in FF^{n+1}, then $\exists x_{n+1}\varphi$ and $\forall x_{n+1}\varphi$ are in FF^n.*

We set $\mathrm{FF} = \bigcup_{n \in \mathbb{N}} \mathrm{FF}^n$.

In contrast to the other logics we have seen, FL and FF do not allow the primitive equality symbol. It can be seen by a simple formula induction that every formula in FF^k can be expressed by a formula in FF^n for every $n > k$; it follows easily that FF can express arbitrary Boolean combinations of its formulas. However, FL cannot: $P(x_1)$ and $P(x_2)$ are in FL, but $P(x_1) \wedge P(x_2)$ is not expressible in FL. Although FF contains formulas which are not in FL, it is known that FF and FL are expressively equivalent at the level of sentences [7]. Furthermore, the satisfiability problems for FL and FF are decidable [48,7].

5.1 Natural Extensions of FF

Given a formula φ, we write $gfv(\varphi)$ to denote the greatest $n \in \mathbb{Z}^+$ such that x_n occurs free in φ; if φ is a sentence, then we set $gfv(\varphi) = 0$. We define *forward logics* to capture the notion of a natural extension of FF.

Definition 5.3. *A* forward logic *is an abstract logic L such that*

1. *L can express all infix n-atoms for every* $n \in \mathbb{Z}^+$,
2. *L can express all Boolean combinations of its formulas, and*
3. *L can express* $\exists x_n \varphi$ *and* $\forall x_n \varphi$ *whenever L can express* φ *and* $n = gfv(\varphi)$.

We refer to the last property of a forward logic as *expressibility of ordered quantification.* Clearly FF is a forward logic, and every forward logic extends FF.

5.2 Finding the Minimal Extension of FF with CIP

Unlike the other fragments we have seen, one peculiar property of FF is that the logic is not closed under variable substitutions. This can be seen simply by considering relational atoms: for a 3-ary relational symbol R, the formula $R(x_1, x_2, x_3)$ is in FF, but the formula $R(x_3, x_1, x_2)$ is not. Before proving our main theorem, we prove the following lemma asserting that whenever a formula is expressible in a forward logic L satisfying CIP, the result of making arbitrary substitutions for the free variables of the formula is also expressible in L.

Lemma 5.1. *Let L be a forward logic satisfying CIP, and let* $\varphi(x_{i_1}, \dots, x_{i_k})$ *be a formula of first-order logic expressible in L, where* $x_{i_1}, \dots, x_{i_k}$ *is not necessarily a contiguous subsequence of variables. Then for every map*

$$\pi : \{i_1, \dots, i_k\} \to \mathbb{Z}^+,$$

we have that L can also express $\varphi(x_{\pi(i_1)}, \dots, x_{\pi(i_k)})$. *In other words, L is closed under renamings of free variables.*

Proof. For brevity, let $\overline{x} = x_{i_1}, \dots, x_{i_k}$, and let $\pi(\overline{x}) = x_{\pi(i_1)}, \dots, x_{\pi(i_k)}$. Without loss of generality, assume that $i_1 \leq \dots \leq i_k$ (we can do this since the notation $\varphi(x_{i_1}, \dots, x_{i_k})$ only indicates that the variables occur free, but says nothing about where or in what order they occur in the formula). Since L can express $\varphi(\overline{x})$, it can evidently express the following formulas, by the definition of a forward logic:

$$\gamma(\overline{x}) := \bigwedge_{m \leq k} G_m(x_{i_m}) \wedge \forall x_{i_1} \dots \forall x_{i_k} \left(\bigwedge_{m \leq k} G_m(x_{\pi(i_m)}) \to \varphi(\overline{x}) \right)$$

$$\chi(\overline{x}) := \bigwedge_{m \leq k} P_m(x_{i_m}) \to \exists x_{i_1} \dots \exists x_{i_k} \left(\varphi(\overline{x}) \wedge \bigwedge_{m \leq k} P_m(x_{\pi(i_m)}) \right)$$

Clearly $\gamma \models \chi$, and so there exists an interpolant ϑ. Hence

$$\exists G_1 \dots G_k \gamma \models \vartheta \models \forall P_1 \dots P_k \chi$$

is a valid second-order entailment. Furthermore, it is easy to see that

$$\exists G_1 \dots G_k \gamma \equiv \forall P_1 \dots P_k \chi \equiv \varphi.$$

Therefore, $\varphi(x_{\pi(i_1)}, \dots, x_{\pi(i_k)})$ is expressible in L. □

We now prove our main theorem, which follows easily from Lemma 5.1.

Theorem 5.1. *Let L be a forward logic satisfying CIP. Then* FO $\preceq L$.

Proof. We proceed by formula induction on FO-formulas φ. For the base case, clearly L can express all atomic FO-formulas by applying Lemma 5.1 to an appropriate infix atom. For the inductive step, the Boolean cases are immediate since L can express all Boolean combinations. Hence the only interesting case is when $\varphi := \exists x_k \psi$ for some formula ψ. By the inductive hypothesis, L can express ψ. Applying Lemma 5.1, L can also express φ', the result of substituting x_{n+1} for all free occurrences of x_k, where $n = gfv(\varphi)$, and leaving all other free variables the same. Then by expressibility of ordered quantification, L can express $\exists x_{n+1} \varphi'$, which is equivalent to φ. □

6 Undecidability of Extensions of FO^2 and FL with CIP

In Section 3, we showed that every strong extension of FO^2 with CIP can express all sentences of FO, and in Section 5, we showed that every forward logic with CIP can express all formulas of FO. These results suggest the undecidability of the satisfiability problems for such logics. In this section, we formalize this idea, showing that extensions of FO^2 and FL with CIP and satisfying very limited expressive assumptions are undecidable. These results rely primarily on known results on the undecidability of FO^2 and FL with additional transitive relations.

Proposition 6.1. *Every abstract logic L with CIP extending* FO^2 *or* FL *can express the following formulas:*

$$\psi_0(x_1) := \forall x_2 \forall x_3 (R(x_1, x_2) \land R(x_2, x_3) \to R(x_1, x_3)), \text{ and}$$
$$\psi_1 := \neg \forall x_1 \forall x_2 \forall x_3 (R(x_1, x_2) \land R(x_2, x_3) \to R(x_1, x_3)).$$

The proof of Proposition 6.1 can be found in the full version of this paper [20]. We also need two additional definitions. First, an *effective translation* from a logic L to a logic L' is a computable function which takes formula of $\varphi \in L$ as input and outputs an equivalent formula $\varphi' \in L'$. Second, we say that a logic L has *effective conjunction* if there is a computable function taking formulas $\varphi, \psi \in L$ as input and outputting a formula $\chi \in L$ which is equivalent to $\varphi \land \psi$.

Theorem 6.1. *Let L be an extension of* FL *which satisfies CIP. Suppose further that there is an effective translation from* FL *to L, and L has effective conjunction. The satisfiability problem for L is undecidable if either*

1. *L can express ordered quantification, or*
2. *L can express negation.*

Proof. Let χ be the sentence asserting the transitivity of the relation R. Since L has CIP and extends FL, it can express both $\psi_0(x_1)$ and ψ_1 by Proposition 6.1. If L can express ordered quantification, it can express $\forall x_1 \psi_0(x_1)$, which is equivalent to χ. If L can express negation, then it can express $\neg\psi_1$, which is also equivalent to χ. Since L, as an abstract logic, can express χ and is closed under predicate renamings, it can express that any number of binary relations are transitive. Let χ_1, χ_2, and χ_3 be sentences expressing transitivity of binary relation symbols R_1, R_2, and R_3, respectively. Let tr be an effective translation from FL to L. Then a formula φ of FL with three designated transitive relations is satisfiable if and only if $tr(\varphi) \wedge \chi_1 \wedge \chi_2 \wedge \chi_3$ is satisfiable. Since tr is computable and L is effectively closed under conjunction, this reduction is computable. Since the satisfiability problem for FL with three transitive relations is undecidable [46], the satisfiability problem for L is undecidable. □

It is also known that satisfiability is undecidable for FO^2-formulas with two transitive relations [36]. Using this fact, along with Proposition 6.1, we obtain the following theorem, by a similar proof to that of Theorem 6.1.

Theorem 6.2. *Let L be an extension of* FO^2 *which satisfies CIP. Suppose further that there is an effective translation from* FO^2 *to L, and L has effective conjunction. The satisfiability problem for L is undecidable if either*

1. *L can express universal quantification, or*
2. *L can express negation.*

We remark that all forward logics and strong extensions of FO^2 with CIP, assuming appropriate effective translations and effective conjunction, meet the requirements of Theorems 6.1 and 6.2, and hence are undecidable.

7 Discussion

In the introduction, we mentioned several results indicating the failure of CIP among many natural proper extension of FO. In [14], van Benthem points out that there is a similar scarcity among FO-fragments as well. Our results in Sections 3 and 5 may be interpreted as additional confirmation of this observation. Furthermore, one tends to study proper fragments of FO for their desirable computational properties, and so our broader undecidability results show that CIP fails for large swaths of *decidable* FO-fragments. However, there are a few notable fragments for which the determination of a minimal extension satisfying CIP is still open, such as FL and the quantifier prefix fragments.

One limitation of our methodology and results is their dependence on a definition of Craig interpolation which mandates the existence of interpolants between proper *formulas*, while many practical applications only require CIP for *sentences*. Throughout this paper, we have established expressibility of a formula ϑ in a logic L by induction (and by constructing two formulas φ and ψ such that $\varphi \models \psi$ and arguing that every interpolant is equivalent to ϑ). In general, this method is difficult to apply unless free variables are allowed; it is not clear how to apply this type of inductive argument if we were only concerned with the existence of interpolants for sentences of the logic.

There are several well-studied properties strictly weaker than CIP. The Δ-interpolation property (also known as Suslin-Kleene interpolation) holds for a logic L if, whenever $\varphi \models \psi$, and (intuitively speaking) there is only one possible interpolant ϑ up to logical equivalence for this entailment, then L contains a formula equivalent to ϑ [5]. It is not hard to see that, unlike the Craig interpolation property, every logic L has a unique extension, denoted $\Delta(L)$, satisfying the Δ-interpolation property. In fact, in our proofs we only rely on Δ-interpolation; every application of the assumption that some abstract logic L satisfies CIP yields a provably unique interpolant, up to logical equivalence. Therefore, all of our results hold also when CIP is replaced by Δ-interpolation.

Two additional weakenings of CIP are the projective and non-projective Beth definability properties. The projective Beth property states, roughly, that whenever a $\sigma \cup \tau \cup \{R\}$-theory Σ implicitly defines a relation R in terms of the relations in σ, then Σ entails an explicit definition of R in terms of σ (the non-projective Beth property being the special case for $\tau = \emptyset$). Many practical applications of CIP in database theory and knowledge representation require only the projective Beth property. It is not immediately clear how to extend our methodology to a systematic study of the (projective) Beth property among decidable FO-fragments. Indeed, GFO already satisfies the non-projective Beth property [33]. Given their applications, an interesting avenue of future work is to map the landscape of FO-fragments satisfying these properties. In the other direction, minimal extensions of logics with *uniform* interpolation (a strengthening of CIP) were studied in [25], although with limited results so far (cf. [25, Thm. 14]). Some of the minimal extensions of PLTL fragments with CIP identified in [28], however, do satisfy uniform interpolation.

Acknowledgements. We thank Jean Jung, Frank Wolter, and Malvin Gattinger for feedback on an earlier draft, and we thank Ian Pratt-Hartmann and Michael Benedikt for helpful remarks during a related workshop presentation. Balder ten Cate is supported by EU Horizon 2020 grant MSCA-101031081.

References

1. Andréka, H., Németi, I., van Benthem, J.: Modal languages and bounded fragments of predicate logic. Journal of Philosophical Logic **27** (06 1998). `https://doi.org/10.1023/A:1004275029985`

2. Areces, C., Blackburn, P., Marx, M.: Repairing the interpolation theorem in quantified modal logic. Annals of Pure and Applied Logic **124**(1), 287–299 (2003). https://doi.org/10.1016/S0168-0072(03)00059-9
3. Bárány, V., Benedikt, M., ten Cate, B.: Rewriting guarded negation queries. In: Proceedings of MFCS 2013. pp. 98–110. Springer Berlin Heidelberg, Berlin, Heidelberg (2013)
4. Barany, V., ten Cate, B., Segoufin, L.: Guarded negation. Journal of the ACM **62**(3), 22.1–22:26 (2015)
5. Barwise, J., Feferman, S. (eds.): Model-Theoretic Logics, Perspectives in Logic, vol. 8. Cambridge University Press (2017)
6. Bednarczyk, B.: Exploiting forwardness: Satisfiability and query-entailment in forward guarded fragment. In: Logics in Artificial Intelligence: 17th European Conference, JELIA 2021, Virtual Event, May 17–20, 2021, Proceedings. p. 179–193. Springer-Verlag, Berlin, Heidelberg (2021). https://doi.org/10.1007/978-3-030-75775-5_13
7. Bednarczyk, B., Jaakkola, R.: Towards a Model Theory of Ordered Logics: Expressivity and Interpolation. In: Szeider, S., Ganian, R., Silva, A. (eds.) 47th International Symposium on Mathematical Foundations of Computer Science (MFCS 2022). Leibniz International Proceedings in Informatics (LIPIcs), vol. 241, pp. 15:1–15:14. Schloss Dagstuhl – Leibniz-Zentrum für Informatik, Dagstuhl, Germany (2022). https://doi.org/10.4230/LIPIcs.MFCS.2022.15
8. Benedikt, M., Bourhis, P., Boom, M.V.: Definability and Interpolation within Decidable Fixpoint Logics. Logical Methods in Computer Science **Volume 15, Issue 3** (Sep 2019). https://doi.org/10.23638/LMCS-15(3:29)2019
9. Benedikt, M., ten Cate, B., Boom, M.V.: Interpolation with decidable fixpoint logics. In: LICS. pp. 378–389 (2015). https://doi.org/10.1109/LICS.2015.43
10. Benedikt, M., ten Cate, B., Tsamoura, E.: Generating plans from proofs. ACM Trans. Database Syst. **40**(4), 22:1–22:45 (2016). https://doi.org/10.1145/2847523
11. Benedikt, M., Cate, B.ten., Boom, M.V.: Effective interpolation and preservation in guarded logics. ACM Trans. Comput. Logic **17**(2) (2015). https://doi.org/10.1145/2814570
12. Benedikt, M., Leblay, J., ten Cate, B., Tsamoura, E.: Generating plans from proofs : the interpolation-based approach to query reformulation. Synthesis Lectures on Data Management, Morgan & Claypool (2016)
13. van Benthem, J.: A new modal lindström theorem. Logica Universalis **1**(1), 125–138 (2007). https://doi.org/10.1007/s11787-006-0006-3
14. van Benthem, J.: The many faces of interpolation. Synthese **164**(3), 451–460 (2008), http://www.jstor.org/stable/40271083
15. van Benthem, J., ten Cate, B., Väänänen, J.A.: Lindström theorems for fragments of first-order logic. Log. Methods Comput. Sci. **5**(3) (2009), http://arxiv.org/abs/0905.3668
16. Börger, E., Grädel, E., Gurevich, Y.: The Classical Decision Problem. Perspectives in Mathematical Logic, Springer (1997)
17. Caicedo, X.: Failure of interpolation for quantifiers of monadic type. In: Di Prisco, C.A. (ed.) Methods in Mathematical Logic. pp. 1–12. Springer Berlin Heidelberg, Berlin, Heidelberg (1985)
18. Calvanese, D., Ghilardi, S., Gianola, A., Montali, M., Rivkin, A.: Combined covers and beth definability. In: Proceedings of the 10th International Joint Conference on Automated Reasoning, Part I, IJCAR 2020. pp. 181–200. Springer (2020). https://doi.org/10.1007/978-3-030-51074-9_11

19. ten Cate, B.: Interpolation for extended modal languages. The Journal of Symbolic Logic **70**(1), 223–234 (2005), http://www.jstor.org/stable/27588355
20. ten Cate, B., Comer, J.: Craig interpolation for decidable first-order fragments. arXiv preprint arXiv:2310.08689 (2023), https://arxiv.org/abs/2310.08689
21. ten Cate, B., Franconi, E., Seylan, I.: Beth definability in expressive description logics. J. Artif. Int. Res. **48**(1), 347–414 (oct 2013)
22. ten Cate, B., Segoufin, L.: Unary negation. Logical Methods in Computer Science **Volume 9, Issue 3** (Sep 2013). https://doi.org/10.2168/LMCS-9(3:25)2013
23. Comer, S.D.: Classes without the amalgamation property. Pacific Journal of Mathematics **28**, 309–318 (1969)
24. Craig, W.: Three uses of the herbrand-gentzen theorem in relating model theory and proof theory. Journal of Symbolic Logic **22**(3), 269–285 (1957). https://doi.org/10.2307/2963594
25. D'Agostino, G., Lenzi, G., French, T.: μ-programs, uniform interpolation and bisimulation quantifiers for modal logics. Journal of Applied Non-Classical Logics **16**(3-4), 297–309 (2006). https://doi.org/10.3166/jancl.16.297-309
26. Friedman, H.: Beth's theorem in cardinality logics. Israel Journal of Mathematics **14**(2), 205–212 (1973)
27. Garcá-Matos, M.: Abstract model theory without negation. Ph.D. thesis, University of Helsinki (2005)
28. Gheerbrant, A., ten Cate, B.: Craig interpolation for linear temporal languages. In: Grädel, E., Kahle, R. (eds.) Computer Science Logic. pp. 287–301. Springer Berlin Heidelberg, Berlin, Heidelberg (2009)
29. Graedel, E., Otto, M., Rosen, E.: Two-variable logic with counting is decidable. In: Proceedings of LICS 1997. p. 306 (1997)
30. Grädel, E.: On the restraining power of guards. The Journal of Symbolic Logic **64**(4), 1719–1742 (1999), http://www.jstor.org/stable/2586808
31. Hoder, K., Holzer, A., Kovács, L., Voronkov, A.: Vinter: A Vampire-based tool for interpolation. In: Jhala, R., Igarashi, A. (eds.) Programming Languages and Systems - 10th Asian Symposium, APLAS 2012, Kyoto, Japan, December 11-13, 2012. Proceedings. Lecture Notes in Computer Science, vol. 7705, pp. 148–156. Springer (2012). https://doi.org/10.1007/978-3-642-35182-2_11
32. Hoogland, E.: Definability and interpolation: model-theoretic investigations. Ph.D. thesis, University of Amsterdam (2000)
33. Hoogland, E., Marx, M.: Interpolation and definability in guarded fragments. Studia Logica **70**(3), 373–409 (2002), http://www.jstor.org/stable/20016403
34. Hustadt, U., Schmidt, R., Georgieva, L.: A survey of decidable first-order fragments and description logics. Journal on Relational Methods in Computer Science **1**, 251–276 (01 2004)
35. Jung, J.C., Wolter, F.: Living without beth and craig: Definitions and interpolants in the guarded and two-variable fragments. In: Proceedings of LICS 2021. pp. 1–14. IEEE Computer Society (jul 2021). https://doi.org/10.1109/LICS52264.2021.9470585
36. Kieroński, E.: Results on the guarded fragment with equivalence or transitive relations. In: Computer Science Logic. Lecture Notes in Computer Science, vol. 3634, pp. 309–324. Springer Verlag (2005)
37. Koopmann, P., Schmidt, R.A.: Uniform interpolation and forgetting for $\mathcal{ALC}$ ontologies with ABoxes. In: Proceedings of the 29th AAAI Conference on Artificial Intelligence, AAAI 2015. pp. 175–181. AAAI Press (2015), http://www.aaai.org/ocs/index.php/AAAI/AAAI15/paper/view/9981

38. Lindström, P.: On extensions of elementary logic. Theoria **35**(1) (1969)
39. Lutz, C., Wolter, F.: Foundations for uniform interpolation and forgetting in expressive description logics. In: Proceedings of the 22nd International Joint Conference on Artificial Intelligence, IJCAI 2011. pp. 989–995. IJCAI/AAAI (2011). https://doi.org/10.5591/978-1-57735-516-8/IJCAI11-170
40. Löwenheim, L.: Über möglichkeiten im relativkalkül. Mathematische Annalen **76**, 447–470 (1915), http://eudml.org/doc/158703
41. McMillan, K.L.: Interpolation and model checking. In: Clarke, E.M., Henzinger, T.A., Veith, H., Bloem, R. (eds.) Handbook of Model Checking, pp. 421–446. Springer (2018). https://doi.org/10.1007/978-3-319-10575-8_14
42. Mortimer, M.: On languages with two variables. Math. Log. Q. **21**, 135–140 (1975)
43. Otto, M.: An interpolation theorem. The Bulletin of Symbolic Logic **6**(4), 447–462 (2000), http://www.jstor.org/stable/420966
44. Pratt-Hartman, I., Szwast, W., Tendera, L.: The fluted fragment revisited. The Journal of Symbolic Logic **84**(3), 1020–1048 (2019). https://doi.org/10.1017/jsl.2019.33
45. Pratt-Hartmann, I., Szwast, W., Tendera, L.: Quine's fluted fragment is non-elementary. In: Regnier, L., Talbot, J. (eds.) 25th EACSL Annual Conference on Computer Science Logic. 25th EACSL Annual Conference on Computer Science Logic (CSL 2016), Schloss Dagstuhl–Leibniz-Zentrum fuer Informatik (Jun 2016). https://doi.org/10.4230/LIPIcs.CSL.2016.39
46. Pratt-Hartmann, I., Tendera, L.: The fluted fragment with transitive relations. Annals of Pure and Applied Logic **173**(1), 103042 (2022). https://doi.org/10.1016/j.apal.2021.103042
47. Purdy, W.C.: Decidability of Fluted Logic with Identity. Notre Dame Journal of Formal Logic **37**(1), 84 – 104 (1996). https://doi.org/10.1305/ndjfl/1040067318
48. Purdy, W.C.: Fluted formulas and the limits of decidability. The Journal of Symbolic Logic **61**(2), 608–620 (1996). https://doi.org/10.2307/2275678
49. Purdy, W.C.: Quine's 'limits of decision'. The Journal of Symbolic Logic **64**(4), 1439–1466 (1999). https://doi.org/10.2307/2586789
50. Purdy, W.C.: Complexity and nicety of fluted logic. Studia Logica **71**, 177–198 (2002)
51. Quine, W.V.: On the limits of decision. 14th International Congress for Philosophy **3**, 57–62 (1969)
52. Skolem, T.: Logisch-Kombinatorische Untersuchungen über die Erfüllbarkeit oder Bewiesbarkeit mathematischer Sätze nebst einem Theorem über dichte Mengen. I. Matematisk-naturvidenskabelig Klasse 4, 1-36, Videnskapsselskapet Skrifter (1920)
53. Toman, D., Weddell, G.E.: Fundamentals of Physical Design and Query Compilation. Synthesis Lectures on Data Management, Morgan & Claypool Publishers (2011)
54. Väänänen, J.: The craig interpolation theorem in abstract model theory. Synthese **164**(3), 401–420 (2008)
55. Vardi, M.Y.: Why is modal logic so robustly decidable? In: Immerman, N., Kolaitis, P.G. (eds.) Descriptive Complexity and Finite Models. DIMACS, vol. 31, pp. 149–183 (1996). https://doi.org/10.1090/dimacs/031/05

Clones, closed categories, and combinatory logic*

Philip Saville(✉)

Department of Computer Science, University of Oxford, Oxford, UK
philip.saville@cs.ox.ac.uk
http://www.philipsaville.co.uk/

Abstract. We explain how to recast the semantics of the simply-typed λ-calculus, and its linear and ordered variants, using multi-ary structures. We define universal properties for multicategories, and use these to derive familiar rules for products, tensors, and exponentials. Finally we outline how to recover both the category-theoretic syntactic model and its semantic interpretation from the multi-ary framework. We then use these ideas to study the semantic interpretation of combinatory logic and the simply-typed λ-calculus without products. We introduce *extensional SK-clones* and show these are sound and complete for both combinatory logic with extensional weak equality and the simply-typed λ-calculus without products. We then show such SK-clones are equivalent to a variant of closed categories called *SK-categories*, so the simply-typed λ-calculus without products is the internal language of SK-categories.

Keywords: categorical semantics · abstract clones · lambda calculus · combinatory logic · closed categories · cartesian closed categories

1 Introduction

Lambek's correspondence between cartesian closed categories and the simply-typed λ-calculus is one of the central pillars of categorical semantics. One way of stating it categorically is to say that the syntax of typed λ-terms over a *signature* of base types and constants forms the free cartesian closed category (for a readable overview, see [27,9]). The existence of this *syntactic model* gives *completeness*: if an equation holds in every model, it holds in the free one, and hence in the syntax. The free property then gives *soundness*: for any interpretation of basic types and constants in a cartesian closed category $(\mathcal{C}, \Pi, \Rightarrow)$ one has a functor $[\![-]\!]$ from the syntactic model to $\mathcal{C}$, which is exactly the semantic interpretation of λ-terms. The fact this functor is required to preserve cartesian closed structure amounts to showing that the semantic interpretation is sound with respect to the usual $\beta\eta$-laws. All this justifies calling the simply-typed λ-calculus the *internal language* of cartesian closed categories.

This framework is powerful, but hides a fundamental mismatch: morphisms $A \to B$ in a category are *unary*—they have just one input—but terms-in-context

* Supported by the Air Force Office of Scientific Research under award number FA9550-21-1-0038.

N. Kobayashi and J. Worrell (Eds.): FoSSaCS 2024, LNCS 14575, pp. 160–181, 2024.
https://doi.org/10.1007/978-3-031-57231-9_8

such as $x_1 : A_1, \ldots, x_n : A_n \vdash t : B$ can have many inputs. The standard solution (*e.g.* [9,23]) is to use categorical products to model contexts, so a term t as above corresponds to a map $\prod_{i=1}^{n} A_i \to B$ out of the product.

Despite its evident success, this solution remains somewhat unsatisfactory, in two ways (see also [21]). First, it forces us to conflate two different syntactic classes, namely contexts and product types. As a result, some encoding is required to construct the syntactic model: the interpretation of $x : A, y : B \vdash t : C$ is a term in context $p : A \times B$. This adds complexity to the construction, and results in the somewhat unintuitive fact that the semantic interpretation of a term t in the syntactic model may not be just t itself. In turn, this complicates the proof of completeness.

Second, we are forced to include products in our type theory if we want a category-theoretic internal language—even though the calculus without products likely has a stronger claim to being called 'the' simply-typed λ-calculus (*e.g.* see Church's original definition [8]). This raises the question: what categorical structure has the simply-typed λ-calculus without products as its internal language?

This paper. This paper has three main aims. First, to explain how removing the mismatch between terms-in-context and morphisms outlined above clarifies the semantic interpretation of simply-typed λ-calculi. To achieve this, one needs to move from the unary setting of categories to a *multi-ary* setting, in which we have *multimaps* $A_1, \ldots, A_n \to B$. These ideas are not new, but are under-appreciated, and I hope this will provide self-contained introduction for a wider audience. Second, to initiate a multi-ary investigation of the semantics of (cartesian) combinatory logic, in the style of Hyland's investigation of similar ideas for the untyped λ-calculus ([18,19]). Finally, to use these results to define a categorical semantics for the simply-typed λ-calculus without products.

Outline. In Sections 2 to 6, we explain how the multi-ary perspective yields a slick way to derive the unary semantic interpretation and syntactic model, together with soundness and completeness results (Section 4.2). We also show how important type-theoretic constructions such as products and exponentials can be derived from the semantics. This framework accommodates different choices of structural rules, such as whether the language is ordered, linear, or cartesian.

The idea of using multi-ary constructions goes back to Lambek ([25,26]), and has recently been exploited to great effect in a very general setting by Shulman [40]. Particular cases can also be found in the works of Hyland ([18,19]), Hyland & de Paiva [20] and Blanco & Zeilberger [7]. A reader familiar with these approaches will likely be unsurprised by the technical development below. However, we believe these ideas deserve to be more widely known, so spend time making them explicit in a concrete setting.

In Section 7 we introduce a multi-ary model of (cartesian) combinatory logic, called *SK-clones*, and prove that the sub-category of *extensional* SK-clones is equivalent to the category of *closed clones* modelling simply-typed λ-calculus without products. This provides a categorical statement of the classical correspondence between λ-calculus and extensional combinatory logic (*e.g.* [5,15]).

Finally, in Section 8 we introduce a version of Eilenberg & Kelly's closed categories ([11,10]), called *SK-categories*, and show that the category of SK-categories is equivalent to the category of extensional SK-clones, and so to the category of closed clones. Hence, SK-categories are a categorical model for the simply-typed λ-calculus without products. SK-categories are a cartesian version of the prounital-closed categories of Uustalu, Veltri & Zeilberger ([43,44]), which in turn are closely related to an (incomplete) suggestion of Shulman's [39].

Jacobs has also isolated a structure that is sound and complete for simply-typed λ-calculus without products [21]. His approach, which fits into his elegant general framework [22], is also predicated on a careful distinction between contexts and products. His models are certain indexed categories, with the contexts encoded by the indexing: this makes them feel closer to multi-ary structures. In SK-categories, by contrast, contexts are modelled within the category itself by using the closed structure (*cf.* [35, §4.4]). Moreover, unlike other work relating closed categories to multi-ary structures, SK-categories do not force us to include a unit object in the corresponding type theory (*cf.* [31]).

Technical preliminaries. For a set S we write $S^\star$ for the set of finite sequences over S, and use Greek letters $\Gamma, \Delta, \ldots$ to denote elements of $S^\star$. The empty string is denoted $\diamond$, and the length of Γ by $|\Gamma|$. Where the length of a sequence is clear, we write simply $A_\bullet$ for $A_1, \ldots, A_n$. Contexts are assumed to be ordered lists.

We call multimaps of the form $A \to B$ *unary* and a multimap $\diamond \to B$ *nullary.*

We define a *signature* $\mathcal{S}$ to be a set $|\mathcal{S}|$ of *basic sorts* with sets $\mathcal{S}(\Gamma; B)$ of *constants* $c : \Gamma \to B$ for each $(\Gamma, B) \in |\mathcal{S}|^\star \times |\mathcal{S}|$. A *homomorphism of signatures* $f : \mathcal{S} \to \mathcal{S}'$ is a map $|f| : |\mathcal{S}| \to |\mathcal{S}'|$ with maps $\mathcal{S}(A_1, \ldots, A_n; B) \to \mathcal{S}'(fA_1, \ldots, fA_n; fB)$ for each $((A_1, \ldots, A_n), B) \in |\mathcal{S}|^\star \times |\mathcal{S}|$. We write **Sig** for the category of signatures and their homomorphisms. One could also consider versions of higher-order constants, which may use the language's constructs. This extension does not change the theory significantly, and would require introducing multiple categories of signatures, so we do not seek this extra generality here (for an outline of this more general approach, see *e.g.* [38, §5.3.1]).

We assume familiarity with the simply-typed λ-calculus, as in *e.g.* [9]. We denote the simply-typed λ-calculus with constants and base types given by a signature $\mathcal{S}$, and both product and exponential types modulo $\alpha\beta\eta$-equality, by $\Lambda_{\mathcal{S}}^{\times,\to}$. We write $\Lambda_{\mathcal{S}}^{\times}$ and $\Lambda_{\mathcal{S}}^{\to}$ for the fragments with just products and just exponentials, respectively. Here we focus on the *typed* cases: the untyped versions—both in the syntax and the multi-ary models—are recovered by fixing a single base type $\star$ such that $\Theta(\star, \ldots, \star) = \star$ for each type constructor Θ.

We also assume familiarity with the basics of cartesian categories, cartesian closed categories, and monoidal categories, as in *e.g.* [30,27]. To avoid having to treat the unit type as a special case, cartesian categories are assumed to have n-ary products $\prod_n$ for all $n \in \mathbb{N}$. We also work with functors preserving structure *strictly*: this simplifies the exposition without any great loss of generality. Thus, **MonCat**, **SMonCat** and **CartCat** denote the categories of monoidal categories, symmetric monoidal categories, and cartesian categories, respectively, with functors preserving all the data on the nose.

2 Multicategories and clones

We begin with an intuitive overview of the place of multi-ary structures in semantics. A multi-ary structure has *multimaps* $A_1, \ldots, A_n \to B$ with multiple inputs and one output; unlike the morphisms in a category, multimaps correspond directly to terms-in-context. As a result, it is often easier to construct a multi-ary free model than it is to construct a unary one, and the interpretation of a term-in-context t in the free model is given by t itself. Moreover, every multi-ary structure gives rise to a unary one by restricting to multimaps with one input. The multi-ary semantics therefore factors the unary one, as shown:

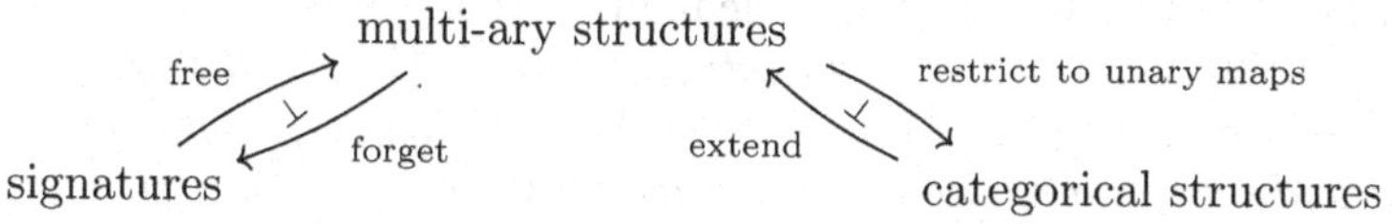

(1)

One can then 'read off' the syntactic category, together with a guarantee that it has the right structure, by restricting the free multi-ary structure to unary maps. Similarly, the usual semantic interpretation in (say) a cartesian closed category $\mathcal{C}$ is exactly the interpretation that arises by extending $\mathcal{C}$ to a multi-ary structure. This gives an algebraic justification for encoding contexts as products: this is how one extends a cartesian closed category to a multi-ary structure. (For the details of these points, see Section 4.2.)

The multi-ary perspective also provides a unifying framework for type theories with different structural rules. The simply-typed λ-calculus is *cartesian*: it admits the structural rules of weakening, contraction, and permutation (as in *e.g.* [9, Fig. 3.2]). The corresponding multi-ary structures are certain *abstract clones*. *Ordered* type theories (*e.g.* [24,36]), also known as *planar* type theories (*e.g.* [2,46]), do not admit weakening, contraction, or permutation, and correspond to certain *multicategories*. *Linear* type theories (*e.g.* [16]), which admit only permutation, correspond to certain *symmetric multicategories* (see also the alternative 'tangled' option in [33]). Since abstract clones and symmetric multicategories may be seen as special cases of multicategories, we can develop a theory of how to add structure to cartesian, linear, and ordered type theories by analysing how to add structure to multicategories.

2.1 Multicategories, clones, and their internal languages

We now introduce multicategories and abstract clones and show how they correspond to certain type theories. An even more general framework for syntax, allowing multi-ary domains and codomains as well as both cartesian and linear contexts, is provided by Shulman's recent work with polycategories [40]. Clones, and their correspondence with syntax, also play a key role in the 'algebraic syntax' programme of Fiore and collaborators initiated in [13] (see *e.g.* [12,3,4]).

Definition 1 ([25]). *A* multicategory $\mathbb{M}$ *consists of a set* $|\mathbb{M}|$ *of* objects *and sets* $\mathbb{M}(\Gamma; B)$ *of* multimaps *for every* $\Gamma \in |\mathbb{M}|^\star$ *and* $B \in |\mathbb{M}|$, *together with*

1. *An* identity *multimap* $\mathrm{Id}_A \in \mathbb{M}(A; A)$ *for every* $A \in |\mathbb{M}|$*;*
2. *For any* $A_1, \ldots, A_n, B \in |\mathbb{M}|$ *and* $(\Delta_i \in |\mathbb{M}|^\star)_{i=1,\ldots,n}$*, a* composition *map*

$$\mathbb{M}(A_1, \ldots, A_n; B) \times \textstyle\prod_{i=1}^{n} \mathbb{M}(\Delta_i; A_i) \to \mathbb{M}(\Delta_1, \ldots, \Delta_n; B)$$
$$(t, (u_1, \ldots, u_n)) \mapsto t \circ \langle u_1, \ldots, u_n \rangle$$

subject to an associativity law and two unit laws (see e.g. [28, p. 35]). A multicategory functor $f : \mathbb{M} \to \mathbb{N}$ *consists of a map* $|f| : |\mathbb{M}| \to |\mathbb{N}|$ *with maps* $f_{A_\bullet, B} : \mathbb{M}(A_1, \ldots, A_n; B) \to \mathbb{N}(fA_1, \ldots fA_n; fB)$ *for every* $A_1, \ldots, A_n, B \in |\mathbb{M}|$*, such that substitution and the identity are preserved (see e.g. [28, p. 39]).*

Definition 2 ([32,20]). *A* symmetric multicategory *consists of a multicategory* $\mathbb{M}$ *together with a symmetric group action: for each* $A_1, \ldots, A_n \in |\mathbb{M}|$ *and* $\sigma \in S_n$ *one has* $(-) \bullet \sigma : \mathbb{M}(A_1, \ldots, A_n; B) \to \mathbb{M}(A_{\sigma 1}, \ldots, A_{\sigma n}; B)$ *compatible with substitution and satisfying unit and associativity laws (e.g. [28, p. 54]). A* symmetric multicategory functor *is a multicategory functor which preserves the action.*

We write **Multicat** (resp. **SMulticat**) for the category of (symmetric) multicategories and their functors, and write $t : \Gamma \to B$ for $t \in \mathbb{M}(\Gamma; B)$.

Example 1. Every monoidal category $(\mathcal{C}, \otimes, I)$ induces a multicategory $\mathcal{T}\mathcal{C}$. The objects are those of $\mathcal{C}$, with multimaps $(\mathcal{T}\mathcal{C})(A_1, \ldots, A_n; B) := \mathcal{C}(\bigotimes_{i=1}^{n} A_i, B)$ for a chosen n-ary bracketing of the tensor product. This determines functors **MonCat** → **Multicat**, and **SMonCat** → **SMulticat** (see *e.g.* [28, p. 39]); we denote both of these by $\mathcal{T}$.

Lambek [25] essentially observed that every multicategory has an internal language, as follows. One identifies multimaps $t : A_1, \ldots, A_n \to B$ with terms $x_1 : A_1, \ldots, x_n : A_n \vdash t : B$, for a fixed ordering of an infinite set of variables $\{x_1, x_2, \ldots\}$. The identity Id_A is identified with the variable $x : A$, and the composition operation becomes a formal substitution operation on the language. Stated in this way, the three axioms become well-known properties of substitution: the unit laws say $x[u] = u$ and $t[x_1, \ldots, x_n] = t$, and the associativity law is a linear version of the so-called Substitution Lemma (*e.g.* [5, Lemma 2.1.16]).

The next result shows this terminology does not differ too much from the notion of internal language in the introduction. For a signature $\mathcal{S}$ and $\Gamma := (x_i : A_i)_{i=1,\ldots,n}$, write $\mathcal{O}_\mathcal{S}$ for the ordered language generated by the two rules on the left below, and $\mathfrak{L}_\mathcal{S}$ for the linear language generated by all three rules:

$$\frac{}{x : A \vdash x : A} \qquad \frac{c \in \mathcal{S}(\Gamma; B) \qquad (\Delta_i \vdash u_i : A_i)_{i=1,\ldots,n}}{\Delta_1, \ldots, \Delta_n \vdash c^{\S}(u_1, \ldots, u_n) : B} \qquad \frac{\Theta, x : A, y : B, \Delta \vdash t : C}{\Theta, y : B, x : A, \Delta \vdash t : C}$$

Substitution is defined as usual, so that the following rule is admissible:

$$\frac{x_1 : A_1, \ldots, x_n : A_n \vdash t : B \qquad (\Delta_i \vdash u_i : A_i)_{i=1,\ldots,n}}{\Delta_1, \ldots, \Delta_n \vdash t[u_1/x_1, \ldots, u_n/x_n] : B} \tag{2}$$

With this rule as composition, $\mathcal{O}_\mathcal{S}$ and $\mathfrak{L}_\mathcal{S}$ define a syntactic multicategory $\mathrm{Syn}(\mathcal{O}_\mathcal{S})$ and a syntactic symmetric multicategory $\mathrm{Syn}(\mathfrak{L}_\mathcal{S})$, respectively. These define left adjoints to the functors **Multicat** $\rightarrow$ **Sig** and **SMulticat** $\rightarrow$ **Sig** sending a (symmetric) multicategory $\mathbb{M}$ to the signature with objects $|\mathbb{M}|$ and constants $\{\mathbb{M}(\Gamma; B)\}_{\Gamma \in |\mathbb{M}|^\star, B \in |\mathbb{M}|}$; we denote both these functors by U.

Lemma 1. $\mathrm{Syn}(\mathcal{O}_\mathcal{S})$ *(resp.* $\mathrm{Syn}(\mathfrak{L}_\mathcal{S})$*) is the free multicategory (resp. symmetric multicategory) on* $\mathcal{S}$.

Thus, the internal language of a symmetric multicategory is the core of Abramsky's linear λ-calculus [1]. To recover a cartesian language, we use *(multi-sorted) abstract clones.* These differ from multicategories in that the result of substituting $(u_i : \Delta \rightarrow A_i)_{i=1,2}$ into $t : A_1, A_2 \rightarrow B$ yields a multimap of type $\Delta \rightarrow B$, not $\Delta, \Delta \rightarrow B$. Abstract clones are equivalently *cartesian multicategories* (see *e.g.* [18]), but this formulation is less natural syntactically: it amounts to adding explicit duplication and deletion operations to the language.

Definition 3. *An* abstract clone $\mathbb{C}$ *consists of a set* $|\mathbb{C}|$ *of* sorts *and sets* $\mathbb{C}(\Gamma; B)$ *of* multimaps *for every* $\Gamma \in |\mathbb{C}|^\star$ *and* $B \in |\mathbb{C}|$*, together with*

1. Projection *multimaps* $\mathsf{p}_i^{A_\bullet} \in \mathbb{C}(A_1, \dots, A_n; A_i)$ *for every* $A_1, \dots, A_n \in |\mathbb{C}|$;
2. *For every* $A_1, \dots, A_n, B \in |\mathbb{C}|$ *and* $\Delta \in |\mathbb{C}|^\star$, *a* substitution *operation*

$$\mathbb{C}(A_1, \dots, A_n; B) \times \textstyle\prod_{i=1}^n \mathbb{C}(\Delta; A_i) \rightarrow \mathbb{C}(\Delta; B)$$
$$(t, (u_1, \dots, u_n)) \mapsto t[u_1, \dots, u_n]$$

subject to an associativity law and two unit laws for any $t \in \mathbb{C}(A_1, \dots, A_n; B)$, $(u_i \in \mathbb{C}(B_1, \dots, B_m; A_i))_{i=1,\dots,n}$ *and* $(v_j \in \mathbb{C}(\Theta; B_j))_{j=1,\dots,m}$:

$$(t[u_\bullet])[v_\bullet] = t[\dots, u_i[v_\bullet], \dots] \ , \ \mathsf{p}_i^{A_\bullet}[u_1, \dots, u_n] = u_i \ , \ t[\mathsf{p}_1^{A_\bullet}, \dots, \mathsf{p}_n^{A_\bullet}] = t$$

A homomorphism of clones $f : \mathbb{C} \rightarrow \mathbb{D}$ *consists of a map* $|f| : |\mathbb{C}| \rightarrow |\mathbb{D}|$ *and maps* $f_{A_\bullet, B} : \mathbb{C}(A_1, \dots, A_n; B) \rightarrow \mathbb{D}(fA_1, \dots fA_n; fB)$ *for every* $A_1, \dots, A_n, B \in |\mathbb{C}|$, *such that* $f(\mathsf{p}_i^{A_\bullet}) = \mathsf{p}_i^{(fA)_\bullet}$ *and* $f(t[u_1, \dots, u_n]) = (ft)[fu_1, \dots, fu_n]$. *We write* **Clone** *for the category of clones and clone homomorphisms.*

Example 2 (cf. Example 1). Any cartesian category $(\mathcal{C}, \Pi)$ determines a clone $\mathcal{PC}$ with sorts the objects of $\mathcal{C}$ and $(\mathcal{PC})(A_1, \dots, A_n; B) := \mathcal{C}(\prod_{i=1}^n A_i; B)$.

We distinguish between clones and multicategories by using $[\dots]$ for a clone's substitution operation and $\langle \dots \rangle$ for a multicategory's composition operation. Every multicategory, and hence every clone, has an underlying category.

Definition 4. *The* nucleus $\overline{\mathbb{M}}$ *of a multicategory or clone* $\mathbb{M}$ *is the category with the same objects and* $\overline{\mathbb{M}}(A, B) := \mathbb{M}(A; B)$. *This defines functors* $\overline{(-)}$: **Multicat** $\rightarrow$ **Cat** *and* $\overline{(-)}$: **Clone** $\rightarrow$ **Cat** *to the category of small categories.*

The internal language of a clone is a cartesian version of that for multicategories. Write $\Lambda_\mathcal{S}$ for the language below; substitution is defined as usual.

$$\frac{(i = 1, \dots, n)}{x_1 : A_1, \dots, x_n : A_n \vdash x_i : A_i} \qquad \frac{c \in \mathcal{S}(\Gamma; B) \qquad (\Delta \vdash u_i : A_i)_{i=1,\dots,n}}{\Delta \vdash c^{\S}(u_1, \dots, u_n) : B}$$

Identifying variables with projections, we get a syntactic clone $\mathrm{Syn}(\Lambda_\mathcal{S})$.

Lemma 2. *The canonical forgetful functor* $\mathrm{U} : \mathbf{Clone} \to \mathbf{Sig}$ *has a left adjoint, and the free clone on* $\mathcal{S}$ *is* $\mathrm{Syn}(\Lambda_{\mathcal{S}})$.

Example 3. The languages $\Lambda_{\mathcal{S}}^{\times}, \Lambda_{\mathcal{S}}^{\rightarrow}$ and $\Lambda_{\mathcal{S}}^{\times,\rightarrow}$ each induce syntactic clones we denote by $\mathrm{Syn}(\Lambda_{\mathcal{S}}^{\times}), \mathrm{Syn}(\Lambda_{\mathcal{S}}^{\rightarrow})$ and $\mathrm{Syn}(\Lambda_{\mathcal{S}}^{\times,\rightarrow})$, respectively.

3 Universal properties for multicategories

In this section we generalise the categorical notion of *universal arrows* (as in *e.g.* [30, §3]) to give a notion of universal property for multicategories. This will provide a uniform way to introduce new connectives to a type theory. One could also define the required conditions directly (see [7,40]), but here we wish to emphasise that they arise from category-theoretic ideas.

Definition 5 (*cf.* **[17]**). *Let* $f : \mathbb{M} \to \mathbb{N}$ *be a multicategory functor.*

1. *A* universal arrow *from* f *to* $Y \in |\mathbb{N}|$ *is a pair* $(R \in |\mathbb{M}|, \rho : fR \to Y)$ *such that for every* $t : fA_1, \ldots, fA_n \to Y$ *there exists a unique multimap* $t^{\#} : A_1, \ldots, A_n \to R$ *such that* $\rho \circ \langle f(t^{\#}) \rangle = t$.
2. *A* universal arrow *from* $X_1, \ldots, X_n \in |\mathbb{N}|$ *to* f *is a pair* $(R \in |\mathbb{M}|, \rho : X_1, \ldots, X_n \to fR)$ *such that for every* $t : X_1, \ldots, X_n \to fB$ *there exists a unique multimap* $t^{\#} : R \to B$ *such that* $f(t^{\#}) \circ \langle \rho \rangle = t$.

We extend this definition—and hence our notion of universal property—to clones by using the next observation (*cf.* the fact a cartesian category is monoidal).

Lemma 3. *There is a faithful functor* $\mathrm{M} : \mathbf{Clone} \to \mathbf{Multicat}$ *sending a clone* $\mathbb{C}$ *to the multicategory with the same objects and hom-sets, and composition given using substitution in* $\mathbb{C}$ *and the projections.*

Definition 5 does not involve 'global' conditions like naturality, so is particularly amenable to a type-theoretic interpretation. As in the categorical setting, however, it can be rephrased using natural isomorphisms (*cf.* [30, §3.2]).

Lemma 4. *Let* $f : \mathbb{M} \to \mathbb{N}$ *be a multicategory functor.*

1. *Giving a universal arrow from* f *to* $X \in |\mathbb{N}|$ *is equivalent to giving* $R \in \mathbb{M}$ *and an isomorphism* $\phi_{A_\bullet} : \mathbb{M}(A_1, \ldots, A_n; R) \xrightarrow{\cong} \mathbb{N}(fA_1, \ldots, fA_n; Y)$, *natural in the sense that the left diagram below commutes for any* $t : A_1, \ldots, A_n \to B$;
2. *Giving a universal arrow from* $X_1, \ldots, X_n \in |\mathbb{N}|$ *to* f *is equivalent to giving* $R \in |\mathbb{M}|$ *and an isomorphism* $\psi_B : \mathbb{M}(R; B) \xrightarrow{\cong} \mathbb{N}(X_1, \ldots, X_n; fB)$, *natural in the sense that the right diagram below commutes for any* $u : B \to C$.

$$\begin{array}{ccc} \mathbb{M}(B;R) & \xrightarrow{\phi_B} & \mathbb{N}(fB;X) \\ {\scriptstyle (-)\circ\langle t\rangle}\downarrow & & \downarrow{\scriptstyle (-)\circ\langle ft\rangle} \\ \mathbb{M}(A_1,\ldots,A_n;R) & \xrightarrow[\phi_{A_\bullet}]{} & \mathbb{N}(fA_1,\ldots,fA_n;X) \end{array} \qquad \begin{array}{ccc} \mathbb{M}(R;B) & \xrightarrow{\psi_B} & \mathbb{N}(X_1,\ldots,X_n;fB) \\ {\scriptstyle u\circ(-)}\downarrow & & \downarrow{\scriptstyle f(u)\circ(-)} \\ \mathbb{M}(R;C) & \xrightarrow[\psi_C]{} & \mathbb{N}X_1,\ldots,X_n;fC) \end{array}$$

A corollary is that giving a right adjoint to a multicategory functor $f : \mathbb{N} \to \mathbb{M}$ in Hermida's 2-category of multicategories [17] is equivalent to giving a mapping $g_0 : |\mathbb{M}| \to |\mathbb{N}|$ and a universal arrow $fg(X) \to X$ from f to X for each $X \in |\mathbb{N}|$.

4 Product structure

We now have enough to define products for multicategories, and hence for clones. An n-ary product is exactly a limit over the discrete category with n objects. Rephrasing in terms of universal arrows (*e.g.* [30, §3]) we get that equipping a category $\mathcal{C}$ with n-ary products is exactly equipping it with a universal arrow from the diagonal functor $\Delta^{(n)} : \mathcal{C} \to \mathcal{C}^{\times n}$ to $(A_1, \ldots, A_n)$ for every $A_1, \ldots, A_n \in \mathcal{C}$.

Since **Multicat** has finite products defined in much the same way as the category of small categories **Cat**, we may make the following definition. The prefix 'cartesian' is already used for multicategories, so we use '**finite-products**'.

Definition 6. *An* fp-multicategory *is a multicategory* $\mathbb{M}$ *equipped with a universal arrow* $\left(\prod_{i=1}^{n} A_i, (\pi_1^{A_\bullet}, \ldots, \pi_n^{A_\bullet})\right)$ *from the diagonal functor* $\Delta^{(n)} : \mathbb{M} \to \mathbb{M}^{\times n}$ *to* $(A_1, \ldots, A_n)$ *for every* $n \in \mathbb{N}$ *and* $A_1, \ldots, A_n \in |\mathbb{M}|$.

Asking for $\mathbb{M}$ to have finite products is equivalent to asking for a product object $\prod_{i=1}^{n} A_i$ and unary multimaps $\left(\pi_i^{A_\bullet} : \prod_{i=1}^{n} A_i \to A_i\right)_{i=1,\ldots,n}$ for each $A_1, \ldots, A_n \in |\mathbb{M}|$, such that composition induces isomorphisms $\mathbb{M}\left(\Gamma; \prod_{i=1}^{n} A_i\right) \cong \prod_{i=1}^{n} \mathbb{M}(\Gamma; A_i)$. In the internal language, this amounts to the following rules:

$$\frac{}{p : \prod_{i=1}^{n} A_i \vdash \pi_i^{A_\bullet}(p) : A_i}\ (i = 1, \ldots, n) \quad , \quad \frac{(\Gamma \vdash t_i : A_i)_{i=1,\ldots,n}}{\Gamma \vdash \langle t, \ldots, t_n \rangle : \prod_{i=1}^{n} A_i} \tag{3}$$

$$\pi_i^{A_\bullet}(p)\big[\langle t_1, \ldots, t_n \rangle\big] = t_i \quad , \quad \big\langle \pi_1^{A_\bullet}(p)[u], \ldots, \pi_n^{A_\bullet}(p)[u] \big\rangle = u$$

We can now derive the rules for & in linear λ-calculus [1]. Indeed, given $\Gamma, x : A_i, \Theta \vdash t : B$, from (3) we get $\Gamma, p : \prod_{i=1}^{n} A_i, \Theta \vdash t[\pi_i^{A_\bullet}(p)/x] : B$. This suggests the following. Let $\mathcal{O}_{\mathcal{S}}^{\&}$ (resp. $\mathfrak{L}_{\mathcal{S}}^{\&}$) be the extension of $\mathcal{O}_{\mathcal{S}}$ (resp. $\mathfrak{L}_{\mathcal{S}}$) with

$$\frac{\Gamma, x_i : A_i, \Theta \vdash t : C \qquad \Delta \vdash u : \&_{i=1}^{n} A_i}{\Gamma, \Delta, \Theta \vdash \mathtt{let}\ x_i\ \mathtt{be}\ \mathtt{p}_i\ \mathtt{of}\ u\ \mathtt{in}\ t : C} \quad , \quad \frac{(\Gamma \vdash t_i : A_i)_{i=1,\ldots,n}}{\Gamma \vdash \langle t_1, \ldots, t_n \rangle : \&_{i=1}^{n} A_i}$$

$$\mathtt{let}\ x_i\ \mathtt{be}\ \mathtt{p}_i\ \mathtt{of}\ \langle u_i \rangle_{i=1}^{n}\ \mathtt{in}\ t = t[u_i/x_i] \quad , \quad \langle \mathtt{let}\ x_i\ \mathtt{be}\ \mathtt{p}_i\ \mathtt{of}\ u\ \mathtt{in}\ x_i \rangle_{i=1}^{n} = u$$

where we write $\langle u_i \rangle_{i=1}^{n}$ for $\langle u_1, \ldots, u_n \rangle$. This syntax defines a free property. To see this, say a multicategory functor f *(strictly) preserves finite products* if it preserves all the data on the nose, so that $f(\prod_{i=1}^{n} A_i) = \prod_{i=1}^{n} f A_i$, $f(\pi_i^{A_\bullet}) = \pi_i^{f A_\bullet}$, and $f(\langle t_1, \ldots, t_n \rangle) = \langle f t_1, \ldots, f t_n \rangle$. Write **fpMulticat** for the category of fp-multicategories and product-preserving functors, and **fpSMulticat** for the subcategory of symmetric multicategories with finite products, with functors preserving both structures.

Lemma 5. *The composite forgetful functor* **fpMulticat** $\to$ **Multicat** $\to$ **Sig** *has a left adjoint, and the free fp-multicategory on* $\mathcal{S}$ *is* $\mathrm{Syn}(\mathcal{O}_{\mathcal{S}}^{\&})$. *This extends to symmetric structure: replace* **fpMulticat** *by* **fpSMulticat** *and* $\mathcal{O}^{\&}$ *by* $\mathfrak{L}^{\&}$.

Returning to the cartesian setting, we define products in a clone using the corresponding structure for multicategories and Lemma 3.

Definition 7. *A* cartesian clone $(\mathbb{C}, \Pi)$ *is a clone* $\mathbb{C}$ *equipped with a choice of finite products on* $\mathcal{M}\mathbb{C}$. *A* (strict) homomorphism of cartesian clones *is a clone homomorphism* f *that strictly preserves all the product structure. We write* **CartClone** *for the category of cartesian clones and strict homomorphisms.*

Writing $\pi_i(t)$ for the multimap $\pi_i^{A_\bullet}[t]$, the rules (3) translate directly to the usual product rules of λ-calculus. So cartesian clones exactly capture $\Lambda^\times$.

Lemma 6. *The composite forgetful functor* **CartClone** → **Clone** → **Sig** *has a left adjoint, and* $\mathrm{Syn}(\Lambda^\times_{\mathcal{S}})$ *is the free cartesian clone on* $\mathcal{S}$.

Using the characterisation of universal arrows in terms of natural isomorphisms we get the following refinement of Example 2.

Example 4. For any cartesian category $(\mathcal{C}, \Pi)$ the induced clone $\mathcal{PC}$ is cartesian, essentially by definition; this extends to a functor $\mathcal{P}$: **CartCat** → **CartClone**. Moreover, if $(\mathbb{C}, \Pi)$ is a cartesian clone, then so is its nucleus $\overline{\mathbb{C}}$. Hence $\overline{(-)}$ restricts to a functor **CartClone** → **CartCat**.

The two functors in this example are actually adjoints, yielding our first version of the schema in (1). The unit is identity-on-objects and sends $t : A_1, \ldots, A_n \to B$ to $t[\pi_1^{A_\bullet}, \ldots, \pi_n^{A_\bullet}] : \prod_{i=1}^n A_i \to B$.

Proposition 1. *The functor* $\overline{(-)}$: **CartClone** → **CartCat** *fits into the following diagram of adjunctions:*

$$\mathbf{Sig} \underset{\mathrm{U}}{\overset{\mathrm{F}}{\rightleftarrows}} \mathbf{CartClone} \underset{\mathcal{P}}{\overset{\overline{(-)}}{\rightleftarrows}} \mathbf{CartCat} \qquad (\mathrm{F} \dashv \mathrm{U},\ \overline{(-)} \dashv \mathcal{P})$$

Moreover, $\mathrm{U} \circ \mathcal{P}$ *is equal to the canonical forgetful functor* **CartCat** → **Sig**. *Hence, the free cartesian category on* $\mathcal{S}$ *is canonically isomorphic to* $\overline{\mathrm{Syn}(\Lambda^\times_{\mathcal{S}})}$.

4.1 Cartesian structure from representability

In the preceding section we defined products using a multi-ary version of the familiar universal property. There is another way to define 'monoidal structure' in a multicategory: Hermida's *representability* [17]. From the perspective of linear logic, the finite product structure explored above corresponds to the additive conjunction &; Hermida's representability will correspond to the multiplicative conjunction ⊗. We shall also see that, for clones, the two are equivalent.

Definition 8. *A* representable multicategory *is a multicategory* $\mathbb{M}$ *equipped with a universal arrow* $\big(\mathrm{T}(X_1, \ldots, X_n), \rho_{X_\bullet} : X_1, \ldots, X_n \to \mathrm{T}(X_1, \ldots, X_n)\big)$ *from* $X_1, \ldots, X_n$ *to the identity* $\mathrm{id}_{\mathbb{M}}$ *for each* $X_1, \ldots, X_n \in |\mathbb{M}|$; *we write* $\mathrm{T}_{i=1}^n X_i$ *for* $\mathrm{T}(X_1, \ldots, X_n)$. *These universal arrows must be closed under composition, so*

$$X_1, \ldots, X_n, Y_1, \ldots, Y_m \xrightarrow{\langle \rho_{X_\bullet}, \rho_{Y_\bullet} \rangle} \mathrm{T}_{i=1}^n X_i, \mathrm{T}_{j=1}^m Y_j \xrightarrow{\rho} \mathrm{T}\big(\mathrm{T}_{i=1}^n X_i, \mathrm{T}_{j=1}^m Y_j\big)$$

must also be universal. A representable multicategory functor f *is a multicategory functor that preserves all the universal arrows, so that* $f(\mathrm{T}_{i=1}^{n} A_i) = \mathrm{T}_{i=1}^{n} fA_i$, $f(\rho_{A_\bullet}) = \rho_{fA_\bullet}$ *and* $f(t^{\#}) = ft^{\#}$. *Write* **RepMulticat** *for the category of representable multicategories, and* **SRepMulticat** *for the category of representable multicategories whose underlying multicategories are also symmetric, with functors preserving both structures.*

Example 5 (cf. Example 1). The multicategory $\mathcal{TC}$ induced by a monoidal category $(\mathcal{C}, \otimes, I)$ is representable. We therefore obtain functors **MonCat** $\to$ **RepMulticat** and **SMonCat** $\to$ **SRepMulticat**; we denote them both $\mathcal{T}$.

A representable multicategory is a multicategory equipped with rules which are dual to those in (3) in the sense that the universal arrow goes the other direction. Indeed, writing $x_1 \otimes \ldots \otimes x_n$ for $\rho_{A_\bullet}$, and $\mathtt{let}\,(x_1, \ldots, x_n)\,\mathtt{be}\,p\,\mathtt{in}\,t$ for $t^{\#}$, and extending this to all terms by

$$u_1 \otimes \ldots \otimes u_n := (x_1 \otimes \ldots \otimes x_n)[u_1/x_1, \ldots, u_n/x_n]$$

$$\mathtt{let}\,(x_1, \ldots, x_n)\,\mathtt{be}\,u\,\mathtt{in}\,t := \big(\mathtt{let}\,(x_1, \ldots, x_n)\,\mathtt{be}\,p\,\mathtt{in}\,t\big)[u/p]$$

we obtain the following rules, where $\Gamma := (x_i : A_i)_{i=1,\ldots,n}$:

$$\frac{(\Delta_i \vdash u_i : A_i)_{i=1,\ldots,n}}{\Delta_1, \ldots, \Delta_n \vdash \otimes_{i=1}^{n} u_i : \bigotimes_{i=1}^{n} A_i}\;,\qquad \frac{\Lambda, \Gamma, \Theta \vdash t : B \qquad \Delta \vdash u : \bigotimes_{i=1}^{n} A_i}{\Lambda, \Delta, \Theta \vdash \mathtt{let}\,(x_1, \ldots, x_n)\,\mathtt{be}\,u\,\mathtt{in}\,t : B} \tag{4}$$

$$\mathtt{let}\,(x_1, \ldots, x_n)\,\mathtt{be}\,p\,\mathtt{in}\,t[\otimes_{i=1}^{n} x_i/p] = t\;,\;\; \mathtt{let}\,(x_1, \ldots, x_n)\,\mathtt{be}\,\otimes_{i=1}^{n} x_i\,\mathtt{in}\,t = t$$

We write $\mathcal{O}_S^{\otimes}$ (resp. $\mathfrak{L}_S^{\otimes}$) for the extension of $\mathcal{O}_S$ (resp. $\mathfrak{L}_S$) with these rules. This is essentially the tensor fragment of Abramsky's linear λ-calculus [1]. The connection with multicategories was already made in by Hyland & de Paiva [20], who showed this type theory arises from Lambek's *monoidal multicategories* [26].

Lemma 7. *The composite forgetful functor* **RepMulticat** $\to$ **Multicat** $\to$ **Sig** *has a left adjoint, and the free representable multicategory on* $\mathcal{S}$ *is the syntactic multicategory* $\mathrm{Syn}(\mathcal{O}_S^{\otimes})$. *The same holds for symmetric structure, if one replaces* **RepMulticat** *by* **SRepMulticat** *and* $\mathcal{O}^{\otimes}$ *by* $\mathfrak{L}^{\otimes}$.

Combining this lemma with Lemma 5, one sees that a multicategory equipped with representable and finite-product structure corresponds to a linear type theory with both $\otimes$ and &.

We can also obtain a linear version of Proposition 1. Hermida [17] showed that the 2-category of representable multicategories is 2-equivalent to the 2-category of monoidal categories, and Weber showed this extends to the symmetric case [45]. From these constructions one can extract functors $\mathcal{T}$: **RepMulticat** $\to$ **MonCat** and $\mathcal{T}_{\mathrm{sym}}$: **SRepMulticat** $\to$ **SMonCat** sending a (symmetric) representable multicategory to a (symmetric) monoidal structure on its nucleus, together with equivalences **RepMulticat** $\simeq$ **MonCat** and **SRepMulticat** $\simeq$ **SMonCat**. So we get the following.

Proposition 2. *The functors $\mathcal{N}$ and $\mathcal{N}_{\text{sym}}$ fit into the following diagram of adjunctions, where in each case the right-hand adjunction is an equivalence:*

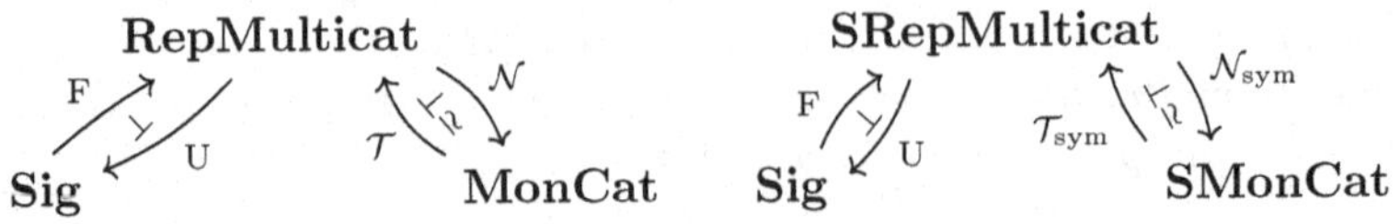

Moreover, $\mathrm{U} \circ \mathcal{T}$ *and* $\mathrm{U} \circ \mathcal{T}_{\text{sym}}$ *are both equal to the canonical forgetful functor to* **Sig**. *Hence, the free monoidal (resp. symmetric monoidal) category on a signature* $\mathcal{S}$ *is canonically isomorphic to* $\mathcal{N}(\mathrm{Syn}(\mathcal{O}_{\mathcal{S}}^{\otimes}))$ *(resp.* $\mathcal{N}(\mathrm{Syn}(\mathfrak{L}_{\mathcal{S}}^{\otimes}))$*).*

We now turn to studying representability in the cartesian setting.

Definition 9. *A* representable clone *is a clone* $\mathbb{C}$ *equipped with a choice of representable structure on* $\mathrm{M}\mathbb{C}$. *A* representable clone homomorphism *is a clone homomorphism which preserves the representable structure as in Definition 8.*

A cartesian clone makes the *projections* primitive (recall (3)), but a representable clone makes the *pairing operation* primitive (recall (4)). It turns out these perspectives are equivalent. In the proof-theoretic setting such ideas are well-studied (*cf.* the equivalence of G-systems and N-systems in [42, §3.3]); the categorical statement has also been made by Pisani [34] and Shulman [40].

Proposition 3. *Equipping a clone* $\mathbb{C}$ *with representable structure is equivalent to equipping* $\mathbb{C}$ *with cartesian structure.*

In Proposition 2 we gave an equivalence of categories but in Proposition 1 we only gave an adjunction. We can now upgrade the latter to an equivalence. Indeed, $\overline{(-)} \circ \mathcal{P}$ is equal to the identity. On the other hand, if $(\mathbb{C}, \Pi)$ is a cartesian clone then by Proposition 3 and Lemma 4 we have a multi-natural isomorphism $\mathbb{C}(A_1, \ldots, A_n; B) \cong \mathbb{C}(\prod_{i=1}^n A_i; B) = \mathcal{P}(\overline{\mathbb{C}})(A_1, \ldots, A_n; B)$.

Corollary 1 ([34]). *The functors* $\mathcal{P}$ *and* $\overline{(-)}$ *of Proposition 1 define an adjoint equivalence* **CartClone** $\simeq$ **CartCat**.

4.2 Recovering the semantic interpretation and syntactic model

We now show how the usual semantic interpretation, syntactic model, and soundness and completeness results can be derived from the multi-ary framework. Although we shall not pursue the point in detail for reasons of space, essentially the same argument holds for all the calculi considered in this paper.

Semantic interpretation and soundness. We recover the usual semantic interpretation of $\Lambda^{\times}$ in a cartesian category by Lemma 6 and Example 4 as follows. Let U : **CartCat** $\to$ **Sig** be the functor sending a cartesian category $(\mathcal{C}, \Pi)$ to the signature with objects those of $\mathcal{C}$ and constants $\{\mathcal{C}(\prod_{i=1}^n A_i, B)\}_{A_1,\ldots,A_n,B \in \mathcal{C}}$. An interpretation $s : \mathcal{S} \to \mathrm{U}\mathcal{C}$ of basic types and constants in $\mathcal{C}$ is exactly an interpretation $s : \mathcal{S} \to \mathrm{U}(\mathcal{PC})$ in the induced cartesian clone. The unique extension $s[\![-]\!] : \mathrm{Syn}(\Lambda^{\times}_{\mathcal{S}}) \to \mathcal{PC}$ sends a term $x_1 : A_1, \ldots, x_n : A_n \vdash t : B$ to a multimap $s[\![x_1 : A_1, \ldots, x_n : A_n]\!] \to s[\![B]\!]$ in $\mathcal{PC}$, which is exactly a map $\prod_{i=1}^{n} s[\![A_i]\!] \to s[\![B]\!]$ in $\mathcal{C}$. It is not hard to show this coincides with the usual, inductively defined semantic interpretation. Unlike with the unary approach, we do not need to prove soundness with respect to $\beta\eta$ as a separate lemma: this holds immediately from the fact $s[\![-]\!]$ is a cartesian clone homomorphism.

Moreover, for any objects $A_1, \ldots, A_n$ in a cartesian clone one can construct a 'multi-isomorphism' $(A_1, \ldots, A_n) \cong \prod_{i=1}^{n} A_i$ (see [38, Lemma 4.2.16]). Hence, in a cartesian simple type theory with products, *contexts must coincide with product types.* Together with the preceding, this provides a mathematical explanation for the identification of contexts and product types in the interpretation of $\Lambda^{\times,\to}$.

Syntactic model. We extract the construction from Proposition 1. For a signature $\mathcal{S}$ the cartesian category $\overline{\mathrm{Syn}(\Lambda^{\times}_{\mathcal{S}})}$ has objects the types of $\Lambda^{\times}_{\mathcal{S}}$ and morphisms $A \to B$ given by $\alpha\beta\eta$-equivalence classes of terms $x : A \vdash t : B$ for a fixed variable x. Composition is substitution and the identity on A is the variable $x : A$. The projections are $x : \prod_{i=1}^{n} A_i \vdash \pi_i^{A_\bullet}(x) : A_i$ and the pairing of the maps $(x : C \vdash t_i : A_i)_{i=1,2}$ is $x : C \vdash \langle t_1, t_2 \rangle : A_1 \times A_2$. The usual proofs that this is indeed cartesian (see *e.g.* [9, Chapter 3]) have been replaced by the simple observation of Example 4.

Completeness. Once again, the proof is largely category-theoretic. Note first that the functor $\overline{(-)} : \mathbf{CartClone} \to \mathbf{CartCat}$ is faithful. One can prove this directly using Proposition 3 or infer it from Corollary 1 and the fact any equivalence is fully faithful. In any case, it follows by standard results (*e.g.* [37, Lemma 4.5.13]) that the unit η' of the adjunction $\overline{(-)} \dashv \mathcal{P}$ is monic. Just as in **Cat**, any monomorphism of clones is injective on objects and injective on multimaps. It suffices, therefore, to find a semantic interpretation $\iota[\![-]\!]$ which is equal to a component of η'. This is accomplished by the next lemma.

Lemma 8. *Let* $\mathcal{C} \underset{\mathrm{U}}{\overset{F}{\rightleftarrows}} \mathcal{D} \underset{\mathrm{U}'}{\overset{F'}{\rightleftarrows}} \mathcal{E}$ (with $\bot$) *be adjunctions with units* $\eta : \mathrm{id}_{\mathcal{C}} \Rightarrow \mathrm{U}F$ *and* $\eta' : \mathrm{id}_{\mathcal{D}} \Rightarrow \mathrm{U}'F'$. *Then for any* $C \in \mathcal{C}$, *the unit* $\eta'_{FC} : FC \to \mathrm{U}'F'FC$ *is the unique map* h *such that the following diagram commutes:*

$$\begin{array}{ccc} \mathrm{U}FC & \overset{\mathrm{U}h}{\dashrightarrow} & \mathrm{U}\mathrm{U}'F'FC \\ {\scriptstyle \eta_C}\uparrow & & \uparrow{\scriptstyle \mathrm{U}\eta'_{FC}} \\ C & \xrightarrow[\eta_C]{} & \mathrm{U}FC \end{array}$$

In the setting of Proposition 1 this lemma implies that the component $\eta'_{F\mathcal{S}} : \overline{\mathrm{Syn}(\Lambda^{\times}_{\mathcal{S}})} \to \mathcal{P}(\overline{\mathrm{Syn}(\Lambda^{\times}_{\mathcal{S}})})$ of the unit for the adjunction $\overline{(-)} \dashv \mathcal{P}$ is exactly the unique cartesian clone homomorphism $\iota[\![-]\!]$ extending the obvious interpretation $\iota := \mathcal{S} \hookrightarrow \overline{\mathrm{Syn}(\Lambda^{\times}_{\mathcal{S}})}$ of base types and constants in the free cartesian category. By our preceding discussion, this clone homomorphism is injective on multimaps: so if $\iota[\![t]\!] = \iota[\![t']\!]$ then $t = t'$ in $\mathrm{Syn}(\Lambda^{\times}_{\mathcal{S}})$, hence $t =_{\beta\eta} t'$.

5 Closed structure

To define closed structure, we follow Lambek's definition and simply upgrade the hom-set definition of exponentials to multicategories.

Definition 10 ([26]). *A* closed multicategory *is a multicategory* $\mathbb{M}$ *equipped with an object* $[A, B]$ *and multimap* $\mathrm{eval}_{A,B} : [A, B], A \to B$ *for every* $A, B \in |\mathbb{M}|$*, such that composition induces isomorphisms as shown:*

$$\mathbb{M}(\Gamma, A; B) \underset{\mathrm{eval}_{A,B}\circ\langle(-),\mathrm{Id}_A\rangle}{\overset{\Lambda^A}{\rightleftarrows}} \mathbb{M}(\Gamma; [A, B]) \qquad (5)$$

where the two maps are mutually inverse ($\cong$).

A (strict) closed multicategory functor *is a multicategory functor* f *which preserves all the data:* $f([A, B]) = [fA, fB]$*,* $f(\mathrm{eval}_{A,B}) = \mathrm{eval}_{fA,fB}$ *and* $f(\Lambda t) = \Lambda(ft)$*. We write* **ClMulticat** *for the category of closed multicategories and their functors, and* **ClSMulticat** *for the category of symmetric multicategories with closed structure, and functors preserving both of these.*

Example 6. If $(\mathcal{C}, \otimes, I, [-, =])$ is a closed (symmetric) monoidal category then the induced (symmetric) multicategory $\mathcal{T}\mathcal{C}$ is also closed.

Closed multicategories allow us to model exponentials without requiring a tensor product. Writing out the rules in the internal language, we get the map Λ^A in (5) as the usual abstraction rule, and the evaluation map as the application $f : A \multimap B, x : A \vdash f\,x : B$. We then see that $\Delta, f : A \multimap B, x : A \vdash u[f\,x/y] : C$ whenever $\Delta, y : B \vdash u : C$, so we recover a small adaptation of Abramsky's rules for exponentials. Write $\mathcal{O}_{\mathcal{S}}^{\multimap}$ (resp. $\mathfrak{L}_{\mathcal{S}}^{\multimap}$) for the extension of $\mathcal{O}_{\mathcal{S}}$ (resp. $\mathfrak{L}_{\mathcal{S}}$) with the following rules and the $\beta\eta$-laws familiar from $\Lambda^{\to}$:

$$\frac{\Delta, y : B \vdash u : C \qquad \Theta \vdash t : A \multimap B \qquad \Gamma \vdash v : A}{\Delta, \Theta, \Gamma \vdash u[t\,v/y] : C}\ , \qquad \frac{\Gamma, x : A \vdash t : B}{\Gamma \vdash \lambda x\,.\,t : A \multimap B}$$

Lemma 9 ([20]). *The composite forgetful functor* **ClMulticat** $\to$ **Multicat** $\to$ **Sig** *has a left adjoint, and the free closed multicategory on* $\mathcal{S}$ *is the syntactic multicategory* $\mathrm{Syn}(\mathcal{O}_{\mathcal{S}}^{\multimap})$*. The same holds for symmetric structure, if one replaces* **ClMulticat** *by* **ClSMulticat** *and* $\mathcal{O}^{\multimap}$ *by* $\mathfrak{L}^{\multimap}$*.*

For the cartesian case, we follow the same procedure as in Section 4.

Definition 11. *A* closed clone *is a clone* $\mathbb{C}$ *equipped with a closed structure on* $\mathbb{MC}$*. We write* **ClClone** *for the category of closed clones and clone homomorphisms preserving the closed structure as in Definition 10.*

Example 7. If $(\mathcal{C}, \Pi, \Rightarrow)$ is a cartesian closed category, the clone $\mathcal{P}\mathcal{C}$ is closed.

Definition 11 recovers the usual $\beta\eta$-laws for exponentials in $\Lambda^{\to}$, complete with the weakenings that are usually implicit. Writing $f\,x$ for eval, we get the following equations in the internal language when $\Gamma := (x_i : A_i)_{i=1,\ldots,n}$:

$$(f\,x)\big[(\lambda x\,.\,t)[x_1/x_1,\ldots,x_n/x_n]/f, x/x\big] = t\ ,\ \lambda x\,.\,(f\,x)\big[t[x_1/x_1,\ldots,x_n/x_n]/f\big] = t$$

Lemma 10. *The composite forgetful functor* **ClClone** $\to$ **Clone** $\to$ **Sig** *has a left adjoint, and the free closed clone on* $\mathcal{S}$ *is the syntactic clone* $\mathrm{Syn}(\Lambda_{\mathcal{S}}^{\to})$*.*

6 Cartesian closed structure

The development above makes defining cartesian closed structure straightforward. For reasons of space we restrict ourselves to the cartesian case, but similar remarks apply to the linear and ordered cases.

Definition 12. *A* cartesian closed clone *is a clone equipped with both closed structure and cartesian structure. We write* **CCClone** *for the category of cartesian closed clones and homomorphisms that strictly preserve both structures.*

By Lemmas 6 and 10, we already have a free property .

Lemma 11. *The composite forgetful functor* **CCClone** → **Clone** → **Sig** *has a left adjoint, and* $\mathrm{Syn}(\Lambda_{\mathcal{S}}^{\times,\to})$ *is the free cartesian closed clone on* $\mathcal{S}$*.*

The nucleus of any cartesian closed clone $(\mathbb{C}, \Pi, \Rightarrow)$ is also cartesian closed:

$$\overline{\mathbb{C}}(A \times B, C) = \mathbb{C}(A \times B; C) \cong \mathbb{C}(A, B; C) \cong \mathbb{C}(A; B \Rightarrow C) = \overline{\mathbb{C}}(A, B \Rightarrow C)$$

Similarly, by Examples 4 and 7, for any cartesian closed category $(\mathcal{C}, \Pi, \Rightarrow)$ the induced category $\mathcal{PC}$ is cartesian closed. Proposition 1 then restricts as follows.

Proposition 4. *The functor* $\overline{(-)}$: **CCClone** → **CCCat** *fits into the following diagram, in which the right-hand adjunction is an equivalence:*

$$\mathbf{Sig} \underset{\mathrm{U}}{\overset{\mathrm{F}}{\underset{\longleftarrow}{\overset{\longrightarrow}{\bot}}}} \mathbf{CCClone} \underset{\mathcal{P}}{\overset{\overline{(-)}}{\underset{\longleftarrow}{\overset{\longrightarrow}{\simeq\bot}}}} \mathbf{CCCat}$$

Moreover, $\mathrm{U}\circ\mathcal{P}$ *is equal to the canonical forgetful functor* **CCCat** → **Sig**. *Hence, the free cartesian closed category on* $\mathcal{S}$ *is canonically isomorphic to* $\overline{\mathrm{Syn}(\Lambda_{\mathcal{S}}^{\times,\to})}$.

As in Section 4.2, the preceding two results are enough to recover the sound semantic interpretation of $\Lambda^{\times,\to}$, and the usual syntactic model.

7 Cartesian combinatory logic and SK-clones

In this section we begin a multi-ary investigation of cartesian combinatory logic, and give a categorical statement of the classical correspondence between combinatory logic and $\Lambda^{\to}$ (for which see *e.g.* [15,6]). In Section 8 we shall use this to define *SK-categories* and show they are sound and complete for $\Lambda^{\to}$.

We briefly recapitulate the rules of typed combinatory logic $\mathsf{CL}_{\mathcal{S}}$ over a signature $\mathcal{S}$; for a fuller account see *e.g.* [6]. Types are as in $\Lambda^{\to}$. Terms are given by the grammar $t, u ::= x \mid c \in \mathcal{S}(\Gamma; B) \mid (t\,u) \mid \mathsf{S} \mid \mathsf{K}$: we have variables, constants and an application operation as in $\Lambda^{\to}$ and, for any context Γ and types A, B and C, two *combinators* $\Gamma \vdash \mathsf{S}^{\Gamma}_{A,B,C} : (A \Rightarrow (B \Rightarrow C)) \Rightarrow ((A \Rightarrow C) \Rightarrow (A \Rightarrow C))$ and $\Gamma \vdash \mathsf{K}^{\Gamma}_{A,B} : A \Rightarrow (B \Rightarrow A)$. Substitution is as in $\Lambda^{\to}$, where the combinators $\mathsf{Z} \in \{\mathsf{S}, \mathsf{K}\}$ satisfy $\mathsf{Z}[u_1/x_1, \dots, u_n/x_n] = \mathsf{Z}$ so that Z^{Γ} is the weakening of $\mathsf{Z}^{\diamond}$.

The correlate of β-equality is *weak equality* $=_{\mathrm{w}}$, which is the smallest congruence containing $\mathsf{S}\, x\, y\, z = (x\, z)\, (y\, z)$ and $\mathsf{K}\, x\, y = x$. The correlate of $\beta\eta$-equality is *extensional weak equality* $=_{\mathrm{wext}}$, which extends $=_{\mathrm{w}}$ with the rule

$$\frac{t\, x_1 \cdots x_n = t'\, x_1 \cdots x_n \qquad x_1, \ldots, x_n \text{ not free in } t \text{ or } t'}{t = t'}\ \mathrm{ext} \tag{6}$$

We write CL^{w} for combinatory logic with weak equality and $\mathsf{CL}^{\mathrm{wext}}$ for combinatory logic with extensional weak equality. The usual encoding of CL^{w} in $\Lambda^{\rightarrow}$ sends S and K to $\lambda f \,.\, \lambda g \,.\, \lambda x \,.\, (f\, x)\, (g\, x)$ and $\lambda x \,.\, \lambda y \,.\, x$, respectively.

The next definition may be obtained by seeing that CL^{w} can be presented as an algebraic theory, and that clones are equivalent to algebraic theories (*e.g.* [29,41]). We implicitly bracket application to the left, so $t \cdot u \cdot v := (t \cdot u) \cdot v$. We also write $(-)^{\Delta;\Theta}$ for the weakening map $\mathbb{C}(\Gamma; B) \to \mathbb{C}(\Delta, \Gamma, \Theta; B)$ sending t to $t[\mathsf{p}^{\Delta,\Gamma,\Theta}_{|\Delta|+1}, \ldots, \mathsf{p}^{\Delta,\Gamma,\Theta}_{|\Delta|+|\Gamma|}]$; when Γ is empty we write just $(-)^{\Delta}$.

Definition 13. *An* SK-clone *is a clone* $\mathbb{C}$ *equipped with a mapping* $[-,=] : |\mathbb{C}| \times |\mathbb{C}| \to |\mathbb{C}|$, *nullary multimaps* $\mathsf{S}_{A,B,C} \in \mathbb{C}(\diamond; [[A,[B,C]], [[A,B],[A,C]]])$ *and* $\mathsf{K}_{A,B} \in \mathbb{C}(\diamond; [A,[B,A]])$ *for every* $A, B, C \in |\mathbb{C}|$, *and a binary* application operation $(- \cdot =) : \mathbb{C}(\Gamma; [A,B]) \times \mathbb{C}(\Gamma; A) \to \mathbb{C}(\Gamma; B)$ *for every* $\Gamma \in |\mathbb{C}|^{\star}$ *and* $B \in |\mathbb{C}|$, *such that the following axioms hold whenever they are well-typed:*

$$(t \cdot u)[v_1, \ldots, v_n] = t[v_1, \ldots, v_n] \cdot u[v_1, \ldots, v_n] \quad , \quad (\mathsf{K}_{A,B})^{A,B} \cdot \mathsf{p}_1 \cdot \mathsf{p}_2 = \mathsf{p}_1$$

$$(\mathsf{S}_{A,B,C})^{[A,[B,C]],[A,B],A} \cdot \mathsf{p}_1 \cdot \mathsf{p}_2 \cdot \mathsf{p}_3 = (\mathsf{p}_1 \cdot \mathsf{p}_3) \cdot (\mathsf{p}_2 \cdot \mathsf{p}_3)$$

A homomorphism of SK-clones *is a clone homomorphism that preserves application,* S *and* K: $f(\mathsf{S}_{A,B,C}) = \mathsf{S}_{fA,fB,fC}$, $f(\mathsf{K}_{A,B}) = \mathsf{K}_{fA,fB}$ *and* $f(t \cdot u) = ft \cdot fu$. *We write* **SKClone** *for the category of SK-clones and their homomorphisms.*

Lemma 12. *The composite forgetful functor* **SKClone** $\to$ **Clone** $\to$ **Sig** *has a left adjoint, and the free SK-clone on* $\mathcal{S}$ *is the syntactic clone* $\mathrm{Syn}(\mathsf{CL}^{\mathrm{w}}_{\mathcal{S}})$.

A core feature of the syntax of combinatory logic, which is at the heart of the correspondence between the terms of $\mathsf{CL}^{\mathrm{wext}}$ and $\Lambda^{\rightarrow}$, is the admissibility of *bracket extension* algorithms (see *e.g.* [5, §7.1]). To express this in the typed setting, we use the following notation. For a binary operation $[-,=]$ on a set S we define $[-;=] : S^{\star} \times S \to S$ inductively as follows:

$$[\diamond; B] := B \quad , \quad [A; B] := [A, B] \quad , \quad [\Gamma, A; B] := [\Gamma; [A, B]]$$

With this notation, bracket abstraction amounts to saying that if $\Gamma := (x_i : A_i)_{i=1,\ldots,n}$ and $\Gamma \vdash t : B$ in CL^{w}, there exists a closed term $\diamond \vdash t^{\mathrm{c}} : [\Gamma; B]$ such that $(t^{\mathrm{c}})^{\Gamma}\, x_1 \ldots x_n =_{\mathrm{w}} t$. The extensionality axiom (6) then says that t^{c} is unique: in other words, $t \mapsto t^{\Gamma}\, x_1 \ldots x_n$ is an isomorphism.

We now translate this into clone-theoretic terms. For any SK-clone $\mathbb{C}$ we obtain the operation $t \mapsto t^{\Gamma}\, x_1 \ldots x_n$ as the composite below:

$$i_{\Gamma;B} := \left(\mathbb{C}(\diamond; [\Gamma; B]) \xrightarrow{\mathrm{w}^{\Gamma}} \mathbb{C}(\Gamma; [\Gamma; B]) \xrightarrow{(-) \cdot \mathsf{p}^{\Gamma}_1 \cdots \mathsf{p}^{\Gamma}_{|\Gamma|}} \mathbb{C}(\Gamma; B)\right) \tag{7}$$

For $\Gamma := \diamond$ this is just the identity. The admissibility of bracket abstraction in the syntax of CL^{w} is then captured by the next lemma. Typically bracket abstraction algorithms restrict to closed constants, because an open constant may have no corresponding closed term. We restrict in the same way. Call a signature $\mathcal{S}$ *nullary* if $\mathcal{S}(\Gamma; A) = \varnothing$ whenever $\Gamma \neq \diamond$, and write $\mathbf{Sig}_0 \hookrightarrow \mathbf{Sig}$ for the full subcategory of nullary signatures.

Lemma 13. *Let $\mathcal{S}$ be a nullary signature. Then for any $\Gamma \in |\mathrm{Syn}(\mathsf{CL}^{\mathrm{w}}_{\mathcal{S}})|^\star$ and $B \in |\mathrm{Syn}(\mathsf{CL}^{\mathrm{w}}_{\mathcal{S}})|$ there exists a map $(-)^{\mathrm{c}}$ such that $i_{\Gamma;B} \circ (-)^{\mathrm{c}} = \mathrm{id}_{\mathrm{Syn}(\mathsf{CL}^{\mathrm{w}}_{\mathcal{S}})}$.*

Because bracket abstraction is defined by induction on the syntax, we cannot straightforwardly define it in an arbitrary SK-clone. We can, however, consider the sub-category of SK-clones (= semantic models of CL^{w}) which admit bracket abstraction in the sense that each $i_{\Gamma;B}$ has a retraction. The *extensional* models are then those for which this retract $(-)^{\mathrm{c}}$ also satisfies uniqueness.

Definition 14. *An SK-clone $\mathbb{C}$ is* extensional *if for every $\Gamma \in |\mathbb{C}|^\star$ and $B \in |\mathbb{C}|$ the map $i_{\Gamma;B}$ defined in (7) is invertible. We write $\mathbf{SKClone}_{\mathrm{ext}}$ for the full subcategory of* **SKClone** *consisting of just the extensional SK-clones.*

Lemma 14. *The composite forgetful functor $\mathbf{SKClone}_{\mathrm{ext}} \to \mathbf{Clone} \to \mathbf{Sig}_0$ has a left adjoint, and the free extensional SK-clone on a nullary signature $\mathcal{S}$ is the syntactic clone $\mathrm{Syn}(\mathsf{CL}^{\mathrm{wext}}_{\mathcal{S}})$.*

7.1 Extensional SK-clones are closed clones

In this section we outline why $\mathbf{SKClone}_{\mathrm{ext}}$ is equivalent to **ClClone**, thereby giving a category-theoretic equivalence not just between the syntax of $\mathsf{CL}^{\mathrm{wext}}$ and $\Lambda^{\to}$ but also between their models. The proof uses extensionality or the η-law to pass from arbitrary multimaps to nullary ones, from which one can build a *strict closed* clone. We shall rely heavily on the following simple observation.

Lemma 15. *Let $\mathbb{C}$ be a clone and $X := \{X(\Gamma; B)\}_{\Gamma \in |\mathbb{C}|^\star, B \in |\mathbb{C}|}$ a family of sets together with an isomorphism $\{\nu_{\Gamma;A} : \mathbb{C}(\Gamma; A) \to X(\Gamma; A)\}_{\Gamma,A}$ between X and the hom-sets of $\mathbb{C}$ in the functor category $[|\mathbb{C}|^\star \times |\mathbb{C}|, \mathbf{Set}]$. Then X acquires a canonical clone structure and ν becomes an isomorphism of clones.*

We now introduce strict closed clones.

Definition 15. *A* strict closed clone *is a closed clone $(\mathbb{C}, \Rightarrow, \mathrm{eval})$ such that every $\Lambda^A : \mathbb{C}(\Gamma, A; B) \to \mathbb{C}(\Gamma, A \Rightarrow B)$ is the identity. We write $\iota : \mathbf{ClClone}_{\mathrm{st}} \hookrightarrow$* **ClClone** *for the full subcategory consisting of just the strict closed clones.*

Any closed clone $(\mathbb{C}, \Rightarrow, \mathrm{eval})$ determines a strict closed clone $\mathrm{S}\mathbb{C}$ and a clone isomorphism $\lambda_{\mathbb{C}} : \mathbb{C} \to \mathrm{S}\mathbb{C}$ by applying Lemma 15 to the isomorphisms $\mathbb{C}(\Gamma; B) \cong \mathbb{C}(\diamond; \Gamma \Rightarrow B)$ arising from the closed structure. This extends to a functor $\mathrm{S} : \mathbf{ClClone} \to \mathbf{ClClone}_{\mathrm{st}}$ sending $f : (\mathbb{C}, \Rightarrow, \mathrm{eval}) \to (\mathbb{D}, \Rightarrow, \mathrm{eval})$ to the composite $\lambda_{\mathbb{D}} \circ f \circ \lambda_{\mathbb{C}}^{-1}$. A short calculation shows that the isomorphisms λ make S : **ClClone** $\leftrightarrows$ **ClClone**$_{st}$: ι into an equivalence of categories.

We play a similar game for turning extensional SK-clones into (strict) closed clones. Indeed, for any extensional SK-clone we have isomorphisms $\mathbb{C}(\Gamma; B) \cong \mathbb{C}(\diamond; [\Gamma; B])$ defining a strict closed clone L$\mathbb{C}$ with $(L\mathbb{C})(\Gamma; B) := \mathbb{C}(\diamond; [\Gamma; B])$, and hence a functor L : **SKClone**$_{ext}$ → **ClClone**$_{st}$ in a similar fashion to S.

Finally, for any closed clone $(\mathbb{C}, \Rightarrow, \mathrm{eval})$ we get an extensional SK-clone E$\mathbb{C}$ with the same underlying clone by taking application to be application in $\Lambda^{\rightarrow}$, so $t \cdot u := \mathrm{eval}_{A,B}[t, u]$, and encoding the combinators as usual.

Theorem 1. *There exist equivalences of categories*

$$\mathbf{SKClone}_{\mathrm{ext}} \;\underset{E' := E \circ \iota}{\overset{L}{\underset{\simeq}{\rightleftarrows}}}\; \mathbf{ClClone}_{\mathrm{st}} \;\underset{S}{\overset{\iota}{\underset{\simeq}{\rightleftarrows}}}\; \mathbf{ClClone}.$$

8   A categorical model of $\Lambda^{\rightarrow}$

In Propositions 1 and 4 we recovered a unary semantic interpretation of $\Lambda^{\times}$ and $\Lambda^{\times, \rightarrow}$ from our clone-theoretic ones. But we do not have a corresponding result for $\Lambda^{\rightarrow}$. In this section we fill this gap: we introduce *SK-categories* and show they play the role for $\Lambda^{\rightarrow}$ that cartesian closed categories play for $\Lambda^{\times, \rightarrow}$. Our definition is inspired by *closed categories* ([11,10]), which axiomatise an 'internal' version of the hom-functor $\mathcal{C}(-, =)$ in the form of a functor $[-, =] : \mathcal{C}^{\mathrm{op}} \times \mathcal{C} \to \mathcal{C}$. Closed categories have a unit object, corresponding to requiring a unit type (*cf.* [31]); our definition avoids this (see also [39,43]).

Recall that in the presence of contravariance, *dinaturality* and *extranaturality* are the right replacements for naturality (see *e.g.* [30, '§IX.4]).

Definition 16. *An SK-category consists of a category $\mathcal{C}$ and functors $[-, =]$: $\mathcal{C}^{\mathrm{op}} \times \mathcal{C} \to \mathcal{C}$ and $U : \mathcal{C} \to$ **Set**, together with*

1. *Maps $S_{C,D,E} : [C, [D, E]] \to [[C, D], [C, E]]$ dinatural in C and natural in D and E;*
2. *Maps $K_D^C : D \to [C, D]$ extranatural in C and natural in D;*
3. *Maps $\varepsilon_{C,D} : U[C, D] \times UC \to UD$ extranatural in C and natural in D;*

*This data is subject to the condition that $U \circ [-, =] = \mathcal{C}(-, =) : \mathcal{C}^{\mathrm{op}} \times \mathcal{C} \to$ **Set** and the 7 axioms of Figure 1a. An SK-functor (F, ϕ, ψ) is a functor $F : \mathcal{C} \to \mathcal{D}$ with natural transformations as below, such that the axioms of Figure 1b hold.*

$$
\begin{array}{ccc}
\mathcal{C}^{\mathrm{op}} \times \mathcal{C} & \xrightarrow{F^{\mathrm{op}} \times F} & \mathcal{D}^{\mathrm{op}} \times \mathcal{C} \\
{\scriptstyle [-,=]}\downarrow & \overset{\phi}{\Longrightarrow} & \downarrow{\scriptstyle [-,=]} \\
\mathcal{C} & \xrightarrow{F} & \mathcal{D}
\end{array}
\qquad\qquad
\begin{array}{ccc}
\mathcal{C} & \xrightarrow{\quad F \quad} & \mathcal{D} \\
{\scriptstyle U}\searrow & \overset{\psi}{\Longrightarrow} & \swarrow{\scriptstyle U} \\
& \mathbf{Set} &
\end{array}
$$

*We call (F, ϕ, ψ) strict if ϕ is the identity, and write **SKCat** for the category of SK-categories and strict SK-functors.*

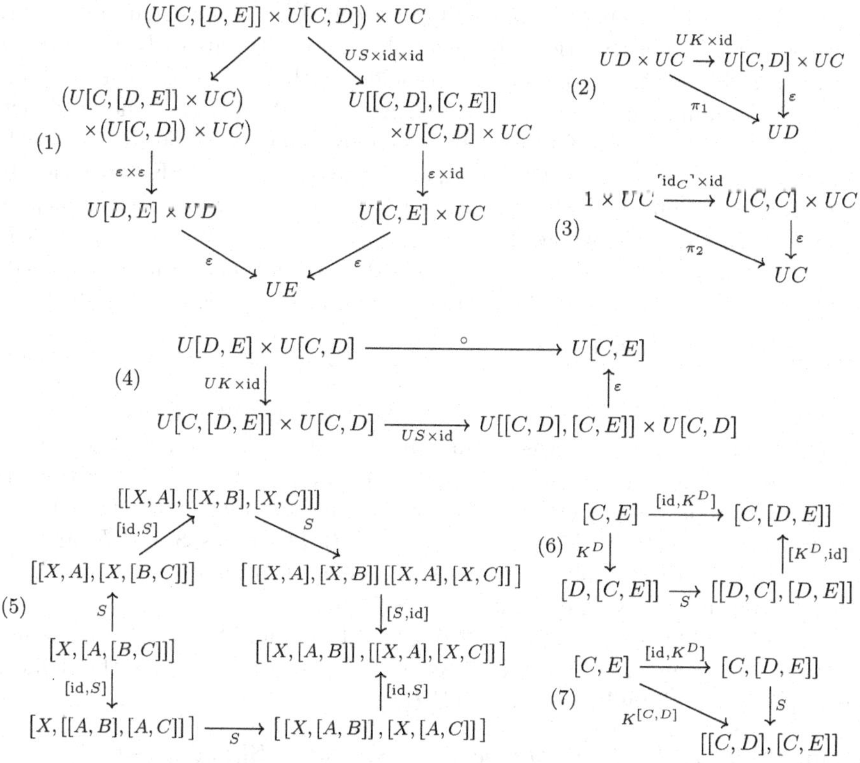

(a) Axioms for an SK-category. In (1) the unlabelled arrow is the canonical map $\langle\langle \pi_1\pi_1, \pi_2\rangle, \langle \pi_2\pi_1, \pi_2\rangle\rangle : (X \times Y) \times Z \to (X \times Z) \times (X \times Z)$. In (3) we write $\ulcorner \mathrm{id}_C \urcorner$ for the set map $* \mapsto \mathrm{id}_C : 1 \to \mathrm{U}[C, C]$.

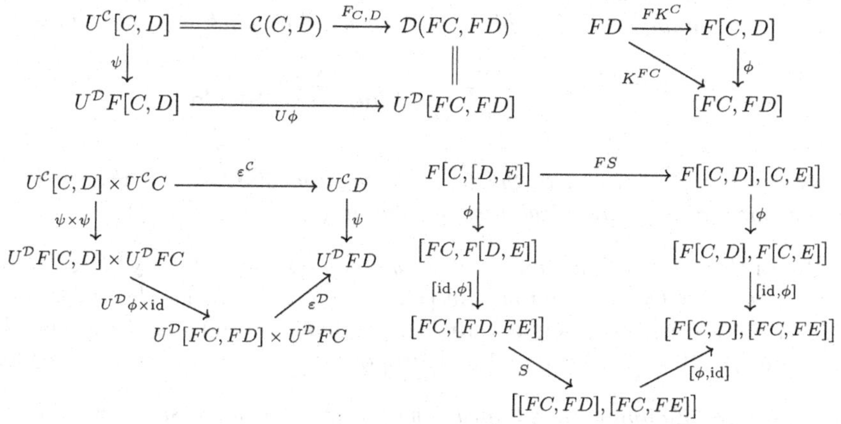

(b) Axioms for an SK-functor

Fig. 1: Extra axioms for Definition 16

We think of UC as the set of multimaps $\diamond \to C$ and ε as a formal application operation $(- \cdot =)$. Axioms (1) and (2) are the weak equality laws from CL. Axioms (3) and (4) ensure compatibility between the category structure and the corresponding CL constructions: for example, axiom (3) implies $U(f)(x) = f \cdot x$, and axiom (4) says that composition coincides with $\mathsf{S}\,(\mathsf{K}\,-)\,(=)$, corresponding to the weak equality $\mathsf{S}\,(\mathsf{K}\,f)\,g\,x = f\,(g\,x)$. Axioms (5) – (7) are coherence laws.

Every extensional SK-clone determines an SK-category. Because we follow [11] and ask for an *equality* $U[A, B] = \mathcal{C}(A, B)$ in the definition of SK-categories, but in general an extensional SK-clone $(\mathbb{C}, [-, =], \mathsf{S}, \mathsf{K}, \cdot)$ only has an *isomorphism* $\mathbb{C}(A; B) \cong \mathbb{C}(\diamond; [A, B])$, we need to strictify in the same manner as Section 7.1. As a notational shorthand, we write I, B and B' for the closed multimaps satisfying the equations below in the internal language of $\mathbb{C}$ (see *e.g.* [15,6]):

$$\mathsf{I}^A \cdot x = x\ ,\ \mathsf{B}^{B \Rightarrow C, A \Rightarrow B, A} \cdot x \cdot y \cdot z = x \cdot (y \cdot z)\ ,\ (\mathsf{B}')^{A \Rightarrow B, B \Rightarrow C, A} \cdot x \cdot y \cdot z = y \cdot (x \cdot z)$$

The category $\mathrm{N}\mathbb{C}$ has objects $|\mathbb{C}|$ and hom-sets $(\mathrm{N}\mathbb{C})(A, B) := \mathbb{C}(\diamond; [A, B])$ (*cf.* [14]). The identity on A is I_A and the composite of t and t' is $\mathsf{B} \cdot t \cdot t'$. For U we take $UA := \mathbb{C}(\diamond; A)$ with the action on maps given by application. For $[-, =]$ the action on objects is given by the SK-structure, with the action on maps given by $[X, t] := \mathsf{B} \cdot t$ and $[t, X] := \mathsf{B}' \cdot t$. The maps S and K are given by the corresponding combinators, and ε is the application operation in $\mathbb{C}$. This extends to a functor $\mathrm{N} : \mathbf{SKClone}_{\mathrm{ext}} \to \mathbf{SKCat}$.

The internal language of SK-categories is $\mathsf{CL}^{\mathrm{wext}}$, and hence $\Lambda^{\to}$. We write U for the functor which sends an SK-category $(\mathcal{C}, U, [-, =], S, K, \varepsilon)$ to the signature with base types $|\mathcal{C}|$ and constants $U[\Gamma, B]$.

Proposition 5. *The forgetful functor* $\mathrm{U} : \mathbf{SKCat} \to \mathbf{Sig}$ *has a left adjoint, and the free SK-category on* $\mathcal{S}$ *is* $\mathrm{N}\big(\mathrm{Syn}(\mathsf{CL}_{\mathcal{S}}^{\mathrm{wext}})\big) \cong (\mathrm{N} \circ \mathrm{E})\big(\mathrm{Syn}(\Lambda_{\mathcal{S}}^{\to})\big)$.

Using Theorem 1, we now obtain a version of Propositions 1 and 4 for $\Lambda^{\to}$.

Theorem 2. *The composite* $\mathrm{N} \circ \iota : \mathbf{ClClone}_{\mathrm{st}} \to \mathbf{SKCat}$ *is invertible; hence we get the diagram below, in which the right-hand adjunction is an equivalence:*

$$\mathbf{Sig} \underset{\mathrm{U}}{\overset{\mathrm{F}}{\underset{\longleftarrow}{\overset{\longrightarrow}{\bot}}}} \mathbf{ClClone} \underset{\mathrm{Cl}}{\overset{\mathrm{N} \circ \mathrm{E}}{\underset{\longleftarrow}{\overset{\longrightarrow}{\simeq\bot}}}} \mathbf{SKCat}$$

Moreover, $\mathrm{U} \circ \mathrm{Cl}$ *is equal to the forgetful functor* $\mathbf{SKCat} \to \mathbf{Sig}$*, so the free SK-category on* $\mathcal{S}$ *is canonically isomorphic to* $(\mathrm{N} \circ \mathrm{E})(\mathrm{Syn}(\Lambda_{\mathcal{S}}^{\to}))$.

Recall that a *closed monoidal category* is a monoidal category $(\mathcal{D}, \otimes, I)$ such that every $(-) \otimes D$ has a right adjoint $[D, -]$, and that in a closed category $\mathcal{C}$ giving every $[C, -]$ a $\mathcal{C}$-enriched left adjoint is equivalent to giving closed monoidal structure ([11,10,43]). Theorem 2 and Proposition 4 imply a cartesian version.

Corollary 2. *Equipping a category* $\mathcal{C}$ *with cartesian closed structure is equivalent to equipping* $\mathcal{C}$ *with SK-structure and natural isomorphisms* $\mathcal{C}(I, [C, D]) \cong \mathcal{C}(C, D)$ *and* $\mathcal{C}(C \otimes D, E) \cong \mathcal{C}(C, [D, E])$ *for every* $C, D, E \in \mathcal{C}$.

Acknowledgements. I thank Nathanael Arkor and Dylan McDermott for useful discussions on early drafts of this paper, and the reviewers for their many useful comments. I am grateful to Nayan Rajesh for pointing out the adjunctions between cartesian categories and cartesian clones, and between cartesian closed categories and cartesian closed clones, are in fact equivalences. Finally, I thank Marcelo Fiore for introducing me to clones.

References

1. Abramsky, S.: Computational interpretations of linear logic. Theoretical Computer Science **111**(1-2), 3–57 (1993). https://doi.org/10.1016/0304-3975(93)90181-r
2. Abramsky, S.: Temperley–Lieb algebra: From knot theory to logic and computation via quantum mechanics. In: Mathematics of Quantum Computation and Quantum Technology. Chapman and Hall/CRC (2007)
3. Arkor, N., Fiore, M.: Algebraic models of simple type theories. In: Proceedings of the 35th Annual ACM/IEEE Symposium on Logic in Computer Science. ACM (2020). https://doi.org/10.1145/3373718.3394771
4. Arkor, N., McDermott, D.: Abstract clones for abstract syntax. In: Kobayashi, N. (ed.) 6th International Conference on Formal Structures for Computation and Deduction, FSCD 2021, July 17-24, 2021, Buenos Aires, Argentina (Virtual Conference). LIPIcs, vol. 195, pp. 30:1–30:19. Schloss Dagstuhl - Leibniz-Zentrum für Informatik (2021). https://doi.org/10.4230/LIPIcs.FSCD.2021.30
5. Barendregt, H.P.: The lambda calculus: its syntax and semantics, Studies in Logic and the Foundations of Mathematics), vol. 103. North-Holland (1985), revised edition
6. Bimbó, K.: Combinatory logic pure, applied, and typed. Taylor & Francis (2012)
7. Blanco, N., Zeilberger, N.: Bifibrations of polycategories and classical linear logic. Electronic Notes in Theoretical Computer Science **352**, 29–52 (Oct 2020). https://doi.org/10.1016/j.entcs.2020.09.003
8. Church, A.: A formulation of the simple theory of types. The Journal of Symbolic Logic **5**(2), 56–68 (1940), http://www.jstor.org/stable/2266170
9. Crole, R.L.: Categories for Types. Cambridge University Press (1994). https://doi.org/10.1017/CBO9781139172707
10. Day, B.J., Laplaza, M.L.: On embedding closed categories. Bulletin of the Australian Mathematical Society **18**(3), 357–371 (1978). https://doi.org/10.1017/s0004972700008236
11. Eilenberg, S., Kelly, G.M.: Closed categories. In: Proceedings of the Conference on Categorical Algebra, pp. 421–562. Springer Berlin Heidelberg (1966). https://doi.org/10.1007/978-3-642-99902-4_22
12. Fiore, M., Mahmoud, O.: Second-order algebraic theories. In: Mathematical Foundations of Computer Science 2010, pp. 368–380. Springer Berlin Heidelberg (2010). https://doi.org/10.1007/978-3-642-15155-2_33
13. Fiore, M., Plotkin, G., Turi, D.: Abstract syntax and variable binding. In: Proceedings of the 14th Annual IEEE Symposium on Logic in Computer Science. pp. 193–. LICS '99, IEEE Computer Society, Washington, DC, USA (1999), http://dl.acm.org/citation.cfm?id=788021.788948
14. Fox, T.: Combinatory logic and cartesian closed categories. Master's thesis, McGill University (1971), https://escholarship.mcgill.ca/concern/theses/6h440t871

15. Gilezan, S.: A note on typed combinators and typed lambda terms. Novi Sad Journal of Mathematics **23**(1), 319–329 (1993), https://sites.dmi.uns.ac.rs/nsjom/Papers/23_1/NSJOM_23_1_319_329.pdf
16. Girard, J.Y., Taylor, P., Lafont, Y.: Proofs and Types. Cambridge University Press, New York, NY, USA (1989), http://www.paultaylor.eu/stable/Proofs+Types.html
17. Hermida, C.: Representable multicategories. Advances in Mathematics **151**(2), 164–225 (2000). https://doi.org/https://doi.org/10.1006/aima.1999.1877
18. Hyland, M.: Towards a notion of lambda monoid. Electronic Notes in Theoretical Computer Science **303**, 59–77 (2014). https://doi.org/10.1016/j.entcs.2014.02.004
19. Hyland, M.: Classical lambda calculus in modern dress. Mathematical Structures in Computer Science **27**(5), 762–781 (2015). https://doi.org/10.1017/s0960129515000377
20. Hyland, M., de Paiva, V.: Full intuitionistic linear logic (extended abstract). presented at the 9th International Congress of Logic, Methodology and Philosophy of Science held in Uppsala, Sweden, August 7-14, 1991. Annals of Pure and Applied Logic **64**(3), 273–291 (1993). https://doi.org/10.1016/0168-0072(93)90146-5
21. Jacobs, B.: Simply typed and untyped lambda calculus revisited. In: Applications of Categories in Computer Science, pp. 119–142. Cambridge University Press (1992). https://doi.org/10.1017/cbo9780511525902.008
22. Jacobs, B.: Categorical logic and type theory. Elsevier Science (1999)
23. Johnstone, P.T.: Sketches of an Elephant: A Topos Theory Compendium Volume 2 (Oxford Logic Guides). Clarendon Press (2002)
24. Lambek, J.: The mathematics of sentence structure. The American Mathematical Monthly **65**(3), 154 (1958). https://doi.org/10.2307/2310058
25. Lambek, J.: Deductive systems and categories II: Standard constructions and closed categories. In: Category theory, homology theory and their applications I, pp. 76–122. Springer (1969). https://doi.org/10.1007/BFb0079385
26. Lambek, J.: Multicategories revisited. In: Gray, J.W., Scedrov, A. (eds.) Categories in Computer Science and Logic: Proceedings of the AMS-IMS-SIAM Joint Summer Research Conference Held June 14-20, 1987 with Support from the National Science Foundation, vol. 92, pp. 217–240. American Mathematical Society (1989). https://doi.org/10.1090/conm/092
27. Lambek, J., Scott, P.J.: Introduction to Higher Order Categorical Logic. Cambridge University Press, New York, NY, USA (1986)
28. Leinster, T.: Higher operads, higher categories. No. 298 in London Mathematical Society Lecture Note Series, Cambridge University Press (2004). https://doi.org/10.1017/CBO9780511525896
29. Linton, F.E.J.: Some aspects of equational categories. In: Proceedings of the Conference on Categorical Algebra, pp. 84–94. Springer Berlin Heidelberg (1966). https://doi.org/10.1007/978-3-642-99902-4_3
30. Mac Lane, S.: Categories for the Working Mathematician, Graduate Texts in Mathematics, vol. 5. Springer-Verlag New York, second edn. (1998). https://doi.org/10.1007/978-1-4757-4721-8
31. Manzyuk, O.: Closed categories vs. closed multicategories. Theory and Applications of Categories **26**(5), 132–175 (2012), http://www.tac.mta.ca/tac/volumes/26/5/26-05.pdf
32. May, J.P.: The Geometry of Iterated Loop Spaces. Springer Berlin Heidelberg (1972). https://doi.org/10.1007/bfb0067491
33. Melliès, P.A.: Ribbon tensorial logic. In: Proceedings of the 33rd Annual ACM/IEEE Symposium on Logic in Computer Science. LICS '18, ACM (Jul 2018). https://doi.org/10.1145/3209108.3209129

34. Pisani, C.: Sequential multicategories. Theory and Applications of Categories **29**(19), 496—541 (2014). https://doi.org/http://www.tac.mta.ca/tac/volumes/29/19/29-19.pdf
35. Pitts, A.M.: Categorical logic. In: Handbook of Logic in Computer Science, chap. 2, pp. 39–123. Oxford University Press, Oxford, UK (2000)
36. Polakow, J., Pfenning, F.: Natural deduction for intuitionistic non-commutative linear logic. In: Lecture Notes in Computer Science, pp. 295–309. Springer Berlin Heidelberg (1999). https://doi.org/10.1007/3-540-48959-2_21
37. Riehl, E.: Category Theory in Context. Dover Publications, Incorporated (2016), https://math.jhu.edu/~eriehl/context.pdf
38. Saville, P.: Cartesian closed bicategories: type theory and coherence. Ph.D. thesis, University of Cambridge (2020). https://doi.org/10.17863/CAM.55080
39. Shulman, M.: Closed category, https://ncatlab.org/nlab/show/closed+category, revision 49 (May 2018)
40. Shulman, M.: LNL polycategories and doctrines of linear logic. Logical Methods in Computer Science **19**(2) (2023). https://doi.org/10.46298/lmcs-19(2:1)2023
41. Taylor, W.: Characterizing Mal'cev conditions. Algebra Universalis **3**(1), 351 (Dec 1973). https://doi.org/10.1007/BF02945141
42. Troelstra, A.S., Schwichtenberg, H.: Basic proof theory. No. 43 in Cambridge Tracts in Theoretical Computer Science, Cambridge University Press, second edn. (2000)
43. Uustalu, T., Veltri, N., Zeilberger, N.: Eilenberg-Kelly reloaded. Electronic Notes in Theoretical Computer Science **352**, 233–256 (Oct 2020). https://doi.org/10.1016/j.entcs.2020.09.012
44. Uustalu, T., Veltri, N., Zeilberger, N.: Deductive systems and coherence for skew prounital closed categories. In: Sacerdoti Coen, C., Tiu, A. (eds.) Proceedings Fifteenth Workshop on Logical Frameworks and Meta-Languages: Theory and Practice, Paris, France, 29th June 2020. Electronic Proceedings in Theoretical Computer Science, vol. 332, pp. 35–53. Open Publishing Association (2021). https://doi.org/10.4204/EPTCS.332.3
45. Weber, M.: Free products of higher operad algebras. Theory and Applications of Categories **28**(2), 24–65 (2013), http://www.tac.mta.ca/tac/volumes/28/2/28-02.pdf
46. Zeilberger, N., Giorgetti, A.: A correspondence between rooted planar maps and normal planar lambda terms. Logical Methods in Computer Science **11** (2015). https://doi.org/10.2168/lmcs-11(3:22)2015

Infinite-State Systems

Reachability in Fixed VASS: Expressiveness and Lower Bounds

Andrei Draghici(✉), Christoph Haase, and Andrew Ryzhikov

Department of Computer Science, University of Oxford, Oxford, UK
andrei.draghici@stcatz.ox.ac.uk

Abstract. The recent years have seen remarkable progress in establishing the complexity of the reachability problem for vector addition systems with states (VASS), equivalently known as Petri nets. Existing work primarily considers the case in which both the VASS as well as the initial and target configurations are part of the input. In this paper, we investigate the reachability problem in the setting where the VASS and the final configuration are fixed and only the initial configuration is variable. We show that fixed VASS fully express arithmetic with counting on initial segments of the natural numbers. It follows that there is a very weak reduction from any fixed such number-theoretic predicate (e.g. square-freeness or "N_1 is the number of primes smaller than N_2") to reachability in fixed VASS where configurations are presented in unary. If configurations are given in binary, we show that there is a fixed VASS with five counters whose reachability problem is PSPACE-hard.

1 Introduction

Vector addition systems with states (VASS), equivalently known as Petri nets, are a fundamental model of computation. A VASS comprises a finite-state controller with a finite number of counters ranging over the non-negative integers. When a transition is taken, counters can be updated by adding an integer, provided that the resulting counter values are all non-negative; otherwise the transition blocks. Given two configurations of a VASS, each consisting of a control state and an assignment of values to the counters, the reachability problem asks whether there is a path connecting the two configurations in the infinite transition system induced by the VASS. The VASS reachability problem has been one of the most intriguing problems in theoretical computer science and studied for more than fifty years. In the 1970s, Lipton showed this problem EXPSPACE-hard [18]. Ever since the 1980s [19, 14, 16], the reachability problem has been known to be decidable, albeit with non-elementary complexity. This wide gap between the EXPSPACE lower bound and a non-elementary upper bound persisted for many years, until a recent series of papers established various non-elementary lower bounds [5, 6, 15], and resulted in matching a recently established upper bound [17], showing the VASS reachability problem Ackermann-complete. The lower bounds for this result require an unbounded number of counters, but even for a fixed number of counters, the Petri net reachability problem requires non-elementary time [6, 7, 15].

N. Kobayashi and J. Worrell (Eds.): FoSSaCS 2024, LNCS 14575, pp. 185–205, 2024.
https://doi.org/10.1007/978-3-031-57231-9_9

Main results. The main focus of this paper is to investigate the reachability problem for *fixed* VASS, where the VASS under consideration and the final configuration are fixed and only the initial configuration forms the input to a reachability query. Here, it is crucial to distinguish between the encoding of numbers used to represent counter values in configurations: in *unary encoding*, the representation length of a natural number $n \in \mathbb{N}$ is its magnitude n whereas in *binary encoding* the bit length of $n \in \mathbb{N}$ is $\lceil \log n \rceil + 1$. It turns out that establishing meaningful lower bounds under unary encoding of configurations is a rather delicate issue; a full discussion is deferred to Section 4. As a first step, we establish a tight correspondence between reachability in VASS and the first-order theory of initial segments of $\mathbb{N}$ with the arithmetical relations addition $(+)$, multiplication $(\times)$ and counting quantifiers. An initial segment in $\mathbb{N}$ is a set $\underline{N} = \{0, \ldots, N\}$ for some arbitrary but fixed $N \in \mathbb{N} \setminus \{0\}$. Relations definable in this family of structures are known as *rudimentary relations* and contain many important number-theoretic relations, cf. [9] and the references therein. For instance, the fixed formula $\mathrm{PRIME}(x) \equiv \neg(x = 0) \wedge \neg(x = 1) \wedge \forall y < x \, \forall z < x \, \neg(x = y \times z)$ evaluates to true in $\underline{N}$ precisely for all prime numbers up to N. The formula $\exists^{=z} y \; (y < x) \wedge \mathrm{PRIME}(y)$ evaluates to true if and only if there exist exactly z prime numbers smaller than x.

Given a fixed rudimentary relation $\Phi(x_1, \ldots, x_k)$, we show how to construct a fixed VASS $\mathcal{V}$ and fixed polynomials $p_1, \ldots, p_m$ such that $\Phi(n_1, \ldots, n_k)$ evaluates to true in $\underline{N}$ if and only if there is a run in $\mathcal{V}$ starting in $(p_1(N, n_1, \ldots, n_k), \ldots, p_m(N, n_1, \ldots, n_k))$ and ending in a zero vector. It thus follows that reachability in fixed VASS under unary encoding of configurations is at least as hard as evaluating any rudimentary relation under unary encoding of numbers. Hence, reachability queries in fixed VASS can, e.g., determine primality and square-freeness of a number given in unary. From those developments, it is already possible to infer that reachability in fixed VASS with configurations encoded in binary is hard for every level of the polynomial hierarchy by a reduction from the validity problem for short Presburger arithmetic [21]. In fact, we can establish a PSPACE lower bound for reachability in a fixed VASS with five counters with configurations encoded in binary, by a generic reduction allowing to simulate space-bounded computations of arbitrary Turing machines encoded as natural numbers. A recent conjecture of Jecker [13] states that for every VASS $\mathcal{V}$, there exists a fixed constant C such that if a target configuration is reachable from an initial configuration, then there exists a witnessing path whose length is bounded by $C \cdot m$, where m is the maximum constant appearing in the initial and final configurations. Thus, assuming Jecker's conjecture, reachability in fixed VASS under binary encoding of configurations would be PSPACE-complete. In the course of our work, we were not able to find any evidence that this conjecture is false. It is also worth noting that while all our results assume that the final configuration is fixed to a zero vector, we did not find any stronger lower bounds for the case where the final configuration is variable, and only the VASS is fixed.

Related work. To the best of our knowledge, the reachability problem for fixed VASS has not yet been systematically explored. Closest to the topics of this paper is the work by Rosier and Yen [22], who conducted a multi-parameter analysis of the complexity of the boundedness and coverability problems for VASS.

However, the study of the computation power of other fixed machines has a long history in the theory of computation. The two classical decision problems for a computation model are *membership* (also called the *word problem*) and *reachability*. Membership asks whether a given machine accepts a given input; the (generic) reachability problem asks whether given an initial and a target configuration, there is a path in the transition system induced by a given machine from the initial configuration to the target configuration. The most prominent example of a reachability problem is the halting problem for different kinds of machines. Classically, the computational complexity of such problems assumes that both the computational model and its input word (for membership) or configurations (for reachability) are part of the input. However, these are two separate parameters. For example, in database theory, the database size and the query size are often considered separately, since the complexity of algorithms may depend very differently on these two parameters, and the sizes of these two parameters in applications can also vary a lot [26]. One approach to study such phenomena is to fix either the database or the query. More generally, the field of parameterised complexity studies the computational difficulty of a problem with respect to multiple parameters of the input.

Returning to our setting, this means fixing either the machine or its input. In this paper, we concentrate on the former. The question can then be seen as follows: in relation to a problem such as membership or reachability, which machine is the hardest one in the given computation model? For some models, the answer easily follows from the existence of universal machines, i.e., machines which are able to simulate any other machine from their class. A classical example here is a universal Turing machine. Sometimes the ability to simulate all other machines has to be relaxed, for example as for Greibach's hardest context-free language [11]. Greibach showed that there exists a fixed context-free grammar such that a membership query for any other context-free grammar can be efficiently reduced to a membership query for this grammar. Similar results are known for two-way non-deterministic pushdown languages [23, 4].

2 Preliminaries

We denote by $\mathbb{Z}$ and $\mathbb{N}$ the set of integers and non-negative integers, respectively. For $N \in \mathbb{N}$ we write $\underline{N}$ to denote the set $\{0, \ldots, N\}$. By $[n, m]$ we define the set of integers between n and m: $[n, m] = \{k \in \mathbb{Z} \mid n \leq k \leq m\}$. By $\mathbf{0}$ we denote the zero vector $(0, 0 \ldots, 0)$ whose dimension is clear from the context.

Counter automata. A *d-counter automaton* is a tuple $\mathcal{A} = (Q, \Delta, \zeta, q_0, q_f)$, where Q is a finite set of states, $\Delta \subseteq Q \times \mathbb{Z}^d \times Q$ is the transition relation, $\zeta : \Delta \to [1, d] \cup \{\top\}$ is a function indicating which counter is tested for zero along a transition ($\top$ meaning no counter is tested), $q_0 \in Q$ is the initial state, and $q_f \in Q$ is the final state. We assume that q_f does not have outgoing transitions.

The set of configurations of $\mathcal{A}$ is $C(\mathcal{A}) := \{(q, n_1, \ldots, n_d) : q \in Q, n_i \in \mathbb{N}, 1 \leq i \leq n\}$. A run ϱ of a counter automaton $\mathcal{A}$ from a configuration $c_1 \in C(\mathcal{A})$ to $c_{n+1} \in C(\mathcal{A})$ is a sequence of configurations interleaved with transitions

$$\varrho = c_1 \xrightarrow{t_1} c_2 \xrightarrow{t_2} \ldots \xrightarrow{t_n} c_{n+1}$$

such that for all $1 \leq i \leq n$, $c_i = (q, m_1, \ldots, m_d)$ and $c_{i+1} = (r, m'_1, \ldots, m'_d)$,

- $t_i = (q, (z_1, \ldots, z_d), r)$ with $m'_j = m_j + z_j$ for all $1 \leq j \leq d$; and
- $m_j = 0$ if $\zeta(t_i) = j$.

Observe that we can without loss of generality assume that each transition $t \in \Delta$ is of one of the two types:

- either no counter is tested for zero along t, that is, $\zeta(t) = \top$, in which case we call it *an update transition*;
- or t does not change the values of the counters, that is, $\zeta(t) = j$ for some $1 \leq j \leq d$ and $t = (q, \mathbf{0}, r)$, in which case we call it *a zero-test transition.*

We say that $\mathcal{A}$ is a *vector addition system with states of dimension d (d-VASS)* if $\mathcal{A}$ cannot perform any zero tests, i.e., ζ is the constant function assigning $\top$ to all transitions. We can now formally define the main decision problem we study in this paper.

Problem 1. FIXED VASS ZERO-REACHABILITY
Fixed: d-VASS $\mathcal{A}$.
Input: A vector $\boldsymbol{x} \in \mathbb{N}^d$ of initial values of the counters.
Output: YES if and only if $\mathcal{A}$ has a run from $(q_0, \boldsymbol{x})$ to $(q_f, \mathbf{0})$.

Counter programs. For ease of presentation, we use the notion of counter programs presented e.g. in [5], which are equivalent to VASS, and allow for presenting VASS (and counter automata) in a serialised way. A counter program is a primitive imperative program that executes arithmetic operations on a finite number of counter variables. Formally, a *counter program* consists of a finite set $\mathcal{X}$ of global counter variables (called *counters* subsequently for brevity) ranging over the natural numbers, and a finite sequence $1, \ldots, m$ of line numbers (subsequently *lines* for brevity), each associated with an instruction manipulating the values of the counters or a control flow operation. Each instruction is in of one the following forms:

```
1: goto 2 or 4
2:    x -= 3
3:    goto 1
4: x += 1
5: halt
```

Fig. 1. Example of a counter program.

- x += c (increment counter x by constant $c \in \mathbb{N}$),
- x −= c (decrement counter x by constant $c \in \mathbb{N}$),
- **goto** L_1 **or** L_2 (non-deterministically jump to the instruction labelled by L_1 or L_2),
- **skip** (no operation).

We write **goto** L as an abbreviation for **goto** L **or** L, and also allow statements of the form **goto** L_1 **or** L_2 **or** ... **or** L_k. Moreover, the line with the largest number is a special instruction **halt**. In our examples of counter programs, we usually omit this last line if it is not referenced explicitly.

An example of a counter program is given in Figure 1. This counter program uses a single counter x and consists of five lines. Starting in line 1, the program non-deterministically loops and decrements the counter x by three every time, until it increments x by one and terminates.

To be able to compose counter programs, we describe the operation of substitution, which substitutes a given line (which we always assume to have a **skip** instruction) of a counter program with the "code" of another counter program. Formally, let C_1, C_2 be counter programs with m_1 and m_2 lines respectively. The result of substituting line k, $1 \leq k \leq m_1 - 1$, of C_1 with C_2 is a counter program C_1' with $m_1 + m_2 - 1$ lines obtained, intuitively, by calling C_2 as a sub-routine in this line and when it halts returning control back to C_1. Formally, the instruction corresponding to a line L, $1 \leq L < m_1 + m_2$, is defined as follows:

- if $L < k$, it is the instruction of line L in C_1,
- if $k \leq L < m_2 + k - 1$, it is the instruction of line $L - k + 1$ in C_2,
- if $L = m_2 + k - 1$, it is the instruction **skip**,
- if $m_2 + k \leq L$, it is the instruction of line $L - m_2$ in C_1.

The line numbers in **goto** instructions are changed accordingly. We also consider a substitution of several counter programs. When specifying counter programs, to denote substitution of another counter program we just write its name instead of an instruction in a line. Also, we write $C_1; C_2$ for

1: C_1
2: C_2

and C_1 **or** C_2 as syntactic sugar for the counter program:

1: **goto** 2 **or** 4
2: C_1
3: **goto** 5
4: C_2

When C is a counter program, we write **loop** C as an abbreviation for the counter program

1: **goto** 2 **or** 4
2: C
3: **goto** 1

Hence, the counter program in Figure 1 corresponds to

1: **loop**
2:     $x \mathrel{-}= 3$
3: $x \mathrel{+}= 1$

We use indentation to mark the scope of the **loop** instruction. We also assume that if several instructions share the same line and are separated by a semicolon, they all belong to the scope of a loop.

Runs of counter programs. Exactly as in the case of VASS, a *configuration* of a counter program is an element $(L, f) \in \mathbb{N} \times \mathbb{N}^{\mathcal{X}}$, where $L \in \mathbb{N}$ is a program line with a corresponding instruction, and $f: \mathcal{X} \to \mathbb{N}$ is a counter valuation. The semantics of counter programs are defined in a natural way: after executing the instructions on the line L, we either non-deterministically go to one of the specified lines (if the instruction on line L is a **goto** instruction), or, otherwise, we go to the line $L + 1$. After executing the last line, we stop.

One can view a counter program as a VASS by treating line numbers as states and defining transitions as specified by the counter program, each labelled with the respective instruction. It is also easy to see how to convert a VASS into a counter program.

A *run* of a counter program is a sequence $\varrho: (L_1, f_1) \to (L_2, f_2) \to \dots \to (L_n, f_n)$ of configurations defined naturally according to the described semantics. For example, $(1, \{x \mapsto 7\}) \to (4, \{x \mapsto 7\}) \to (5, \{x \mapsto 8\})$ is a run of the counter program in Figure 1. Given a run $\varrho: (L_1, f_1) \to (L_2, f_2) \to \dots \to (L_n, f_n)$, we say that ϱ is *terminating* if $L_1 = 1$ and the instruction on line L_n is **halt**, and *zero-terminating* if additionally $f_n(x) = 0$ for all $x \in \mathcal{X}$. We denote by $\mathrm{val}_{\mathrm{end}}(\varrho, x) := f_n(x)$ the value of the counter x at the end of a terminating run. Sometimes, we also want to talk about the value of a counter at a specific point during the execution of a run and define $\mathrm{val}_i(\varrho, x)$ to be the value of the counter x right before we execute the instruction on line i in the run ϱ for the first time, i.e. $\mathrm{val}_i(\varrho, x) := f_k(x)$, where k is the smallest index such that $L_k = i$. For instance, in the example above, we have $\mathrm{val}_{\mathrm{end}}(\varrho, x) = 8$ and $\mathrm{val}_4(\varrho, x) = 7$. We often construct counter programs that admit exactly one run ϱ from a given initial configuration to a target configuration. In such a setting, we may omit the reference to ϱ and simply write $\mathrm{val}_{\mathrm{end}}(x)$ and $\mathrm{val}_i(x)$. The effect $\mathrm{eff}(\varrho): \mathcal{X} \to \mathbb{Z}$ of a run ϱ starting in $(1, f_1)$ and ending in (n, f_n) is a map such that $\mathrm{eff}(\varrho, x) = f_n(x) - f_1(x)$ for all $x \in \mathcal{X}$.

For counter programs, the zero-reachability problem is as follows.

Problem 2. FIXED COUNTER PROGRAM ZERO-REACHABILITY
Fixed: Counter program $\mathcal{C}$.
Input: A vector $\boldsymbol{x} \in \mathbb{N}^d$ of initial values of the counters.
Output: YES if and only if $\mathcal{C}$ has a zero-terminating run from $\boldsymbol{x}$.

3 Implementation of zero tests

The structure of runs in arbitrary counter programs is very complicated and hard to analyse, and hence it is difficult to force a counter program to have a prescribed behaviour. One of the common ways to deal with this issue is to introduce some restricted zero tests, that is, some gadgets that guarantee that if a run reaches a certain configuration, then along this run, the values of some counters are zero at prescribed positions. In this section, summarising [5], we describe such a gadget in the case where the values of counters are bounded by a given number. The number of zero tests that can be performed this way is also bounded. For a counter v, we call this gadget **zero-test**(v), and later on we will use it as a single instruction to test that the value of v is zero before executing it.

In Section 4, the assumption that the values of the counters are bounded comes from the the fact that the corresponding values of the variables in rudimentary arithmetic are bounded. In Section 5, we enforce this property for more powerful models of computation and show how to simulate them with VASS.

Let $N \in \mathbb{N}$ be an upper bound on the value of a counter v. Then, we can introduce a counter $\hat{v}$ and enforce the invariant $f(v)+f(\hat{v}) = N$ to hold in all the configurations of any run of our counter program. We achieve this by ensuring that every line containing an instruction of type $v \mathrel{+}= c$ must be followed by a line with a $\hat{v} \mathrel{-}= c$ instruction. From now on, we make the convention that the instruction $v \mathrel{+}= c$ is an abbreviation for $v \mathrel{+}= c; \hat{v} \mathrel{-}= c$. This allows us to remove the hatted counters from our future counter programs whenever it is convenient for us, which will ease readability. So, if we choose an initial configuration in which $f(v) + f(\hat{v}) = N$, we have that this invariant holds whenever the zero-test gadget is invoked.

We introduce auxiliary counters u_1, u_2 that will be tested for zero only in the final configuration, and hence have no hat counterpart. In the following, the instruction **zero-test**(v) denotes the following gadget:

1: **loop**
2: $\quad v \mathrel{+}= 1; \hat{v} \mathrel{-}= 1; u_2 \mathrel{-}= 1$
3: **loop**
4: $\quad v \mathrel{-}= 1; \hat{v} \mathrel{+}= 1; u_2 \mathrel{-}= 1$
5: $u_1 \mathrel{-}= 2$

Consider an initial configuration in which $f(u_1) = 2n$ and $f(u_2) = 2n \cdot N$ for some $n > 0$. Initially, it is true that $f(u_2) = f(u_1) \cdot N$.

Lemma 1 ([5]). *There exists a run of the counter program* **zero-test**(v) *that starts in a configuration with* $f(u_2) \geq 2$, $f(u_2) = f(u_1) \cdot N$, *and ends in a configuration with* $f(u_2) = f(u_1) \cdot N$ *if and only if* $f(v) = 0$ *in the initial configuration.*

Proof. The invariant $f(v)+f(\hat{v}) = N$ ensures that the loops on line 1 and line 3 can each decrease the value of u_2 by at most N. Moreover, this can only happen if $f(v) = 0$ in the initial configuration. □

From a configuration with $f(u_2) = f(u_1) \cdot N$, a run "incorrectly" executing the **zero-test**(v) subroutine can only reach a configuration with $f(u_2) > f(u_1) \cdot N$. Observe that from such a configuration, we can never reach a configuration respecting the invariant $f(u_2) = f(u_1) \cdot N$ if the values of u_1, u_2 are only changed by **zero-test**(v) instructions. Now, consider a counter v and a counter program C that modifies the values of counters u_1 and u_2 only through the **zero-test**(v) instruction. If we start in a configuration in which $f(u_1) = 2n$ and $f(u_2) = 2n \cdot N$ for some $n > 0$, and we are guaranteed that any run of C cannot execute more than n **zero-test**(v) instructions, then after any run of C, we have that $f(u_2) = f(u_1) \cdot N$ only if the value of the counter v was zero at the beginning of every **zero-test**(v) instruction. If all the counters that we are interested in are bounded by the same value N, we can use a single pair of counters u_1, u_2 to perform zero tests on all our counters. We subsequently call the counters u_1 and u_2 *testing counters.* To summarise, using this technique, we can perform n zero tests on counters bounded by N via a reachability query in a VASS.

Given a configuration (L, f), we say that (L, f) is a *valid configuration* if f respects the condition that $f(u_2) = f(u_1) \cdot N$. A *valid run* is a run that starts in a valid configuration and ends in a valid configuration. Also, a counter program *admits* a valid run if there exists a valid run that reaches the terminal instruction **halt**. Observe that in every valid run the **zero-test**() subroutine does not change the value of the counter which is tested for zero, that is, this value remains zero. Only the values of the testing counters are changed.

We now introduce components. Informally, a component is a counter program acting as a subroutine such that, if it is invoked in a configuration fulfilling the invariants required for valid runs, upon returning, those invariants still hold. Formally, a *component* is a counter program such that:

- there is a polynomial p such that every valid run performs at most $p(N)$ calls of **zero-test**() on all counters; and
- the values of u_1 and u_2 are updated only by **zero-test**() instructions.

We conclude this section with Lemma 2, which states that sequential composition and non-deterministic branching of components yields components. We will subsequently implicitly make use of this obvious lemma without referring to it.

Lemma 2. *If C_1, C_2 are components then both $C_1; C_2$ and C_1* **or** *C_2 are also components.*

Remark 1. Let $\mathcal{V}$ be a fixed VASS, and $s = (q_0, \boldsymbol{n}), t = (q_f, \boldsymbol{m})$ be a pair of its configurations. Given s and t, the FIXED VASS COVERABILITY problem asks where there exists a run in $\mathcal{V}$ from s to a configuration $t' = (q_f, \boldsymbol{m}')$ such that $\boldsymbol{m}' \geq \boldsymbol{m}$ componentwise. Note that when simulating zero tests as described above, for each counter x except u_1, u_2, we have a counter $\hat{x}$ such that the sum of the values of x and $\hat{x}$ is always the same and is known in advance. Since the values of u_1, u_2 are never increased, we can introduce in the same way the counters $\hat{u}_1, \hat{u}_2$, initially set to zero, so that $u_i + \hat{u}_i$ is constant for $i = 1, 2$. Hence, by requiring that the final value of $\hat{x}$ is at least the initial value of x, we make sure that the final value of x is equal to zero. Thus, in this setting, reachability queries reduce to coverability queries.

4 Rudimentary arithmetic and unary VASS

In this section, we provide a lower bound for the zero-reachability problem for a VASS when the input configuration is encoded in unary. We observe that there is a close relationship between this problem and deciding validity of a formula of first-order arithmetic with counting, addition, and multiplication on an initial segment of $\mathbb{N}$, also known as rudimentary arithmetic with counting [9].

4.1 Rudimentary arithmetic with counting

For the remainder of this section, all the structures we consider are relational. We denote by $\mathbf{FO}(+, \times)$ the first-order theory of the structure $\langle \mathbb{N}, +, \times \rangle$, where $+$ and $\times$ are the natural ternary addition and multiplication relations. When interpreted over initial segments of $\mathbb{N}$, i.e. sets $\{0, 1, \ldots, N\}$, for some fixed $N \in \mathbb{N}$, the family of the first-order theories is known as rudimentary arithmetic. Note that, in particular, for a predicate $x + y = z$ to hold, all of x, y, z must be at most N. It thus might seem that after we fix N, a formula $\Phi(\boldsymbol{x})$ can only express facts about numbers up to N. However, as discussed in [25] and [9], this can be improved to quantifying over variables up to N^d for any fixed d using $(N+1)$-ary representations of numbers. In other words, for any fixed d and formula $\Phi(\boldsymbol{x})$, there exists a formula $\Phi'(\boldsymbol{x})$ such that for any $N \in \mathbb{N}$ and $\boldsymbol{x} \in \underline{N}^n$, we have that $\langle \underline{N}, +, \times \rangle \models \Phi'(\boldsymbol{x})$ iff $\langle \underline{N}^d, +, \times \rangle \models \Phi(\boldsymbol{x})$.

Rudimentary arithmetic can be extended with counting quantifiers. As described in [25], let rudimentary $\mathbf{FOunC}(+, \times)$ be rudimentary $\mathbf{FO}(+, \times)$ extended with counting quantifiers of the form $\exists^{>x} y\ \varphi(y)$. In this expression, the variable x is free and the variable y is bounded by the quantifier. The semantics of this expression is that there exist more than x different values of y such that the formula $\varphi(y)$ is satisfied. The paper [25] actually uses the counting quantifier $\exists^{=x} y\ \varphi(y)$ to state that the number of such values is exactly x, which can be expressed as $(x = 0 \wedge \neg\exists y\ \varphi(y)) \vee ((\exists^{>x'} y\ \varphi(y)) \wedge (x' + 1 = x) \wedge \neg\exists^{>x} y\ \varphi(y))$.

Moreover, $\mathbf{FOunC}(+, \times)$ can be extended to $\mathbf{FO}k\textbf{-aryC}(+, \times)$, $\mathbf{FO}(+, \times)$ with k-ary counting quantifiers $\exists^{=\boldsymbol{x}} \boldsymbol{y}\ \varphi(\boldsymbol{y})$. In this expression, $\boldsymbol{x}, \boldsymbol{y}$ are vectors of the same dimension, and similarly to the previous case, all the variables of $\boldsymbol{x}$ are free and all the variables of $\boldsymbol{y}$ are bounded by the quantifier. The semantics is that the k-tuple $\boldsymbol{x}$ is the $(N+1)$-ary representation of the number of k-tuples $\boldsymbol{y}$ that satisfy $\varphi(\boldsymbol{y})$. As shown in [3], rudimentary $\mathbf{FOunC}(+, \times)$ and rudimentary $\mathbf{FO}k\textbf{-aryC}(+, \times)$ have the same expressive power. In order to have a meaningful reduction to *fixed* VASS, we are interested in the following decision problem:

Problem 3. FIXED RUDIMENTARY $\mathbf{FO}k\textbf{-aryC}(+, \times)$ VALIDITY
Fixed: $\Phi(\boldsymbol{x}) \in \mathbf{FO}k\textbf{-aryC}(+, \times)$.
Input: $N \in \mathbb{N}$ and $\boldsymbol{x} \in \underline{N}^n$ given in unary.
Output: YES if and only if $\langle \underline{N}, +, \times \rangle \models \Phi(\boldsymbol{x})$.

4.2 Reductions between unary languages

In order to study decision problems whose input is, for some constant k, a k-tuple of numbers presented in unary, and hence to analyse languages corresponding to them, we need a notion of reductions that are weaker compared to the standard ones that are widely used in computational complexity. The reason is that classical problems involving numbers represented in unary, such as UNARY SUBSET SUM [8], have as an input a variable-length sequence of numbers given in unary. Hence, languages of such problems are in fact binary, as we need a delimiter symbol to separate the elements of the sequence. It is not clear how a reasonable reduction from such a language to a language consisting of k-tuples of numbers for a *fixed* k would look like. In particular, note that unary FIXED VASS ZERO-REACHABILITY is not the unary "counterpart" of binary FIXED VASS ZERO-REACHABILITY in the classical sense. Conversely, arithmetic properties of a single number, e.g. primality or square-freeness, require very low computational resources if the input is represented in unary. Hence, the notion of a reduction between such "genuinely unary" languages has to be very weak.

In view of this discussion, we introduce the following kind of reduction. Given $k > 0$, a *k-tuple unary language* is a subset $L \subseteq \mathbb{N}^k$. We say that L is a tuple unary language if L is a k-tuple unary language for some $k > 0$. Let $L \subseteq \mathbb{N}^k$ and $M \subseteq \mathbb{N}^\ell$ be tuple unary languages, we say that L *arithmetically reduces* to M if there are fixed polynomials $p_1, \ldots, p_\ell \colon \mathbb{N}^k \to \mathbb{N}$ such that $(m_1, \ldots, m_k) \in L$ if and only if $(p_1(m_1, \ldots, m_k), \ldots, p_\ell(m_1, \ldots, m_k)) \in M$.

We believe that this reduction is sensible for the following informal reasons. Polynomials can be represented as arithmetic circuits. To the best of our knowledge, there are no known lower bounds for, e.g. comparing the output of two arithmetic circuits with all input gates having value one [1], suggesting that evaluating a polynomial is a computationally weak operation. Moreover, in the light of sets of numbers definable in rudimentary arithmetic, it seems implausible that applying a polynomial transformation makes, e.g. deciding primality of a number substantially easier.

For a formula Φ, let $\mathcal{L}_\Phi$ be the tuple unary language of yes-instances for FIXED RUDIMENTARY **FO**k-**aryC**$(+, \times)$ VALIDITY. Also, for a counter program C, define $\mathcal{L}_C$ as the tuple unary language of yes-instance for the FIXED COUNTER PROGRAM ZERO-REACHABILITY problem. The remainder of this section is devoted to proving the following theorem.

Theorem 1. *For every formula Φ of rudimentary* **FO**k-**aryC**$(+, \times)$*, there exists a counter program C such that $\mathcal{L}_\Phi$ arithmetically reduces to $\mathcal{L}_C$.*

This theorem can be viewed in two different contexts. On the one hand, it relates the computational complexity of the two problems using a very weak reduction as described above. On the other hand, it also relates the expressivity of two formalisms. Namely, the set of satisfying assignments for formulas of rudimentary arithmetic is at most as expressive as the composition of polynomial transformations with the sets of initial configurations for zero-reachable runs in counter programs. In particular, it shows that fixed VASS can, up to a polynomial transformation, decide number-theoretic properties such as primality, square-freeness, see [9] for further examples. Note that by Remark 1, an analogue of Theorem 1 holds for tuple unary languages of yes-instances of FIXED VASS COVERABILITY.

4.3 Components for arithmetic operations

Since there is no straightforward way to model negation with a counter program, we need to provide gadgets for both the predicates $+$ and $\times$ of rudimentary **FO**k-**aryC**$(+, \times)$ and their negations, and hence design a separate component for each literal. However, these components may change the values of the counters representing first-order variables, and since a first-order variable might appear in multiple literals, we first provide a gadget to copy the value of a chosen counter to some auxiliary counter before it can be manipulated.

Copy. We provide a counter program $\textsc{Copy}[x, x']$ with the following properties:

1. it admits a valid run if and only if $\text{val}_{\text{end}}(x') = \text{val}_{\text{end}}(x) = \text{val}_1(x)$; and
2. $\textsc{Copy}[x, x']$ is a component.

We implement $\textsc{Copy}[x, x']$ as follows:

1: **loop**
2: $\quad x' -= 1$
3: **zero-test**(x')
4: **loop**
5: $\quad x -= 1; x' += 1; t += 1$
6: **zero-test**(x)
7: **loop**
8: $\quad t -= 1; x += 1$
9: **zero-test**(t)

The loop on line 1 ensures that $\text{val}_4(x') = 0$. We do not do this for the auxiliary counter t because any valid run sets $\text{val}_{\text{end}}(t) = 0$. Observe that $\textsc{Copy}[x, x']$ admits a valid run if and only if the loop on line 4 is executed $\text{val}_1(x)$ many times and the loop on line 7 is executed $\text{val}_4(t) = \text{val}_1(x)$ many times which happens if and only if $\text{val}_{\text{end}}(x') = \text{val}_{\text{end}}(x) = \text{val}_1(x)$. Moreover, any valid run performs 3 calls to the **zero-test**() subroutine, so $\textsc{Copy}[x, x']$ is a component.

Addition. We define a counter program $\textsc{Addition}[x, y, z]$ that enables us to check whether the value stored in counter z is equal to the sum of the values stored in x, y. Formally, it has following properties:

1. $\textsc{Addition}[x, y, z]$ admits a valid run if and only if $\text{val}_1(x)+\text{val}_1(y) = \text{val}_1(z)$;
2. $\textsc{Addition}[x, y, z]$ is a component; and
3. the effect of $\textsc{Addition}[x, y, z]$ is zero on counters x, y, z.

We implement $\textsc{Addition}[x, y, z]$ as follows:

1: $\textsc{Copy}[x, x']; \textsc{Copy}[y, y']; \textsc{Copy}[z, z']$
2: **loop**
3: $\quad z' -= 1$
4: $\quad x' -= 1$ **or** $y' -= 1$
5: **zero-test**(x'); **zero-test**(y'); **zero-test**(z')

It is easy to see that the first property is fulfilled by the counter program and that $\textsc{Addition}[x, y, z]$ is a component because any run performs exactly 12 class to **zero-test**() (9 calls on line 1, and 3 calls on line 5). The last property is true based on the properties of $\textsc{Copy}$. The component for the negation of the addition predicate is defined similarly.

Multiplication. We now define a counter program $\textsc{Multiplication}[x, y, z]$ with the following properties:

1. it admits a valid run if and only if $\mathrm{val}_1(z) = \mathrm{val}_1(x) \cdot \mathrm{val}_1(y)$;
2. $\textsc{Multiplication}[x, y, z]$ is a component; and
3. the effect of $\textsc{Multiplication}[x, y, z]$ is zero on counters x, y, z.

We implement $\textsc{Multiplication}[x, y, z]$ as follows:

1: $\textsc{Copy}[x, x']; \textsc{Copy}[y, y']; \textsc{Copy}[z, z']$
2: **loop**
3:     **loop**
4:         $x' -= 1; t += 1; z' -= 1$
5:     **zero-test**(x')
6:     **loop**
7:         $x' += 1; t -= 1;$
8:     **zero-test**(t)
9:     $y' -= 1$
10: **zero-test**(y'); **zero-test**(z')

Observe that the loop on line 3 of any valid run must be executed $\mathrm{val}_1(x)$ $\mathrm{val}_1(x)$ many times in order to pass the zero test on line 5. The effect of this loop is then to decrease the value of z' by $\mathrm{val}_1(x)$ and to set the value of t to $\mathrm{val}_1(x)$. Next, the loop on line 6 must be executed $\mathrm{val}_5(t) = \mathrm{val}_1(x)$ many times to pass the zero test on line 8, so the value of x' is set to $\mathrm{val}_1(x)$ and the value of t is set again to zero. Hence, the effect of lines 3-8 is to subtract $\mathrm{val}_1(x)$ from the value of z' without changing the value of x'. Finally, any valid run passes the test on line 10 if and only if the loop on line 2 is executed $\mathrm{val}_1(y)$ many times, which happens if and only if $\mathrm{val}_1(z) = \mathrm{val}_1(x) \cdot \mathrm{val}_1(y)$. Since we argued that the loop on line 2 is executed $\mathrm{val}_1(y)$ many times, we conclude that any valid run of $\textsc{Multiplication}[x, y, z]$ performs at most $2N + 9$ calls to **zero-test**(), so $\textsc{Multiplication}[x, y, z]$ is a component. Again, the last property is ensured by the properties of $\textsc{Copy}$. The definition of $\neg\textsc{Multiplication}[x, y, z]$ is similar.

4.4 Components for quantification

We define the remaining components that we need in order to prove Theorem 1. These components allow us to existentially and universally quantify over variables in a bounded range.

Existential quantifiers. We start with a counter program $\textsc{Exists}[v]$ with the following properties:

1. for every $n \in \underline{N}$, $\textsc{Exists}[v]$ admits a valid run ϱ such that $\mathrm{val}_{\mathrm{end}}(\varrho, v) = n$;
2. $\textsc{Exists}[v]$ is a component.

We define $\textsc{Exists}[v]$ as follows:

1: **loop** $v -= 1$

```
zero-test(v)
loop v += 1
```

It is easy to see that both properties hold, since EXISTS[v] performs exactly one call to the **zero-test**() subroutine.

Universal quantifiers. While the component used for simulating existential quantification can be sequentially composed with a component for a subformula, universal quantification requires directly integrating the component over whose variable we universally quantify. Let $C[v]$ be a component that may access the counter v, test it for zero, and change its value on intermediate steps, but has overall effect zero on counter v. We write FORALL[v] : $C[v]$ for the following counter program:

```
1: loop
2:     v -= 1
3: zero-test(v)
4: loop
5:     C[v]
6:     v += 1
7: zero-test(v̂)
```

The properties of FORALL[v] : $C[v]$ are as follows:

1. it admits a valid run if and only if for all $n \in \underline{N}$, C has a valid run with $\text{val}_1(v) = n$; and
2. FORALL[v] : $C[v]$ is a component.

Notice that the instruction on line 7 tests if $\text{val}_7(v) = N$. Thus, any valid run that passes the test on line 7 must be able to execute $C[v]$ for all values of $v \in \underline{N}$. Moreover, since $C[v]$ is a component, we know that the number of calls to **zero-test**() it makes is polynomial in N. Denote this number by B. Then FORALL[v] : $C[v]$ executes at most $N \cdot B + 1$ many calls to **zero-test**() and it is thus a component.

Counting quantifiers. Finally, we design a component which is an extension of the FORALL[v] : $C[v]$ component, where, as in the case of FORALL, $C[v]$ has overall effect zero on v. Formally, EXISTSC[x, v] : $C[v]$ component has the following properties:

- it admits a valid run if and only if there exist more than $\text{val}_1(x)$ different integers $n \in \underline{N}$ such that C has a valid run with $\text{val}_1(v) = n$
- the overall effect on counter x is zero; and
- EXISTSC[x, v] : $C[v]$ is a component.

We write EXISTSC[x, v] : $C[v]$ for the following counter program:

```
1: loop
2:     v −= 1
3: zero-test(v)
4: Copy[x, x′]
5: goto 6 or 9
6: zero-test(x′)
7: ForAll[v] : C[v]
8: goto 15
9: x += 1
10: loop
11:     v += 1
12:     goto 13 or 10
13:     C[v]; x′ −= 1
14: zero-test(x′)
15: halt
```

The branching on line 5 checks whether $\text{val}_1(x) = N$. If so, $C[v]$ must have a valid run for all values of v, which is checked on line 7. Otherwise, the instructions on line 13 ensure that the value of x' can be decremented if only if $C[v]$ admits at least one valid run with the current value of v. Moreover, the zero test on line 14 is passed if and only if $C[v]$ admitted a valid run for more than $\text{val}_1(x)$ different values. Similarly to the ForAll case, since $C[v]$ is a component, we have that it makes at most a polynomial number of calls to **zero-test**(). If we denote this number by B, the maximum number of calls to **zero-test**() performed by ExistsC$[x, v] : C[v]$ is bounded by $N \cdot B + 5$. Hence, it is indeed a component.

4.5 Putting it all together

Having defined all the building blocks above, we now prove Theorem 1, which is a consequence of the following lemma.

Lemma 3. *For any formula $\Phi(\boldsymbol{x})$ of* **FO***k*-**aryC**$(+, \times)$, *there exists a component C over k counters and polynomials $p_1, \dots, p_k : \mathbb{N} \times \mathbb{N}^n \to \mathbb{N}$ such that for any $N \in \mathbb{N}$ and $\boldsymbol{x} \in \mathbb{N}^n$, $\langle \underline{N}, +, \times \rangle \models \Phi(\boldsymbol{x})$ if and only if C admits a valid run from the initial configuration $(p_1(N, \boldsymbol{x}), \dots, p_k(N, \boldsymbol{x}))$.*

Proof. We prove this statement by structural induction on subformulas of Φ. As shown in [3], rudimentary **FOunC**$(+, \times)$ has the same expressive power as rudimentary **FO***k*-**aryC**$(+, \times)$. Since in our setting the formula is fixed, we can thus assume that $\Phi \in$ **FOunC**$(+, \times)$. Moreover, it is easy to see that we can assume that only $\exists^{>x}$ is used as a counter quantifier, since $\exists^{=x}$ can easily be defined using it as described above. Finally, we can assume that negations appear in Φ only in front of arithmetic predicates. In particular, $\neg\exists^{>x} y\, \varphi(y)$ is equivalent to $(\exists^{>x'} y\, \neg\varphi(y)) \wedge (x + x' = N)$.

The counters of the component C are defined to be:

- a counter in vector $\boldsymbol{x}_C$ corresponding to every free variable of $\Phi(\boldsymbol{x})$;
- a counter in vector $\boldsymbol{y}_C$ corresponding to every quantified variable of $\Phi(\boldsymbol{x})$;
- a counter in vector $\boldsymbol{a}_C$ corresponding to every constant of $\Phi(\boldsymbol{x})$; and
- the auxiliary counters $t_C, \boldsymbol{x}'_C, \boldsymbol{y}'_C, \boldsymbol{c}'_C$ used inside the components for predicates and counting quantifiers described above.

We initialise them as follows:

- $f_1(x_C) = x$ and $f_1(\hat{x}_C) = N - x$ for each counter x_C corresponding to a variable x in $\boldsymbol{x}$;
- $f_1(v) = 0$ and $f_1(\hat{v}) = N$ for all the counters corresponding to quantified variables and constants, and auxiliary counters; and
- for the testing counters, $f_1(u_1) = 2N$ and $f_1(u_2) = 2N \cdot P(N)$, where the polynomial $P(N)$ will be defined later.

Assume first that a subformula φ of Φ consists of a single literal. Then, by using the previously defined components, we can construct a fixed component C' corresponding to this literal. In C', for every valid initial configuration (L, f), there exists a valid run starting in it if and only if φ is true under the assignment of the values of the counters in (L, f) to the corresponding variables in φ. If φ is a Boolean combination of multiple literals, by simulating conjunction via sequential composition and disjunction by non-deterministic branching, we can construct a component C_φ with the same property.

We now need to show how to simulate the quantifiers. Let C be the component constructed for φ. We then take

- for $\exists y\ \varphi$:

  1: EXISTS$[y_C]$

  2: $C[y_C]$
- for $\forall y\ \varphi$:

  1: FORALL$[y_C]$:

  2: $C[y_C]$
- for $\exists^{>x} y\ \varphi$:

  1: EXISTSC$[x_C, y_C]$:

  2: $C[y_C]$

As noted above, to be able to use these components, we need to make sure that $C[y_C]$ has overall zero effect on the value of y_C. This is indeed true, since the only place where the value of a counter y_C is changed by a subroutine is in the component corresponding to the quantifier bounding y.

The counter program C starts with a component C_0 that initialises the counters $\boldsymbol{a}$ corresponding to the constants of $\Phi(\boldsymbol{x})$ by a sequence of instruction of the type $a \mathrel{+}= c$ for a corresponding constant c appearing in $\Phi(\boldsymbol{x})$. Finally, we let $C = C_0; C_1$. By the properties established above, it is clear that C admits a valid run starting with f_1 defined above if and only if $\Phi(\boldsymbol{x})$ is valid. To see that C is a component, it remains to note that at every step of the structural induction the number of calls to **zero-test**() is polynomial in N. Hence, there exists a polynomial $P(N)$ such that the overall number calls to **zero-test**() performed by C is bounded by $P(N)$. We conclude by reminding that we use this polynomial to initialise the value of the testing counter u_2. □

To prove Theorem 1, add a loop repeating zero tests at the end of C, thus setting the values of the testing counters to zero if and only if the invariant described in Section 3 holds. After that, set to zero all the remaining counters (including the hatted counters) by decrementing them in loops. A run in thus constructed counter program is zero-accepting if and only if it is valid.

As proved in [3], rudimentary **FOk-aryC**$(<)$ has the same expressive power as **FOk-aryC**$(+, \times)$. Hence, an alternative proof for Theorem 1 is to express k-ary counting quantifiers without the need for components for addition and multiplication. However, this approach is more technical and less insightful.

5 A universal VASS for polynomial space computations

The goal of this section is to show that there is a fixed 5-VASS whose zero-reachability problem is PSPACE-hard, provided that the initial configuration is encoded in binary. Let us first remark that we can actually use the techniques developed in the previous section to prove that for every i, there exists a fixed VASS $\mathcal{V}_i$ such that deciding zero-reachability for $\mathcal{V}_i$ is Σ_i^{P}-hard. A result by Nguyen and Pak [21] shows that for every i, there is a formula Φ_i of so-called *short Presburger arithmetic* such that deciding Φ_i is Σ_i^{P}-hard. Applying bounds on quantifier elimination established in [27], it can be shown that quantification for formulas of short Presburger arithmetic relativises in a certain sense to an initial segment $\underline{N}$ for some $N \in N$ whose bit length is polynomial in the size of Φ_i. Hence, by combining the results from [21] with Lemma 3, it is possible to show that zero-reachability for fixed binary VASS is hard for the polynomial hierarchy. We do not explore this method further because we can actually construct a fixed binary VASS such that the zero-reachability problem is PSPACE-hard for it and which has a smaller number of counters than the fixed binary VASS obtained from showing NP-hardness via the reduction from short Presburger arithmetic outlined above.

We proceed with our construction as follows. We start with the halting problem for Turing machines (TMs) working in polynomial space and show that this problem is PSPACE-hard even if the space complexity of the TM is bounded by the length of its encoding and its input is empty. In Proposition 2, we then reformulate the halting problem as follows: given the encoding of such a machine as an input to a universal one-tape TM $\mathcal{U}$, does $\mathcal{U}$ accept?

We then use two consecutive simulation. First, we simulate $\mathcal{U}$ with a 3-counter automaton $\mathcal{A}$ (Proposition 3), and then simulate $\mathcal{A}$ with an 5-VASS $\mathcal{V}$ (Theorem 2). To be able to apply the technique described in Section 3, we make sure that the space complexity stays linear in the size of the input throughout these simulations. This implies that both the upper bound on the value of the counters and the required number of zero tests are polynomial in the size of the input, which enables us to establish a polynomial time reduction. As a result we obtain a VASS $\mathcal{V}$ which, in a certain sense, can simulate arbitrary polynomial-space computations.

To provide the reduction, we then show how to transform in polynomial time the input of the problem we started with, the halting problem for polynomial-space TMs, into a zero-reachability query for $\mathcal{V}$.

5.1 The halting problem for space-bounded TMs

The goal of this subsection is to show that there exists a *fixed* polynomial-space TM whose halting problem is PSPACE-complete. Note that using standard arguments, we can assume that $\mathcal{M}$ below always halts.

Proposition 1 ([2, Section 4.2]). *The following problem is PSPACE-complete: given a TM $\mathcal{M}$, an input word w and a number n encoded in unary, decide if $\mathcal{M}$ accepts w in at most n space.*

We fix some way of encoding, using an alphabet of size at least two, of Turing machines and we denote by $|\mathcal{M}|$ the length of the encoding of $\mathcal{M}$, which we call the *size* of $\mathcal{M}$. Given a TM $\mathcal{M}$, we say that it is $|\mathcal{M}|$-space-bounded if on every input it halts using at most $|\mathcal{M}|$ space. Given $\mathcal{M}$, an input word w and a number n encoded in unary, it is easy to construct a $|\mathcal{M}|$-space-bounded TM $\mathcal{M}'$ such that if $\mathcal{M}$ accepts w in space at most n, then $\mathcal{M}'$ accepts on the empty input, otherwise $\mathcal{M}'$ rejects on the empty input. Moreover, the size of $\mathcal{M}'$ is polynomial in $|\mathcal{M}|$, $|w|$ and 2^n.

Indeed, $\mathcal{M}'$ can be constructed as follows. When run on the empty input, it writes w on some tape, and then runs $\mathcal{M}$ treating this tape as the input tape. Additionally, it initialises another tape with n written in unary, and before each step of $\mathcal{M}$ it checks that the space used by the tape where $\mathcal{M}$ is simulated does not exceed n. If it does, it immediately rejects. It is easy to see that such a TM is $|\mathcal{M}'|$-space-bounded and satisfies the required conditions.

Hence we get that the following problem is PSPACE-complete: given a $|\mathcal{M}|$-space-bounded TM $\mathcal{M}$, does $\mathcal{M}$ accept on the empty input? Observe that from the construction above we can assume that $\mathcal{M}$ has a special representation such that the fact that it is $|\mathcal{M}|$-space-bounded can be checked in polynomial time.

Let $\mathcal{U}$ be a one-tape universal TM. This TM has a single read-write tape, which in the beginning contains the input, that is, a description of a TM $\mathcal{M}$ it is going to simulate. If $\mathcal{M}$ is $|\mathcal{M}|$-space-bounded (and represented as mentioned in the previous paragraph), $\mathcal{U}$ simulates $\mathcal{M}$ on the empty input in space linear in $|\mathcal{M}|$ [2, Claim 1.6], otherwise $\mathcal{U}$ rejects. That is, in this space, $\mathcal{U}$ accepts or rejects depending on whether $\mathcal{M}$ accepts or rejects the empty word. Hence we get the following proposition.

Proposition 2. *There exists a fixed linear-space TM $\mathcal{U}$ such that the question whether $\mathcal{U}$ halts on a given input is PSPACE-complete.*

5.2 From TMs to a counter automata

In the previous subsection, we obtained a PSPACE-complete problem which already resembles the form of the reachability problem for a fixed counter program: given a fixed linear-space TM $\mathcal{U}$, does it accept a given input? In this section we show how to simulate $\mathcal{U}$ with a fixed counter automaton $\mathcal{A}$, and in the next section we show how to simulate $\mathcal{A}$ with a fixed binary VASS $\mathcal{V}$.

Let $\mathcal{A}$ be a counter automaton. We say that $\mathcal{A}$ is *deterministic* if for every configuration $(q, n_1, \ldots, n_d)$ there is at most one transition that $\mathcal{A}$ can take from this configuration. Suppose that $\mathcal{A}$ is deterministic, and that its final state q_f does not have any outgoing transitions. Let $\boldsymbol{n} = (n_1, \ldots, n_d) \in \mathbb{N}^d$. We treat $\mathcal{A}$ as an acceptor for such vectors. We say that $\mathcal{A}$ works in time t and space s on $\boldsymbol{n}$ if the unique run starting in the configuration $(q_0, n_1, \ldots, n_d)$ ends in a state without outgoing transitions, has length t, and the bit length of the largest value of a counter along this run is s. If this run ends in q_f, we say that $\mathcal{A}$ *accepts* this vector, otherwise we say that it *rejects* it. In all our constructions we make sure that there are no infinite runs. Note that, as in the case of TMs, we measure space complexity in the bit length of the values of the counters, and not in their actual values.

Let Σ be a finite alphabet. Let us bijectively assign a natural number to each word over Σ as follows. First, assign a natural number between 1 and $|\Sigma|$ to each symbol in Σ. Then w can be considered as a number in base $|\Sigma|+1$, with the least significant digit corresponding to the first letter of w. We denote this number by $\text{num}(w)$.

Let $\mathcal{M}$ be a TM, and w be its input. We can transform w into a vector $(\text{num}(w), 0, \ldots, 0)$, which will be the input of a deterministic counter automaton $\mathcal{A}$. We say that $\mathcal{A}$ *simulates* $\mathcal{M}$ if w is accepted by $\mathcal{M}$ if and only if the corresponding vector is accepted by $\mathcal{A}$. We say that this simulation is *in linear space* if there exists a constant c such that if the space complexity of $\mathcal{M}$ is s on some input, then the space complexity of $\mathcal{A}$ on the corresponding input is cs.

The proof of the following proposition uses the techniques described in the proofs of [10, Theorem 4.3(a)] and [12, Theorem 2.4].

Proposition 3. *For every one-tape TM $\mathcal{M}$, there exists a deterministic 3-counter automaton $\mathcal{A}$ that simulates it in linear space.*

Proof. The idea of the proof is as follows. Two counters of $\mathcal{A}$, call them ℓ and r, represent the content of the tape of $\mathcal{M}$ to the left and to the right of the reading head. They are encoded similarly to the way we encode the input word. Namely, let $w_1 a w_2$, where $w_1, w_2 \in \Sigma^*$ and $a \in \Sigma$, be the content of the tape at some moment of time, with the working head in the position of the letter a. Denote by w_1^R the reversal of the word w. Then ℓ stores $\text{num}(w_1^R)$, r stores $\text{num}(w_2)$, and a is stored in the finite memory of the underlying finite automaton.

Now, to make a step to the left, we do the following. First, we need to add a to the end of the word encoded by the value of r. This is done by multiplying the value of r by $|\Sigma|+1$ and adding $\text{num}(a)$ to it. Next, we need to extract the last letter of the word encoded by the value of ℓ, and remove this letter. To do so, we do the opposite of what we did for r: this letter is the residue of dividing the value of ℓ by $|\Sigma|+1$, and the new value of ℓ is the result of this division.

The reason we need the third counter x is to perform these multiplications and divisions. Namely, to divide the value of a counter ℓ by a constant c, we repeat the following until it is no longer possible: subtract c from the value of ℓ and add one to the value of x. When the value of ℓ becomes smaller than c, we get the result of the division in the counter x, and the remainder in ℓ. Multiplication by a constant is done similarly. Observe that by construction the largest value of a counter of $\mathcal{A}$ at any moment of time is at most $(|\Sigma|+1)^S$, where S is the maximal amount of space $\mathcal{M}$ uses on given input. The bit length of this number is linear in S, hence $\mathcal{A}$ simulates $\mathcal{M}$ in linear space. □

By simulating $\mathcal{U}$ from Proposition 2 with a counter automaton $\mathcal{A}$, we get the following statement.

Corollary 1. *There exists a fixed 3-counter automaton $\mathcal{A}$ working in linear space such that the zero-reachability problem for it is PSPACE-complete.*

For 2-counter automata, no such result is known. Informally speaking, such automata are exponentially slower than 3-counter automata: the known simulation requires storing the values of the three counters x, y, z as $2^x 3^y 5^z$ [20]. They are also less expressive: for example, 2-counter automata cannot compute the function 2^n [24], while for 3-counter automata this is trivial. It is worth noting the developments of the next subsection imply that a lower bound for fixed 2-counter automata translates into a lower bound for fixed 4-VASS.

5.3 From counter automata to VASS

To go from a counter automaton to a VASS, we need to simulate zero tests with a VASS. In general, this is not possible. However, the space complexity of the counter automaton in Corollary 1 is linear, so the values of all its counters are bounded by a polynomial in the bit length of the input. The number of zero tests $\mathcal{A}$ performs does not exceed its time complexity, which is at most exponential in the space complexity. However, this is not a problem, since all the values are provided and stored in binary. The bit length of the number of zero tests is thus polynomial in the input, and hence the testing counters described in Section 3 can be initialised with a polynomial time reduction, hence obtaining PSPACE-hardness of the zero-reachability problem in fixed 8-VASS.

Moreover, a more advanced technique of quadratic pairs described in [7] allows to deduce the same result for 5-VASS. Namely, a slight variation of [7, Lemma 2.7] states that given a 3-counter automaton $\mathcal{A}$ working in linear space, one can construct a 5-VASS $\mathcal{V}$ such that fixed zero-reachability in $\mathcal{A}$ can be reduced in polynomial time to fixed zero-reachability in $\mathcal{V}$. The same reasoning as before shows that we can initialise the counters of $\mathcal{V}$ to account for enough zero tests. Hence we get the main result of this section.

Theorem 2. *There exists a fixed* 5-*VASS such that the* FIXED VASS ZERO-REACHABILITY *problem for it is PSPACE-hard assuming that the input configuration is given in binary.*

By Remark 1 and by further inspecting the construction in [7, Lemma 2.7], together with the PSPACE upper bound for coverability in fixed VASS with configurations given in binary established in [22], we moreover obtain the following corollary.

Corollary 2. *There exists a fixed* 6-*VASS such that the* FIXED VASS COVERABILITY *problem for it is PSPACE-complete assuming that the input configurations are given in binary.*

Acknowledgements. We would like to thank anonymous reviewers for their useful comments on the content and presentation of the paper. This work is part of a project that has received funding from the European Research Council (ERC) under the European Union's Horizon 2020 research and innovation programme (Grant agreement No. 852769, ARiAT).

References

1. Allender, E., Bürgisser, P., Kjeldgaard-Pedersen, J., Miltersen, P.B.: On the complexity of numerical analysis. SIAM Journal on Computing **38**(5), 1987–2006 (2009). https://doi.org/10.1137/070697926
2. Arora, S., Barak, B.: Computational Complexity – A Modern Approach. Cambridge University Press (2009)
3. Barrington, D.A.M., Immerman, N., Straubing, H.: On uniformity within NC^1. Journal of Computer and System Sciences **41**(3), 274–306 (1990). https://doi.org/10.1016/0022-0000(90)90022-D
4. Chistikov, D., Majumdar, R., Schepper, P.: Subcubic certificates for CFL reachability. Proceedings of the ACM on Programming Languages **6**(POPL) (2022). https://doi.org/10.1145/3498702
5. Czerwiński, W., Lasota, S., Lazić, R., Leroux, J., Mazowiecki, F.: The reachability problem for petri nets is not elementary. Journal of the ACM **68**(1), 1–28 (2020). https://doi.org/10.1145/3313276.3316369
6. Czerwinski, W., Orlikowski, L.: Reachability in vector addition systems is Ackermann-complete. In: Annual Symposium on Foundations of Computer Science, FOCS. pp. 1229–1240. IEEE (2021). https://doi.org/10.1109/FOCS52979.2021.00120
7. Czerwinski, W., Orlikowski, L.: Lower bounds for the reachability problem in fixed dimensional VASSes. In: Symposium on Logic in Computer Science, LICS. Association for Computing Machinery, New York, NY, USA (2022). https://doi.org/10.1145/3531130.3533357
8. Elberfeld, M., Jakoby, A., Tantau, T.: Logspace versions of the theorems of Bodlaender and Courcelle. In: Annual Symposium on Foundations of Computer Science, FOCS. pp. 143–152 (2010). https://doi.org/10.1109/FOCS.2010.21
9. Esbelin, H.A., More, M.: Rudimentary relations and primitive recursion: A toolbox. Theoretical Computer Science **193**(1), 129–148 (1998). https://doi.org/https://doi.org/10.1016/S0304-3975(97)00002-9
10. Fischer, P.C., Meyer, A.R., Rosenberg, A.L.: Counter machines and counter languages. Mathematical systems theory **2**, 265–283 (1968). https://doi.org/10.1007/BF01694011
11. Greibach, S.A.: The hardest context-free language. SIAM Journal on Computing **2**(4), 304–310 (1973). https://doi.org/10.1137/0202025
12. Greibach, S.A.: Remarks on the complexity of nondeterministic counter languages. Theoretical Computer Science **1**(4), 269–288 (1976). https://doi.org/10.1016/0304-3975(76)90072-4
13. Jecker, I.: 22.1 complexity of fixed vas reachability. https://autoboz.org/open-problems (2023), accessed: 2023-10-12
14. Kosaraju, S.R.: Decidability of reachability in vector addition systems (preliminary version). In: Symposium on Theory of Computing, STOC. pp. 267–281. ACM (1982). https://doi.org/10.1145/800070.802201
15. Leroux, J.: The reachability problem for petri nets is not primitive recursive. In: Annual Symposium on Foundations of Computer Science, FOCS. pp. 1241–1252. IEEE (2021). https://doi.org/10.1109/FOCS52979.2021.00121
16. Leroux, J., Schmitz, S.: Demystifying reachability in vector addition systems. In: Symposium on Logic in Computer Science, LICS. pp. 56–67. IEEE Computer Society (2015). https://doi.org/10.1109/LICS.2015.16

17. Leroux, J., Schmitz, S.: Reachability in vector addition systems is primitive-recursive in fixed dimension. In: Symposium on Logic in Computer Science (LICS). pp. 1–13 (2019). https://doi.org/10.1109/LICS.2019.8785796
18. Lipton, R.J.: The reachability problem requires exponential space. Research report (Yale University. Department of Computer Science), Department of Computer Science, Yale University (1976)
19. Mayr, E.W.: An algorithm for the general petri net reachability problem. SIAM Journal on Computing **13**(3), 441–460 (1984). https://doi.org/10.1137/0213029
20. Minsky, M.L.: Computation: Finite and Infinite Machines. Prentice-Hall, USA (1967)
21. Nguyen, D., Pak, I.: Short Presburger arithmetic is hard. SIAM Journal on Computing **51**(2), 17:1–30 (2022). https://doi.org/10.1137/17M1151146
22. Rosier, L.E., Yen, H.C.: A multiparameter analysis of the boundedness problem for vector addition systems. Journal of Computer and System Sciences **32**(1), 105–135 (1986). https://doi.org/10.1016/0022-0000(86)90006-1
23. Rytter, W.: A hardest language recognized by two-way nondeterministic pushdown automata. Information Processing Letters **13**(4), 145–146 (1981). https://doi.org/10.1016/0020-0190(81)90045-4
24. Schroeppel, R.: A two counter machine cannot calculate 2^N. Artificial Intelligence Memo 257, Massachusetts Institute of Technology (1972)
25. Schweikardt, N.: Arithmetic, first-order logic, and counting quantifiers. ACM Transactions on Computational Logic (TOCL) **6**(3), 634–671 (2005). https://doi.org/10.1145/1071596.1071602
26. Vardi, M.Y.: The complexity of relational query languages (extended abstract). In: Symposium on Theory of Computing, STOC. pp. 137–146. Association for Computing Machinery, New York, NY, USA (1982). https://doi.org/10.1145/800070.802186
27. Weispfenning, V.: The complexity of almost linear diophantine problems. Journal of Symbolic Computation **10**(5), 395–403 (1990). https://doi.org/10.1016/S0747-7171(08)80051-X

From Innermost to Full Almost-Sure Termination of Probabilistic Term Rewriting*

Jan-Christoph Kassing(✉), Florian Frohn(✉), and Jürgen Giesl(✉)

LuFG Informatik 2, RWTH Aachen University, Aachen, Germany
{kassing,florian.frohn}@cs.rwth-aachen.de,
giesl@informatik.rwth-aachen.de

Abstract. There are many evaluation strategies for term rewrite systems, but proving termination automatically is usually easiest for innermost rewriting. Several syntactic criteria exist when innermost termination implies full termination. We adapt these criteria to the probabilistic setting, e.g., we show when it suffices to analyze almost-sure termination (AST) w.r.t. innermost rewriting to prove full AST of probabilistic term rewrite systems. These criteria also apply to other notions of termination like positive AST. We implemented and evaluated our new contributions in the tool AProVE.

1 Introduction

Termination analysis is one of the main tasks in program verification, and techniques and tools to analyze termination of term rewrite systems (TRSs) automatically have been studied for decades. While a direct application of classical reduction orderings is often too weak, these orderings can be used successfully within the *dependency pair* (DP) framework [3, 20]. This framework allows for modular termination proofs by decomposing the original termination problem into sub-problems whose termination can then be analyzed independently using different techniques. Thus, DPs are used in essentially all current termination tools for TRSs (e.g., AProVE [21], MuTerm [25], NaTT [46], TTT2 [33]). To allow certification of termination proofs with DPs, they have been formalized in several proof assistants and there exist several corresponding certification tools for termination proofs with DPs (e.g., CeTA [43]).

On the other hand, *probabilistic* programs are used to describe randomized algorithms and probability distributions, with applications in many areas, see, e.g., [23]. To use TRSs also for such programs, *probabilistic term rewrite systems* (PTRSs) were introduced in [4, 9, 10]. In the probabilistic setting, there are several notions of "termination". In this paper, we mostly focus on analyzing *almost-sure termination* (AST), i.e., we want to prove automatically that the probability for termination is 1.

* funded by the Deutsche Forschungsgemeinschaft (DFG, German Research Foundation) - 235950644 (Project GI 274/6-2) and the DFG Research Training Group 2236 UnRAVeL

Supplementary Information The online version contains supplementary material available at https://doi.org/10.1007/978-3-031-57231-9_10.

N. Kobayashi and J. Worrell (Eds.): FoSSaCS 2024, LNCS 14575, pp. 206–228, 2024.
https://doi.org/10.1007/978-3-031-57231-9_10

While there exist many automatic approaches to prove (P)AST of imperative programs on numbers (e.g., [2, 5, 11, 16, 22, 26–28, 36–38, 40]), there are only few automatic approaches for programs with complex non-tail recursive structure [8, 12, 13]. The approaches that are also suitable for algorithms on recursive data structures [7, 35, 45] are mostly specialized for specific data structures and cannot easily be adjusted to other (possibly user-defined) ones, or are not yet fully automated.

For innermost AST (i.e., AST restricted to rewrite sequences where one only evaluates at innermost positions), we recently presented an adaption of the DP framework which allows us to benefit from a similar modularity as in the non-probabilistic setting [29, 32]. Unfortunately, there is no such modular powerful approach available for *full* AST (i.e., AST when considering arbitrary rewrite sequences). Up to now, full AST of PTRSs can only be proved via a direct application of orderings [4, 29], but there is no corresponding adaption of dependency pairs. (As explained in [29], a DP framework to analyze full instead of innermost AST would be "considerably more involved".) Indeed, also in the non-probabilistic setting, innermost termination is usually substantially easier to prove than full termination, see, e.g., [3, 20]. To lift innermost termination proofs to full rewriting, in the non-probabilistic setting, there exist several sufficient criteria which ensure that innermost termination implies full termination [24].

Up to now no such results were known in the probabilistic setting. Our paper presents the first sufficient criteria for PTRSs which ensure that AST coincide for full and innermost rewriting, and we also show similar results for other rewrite strategies like *leftmost-innermost* rewriting. We focus on criteria that can be checked automatically, so we can combine our results with the DP framework for proving innermost AST of PTRSs [29, 32]. In this way, we obtain a modular powerful technique that can also prove AST for *full* rewriting automatically.

We will also consider the stronger notion of *positive almost-sure termination* (PAST) [10, 42], which requires that the expected runtime is finite, and show that our criteria for the relationship between full and innermost probabilistic rewriting hold for PAST as well. In contrast to AST, PAST is not modular, i.e., the sequence of two programs that are PAST may yield a program that is not PAST (see, e.g., [27]). Therefore, up to now there is no variant of DPs that allows to prove PAST of PTRSs, but there only exist techniques to apply polynomial or matrix orderings directly [4].

We start with preliminaries on term rewriting in Sect. 2. Then we recapitulate PTRSs based on [4, 10, 14, 15, 29] in Sect. 3. In Sect. 4 we show that the properties of [24] that ensure equivalence of innermost and full termination do not suffice in the probabilistic setting and extend them accordingly. In particular, we show that innermost and full AST coincide for PTRSs that are non-overlapping and linear. This result also holds for PAST, as well as for strategies like leftmost-innermost evaluation. In Sect. 5 we show how to weaken the linearity requirement in order to prove full AST for larger classes of PTRSs. The implementation of our criteria in the tool AProVE is evaluated in Sect. 6. We refer to [30] for all proofs.

2 Preliminaries

We assume familiarity with term rewriting [6] and regard (possibly infinite) TRSs over a (possibly infinite) signature Σ and a set of variables $\mathcal{V}$. Consider the TRS $\mathcal{R}_{\mathsf{d}}$ that doubles a natural number (represented by the terms s and $\mathcal{O}$) with the rewrite rules $\mathsf{d}(\mathsf{s}(x)) \to \mathsf{s}(\mathsf{s}(\mathsf{d}(x)))$ and $\mathsf{d}(\mathcal{O}) \to \mathcal{O}$ as an example. A TRS $\mathcal{R}$ induces a *rewrite relation* $\to_{\mathcal{R}} \subseteq \mathcal{T}(\Sigma, \mathcal{V}) \times \mathcal{T}(\Sigma, \mathcal{V})$ on terms where $s \to_{\mathcal{R}} t$ holds if there is a position π, a rule $\ell \to r \in \mathcal{R}$, and a substitution σ such that $s|_\pi = \ell\sigma$ and $t = s[r\sigma]_\pi$. A rewrite step $s \to_{\mathcal{R}} t$ is an *innermost* rewrite step (denoted $s \xrightarrow{\mathsf{i}}_{\mathcal{R}} t$) if all proper subterms of the used redex $\ell\sigma$ are in normal form w.r.t. $\mathcal{R}$ (i.e., they do not contain redexes themselves and thus, they cannot be reduced with $\to_{\mathcal{R}}$). For example, we have $\mathsf{d}(\mathsf{s}(\mathsf{d}(\mathsf{s}(\mathcal{O})))) \xrightarrow{\mathsf{i}}_{\mathcal{R}_{\mathsf{d}}} \mathsf{d}(\mathsf{s}(\mathsf{s}(\mathsf{s}(\mathsf{d}(\mathcal{O})))))$.

Let $<$ be the prefix ordering on positions and let $\leq$ be its reflexive closure. Then for two parallel positions τ and π we define $\tau \prec \pi$ if we have $i < j$ for the unique i, j such that $\chi.i \leq \tau$ and $\chi.j \leq \pi$, where χ is the longest common prefix of τ and π. An innermost rewrite step $s \xrightarrow{\mathsf{i}}_{\mathcal{R}} t$ at position π is *leftmost* (denoted $s \xrightarrow{\mathsf{li}}_{\mathcal{R}} t$) if there exists no redex at a position τ with $\tau \prec \pi$.

We call a TRS $\mathcal{R}$ *strongly (innermost/leftmost innermost) normalizing* (SN / iSN / liSN) if $\to_{\mathcal{R}}$ ($\xrightarrow{\mathsf{i}}_{\mathcal{R}}$ / $\xrightarrow{\mathsf{li}}_{\mathcal{R}}$) is well founded. SN is also called "*terminating*" and iSN/liSN are called "*innermost/leftmost innermost terminating*". If every term $t \in \mathcal{T}(\Sigma, \mathcal{V})$ has a normal form (i.e., we have $t \to_{\mathcal{R}}^* t'$ where t' is in normal form) then we call $\mathcal{R}$ *weakly normalizing* (WN). Two terms s, t are *joinable* via $\mathcal{R}$ (denoted $s \downarrow_{\mathcal{R}} t$) if there exists a term w such that $s \to_{\mathcal{R}}^* w \leftarrow_{\mathcal{R}}^* t$. Two rules $\ell_1 \to r_1, \ell_2 \to r_2 \in \mathcal{R}$ with renamed variables such that $\mathcal{V}(\ell_1) \cap \mathcal{V}(\ell_2) = \varnothing$ are *overlapping* if there exists a non-variable position π of ℓ_1 such that $\ell_1|_\pi$ and ℓ_2 are unifiable with a mgu σ. If $(\ell_1 \to r_1) = (\ell_2 \to r_2)$, then we require that $\pi \neq \varepsilon$. $\mathcal{R}$ is *non-overlapping* (NO) if it has no overlapping rules. As an example, the TRS $\mathcal{R}_{\mathsf{d}}$ is non-overlapping. A TRS is *left-linear* (LL) (*right-linear*, RL) if every variable occurs at most once in the left-hand side (right-hand side) of a rule. A TRS is *linear* if it is both left- and right-linear. A TRS is *non-erasing* (NE) if in every rule, all variables of the left-hand side also occur in the right-hand side.

Next, we recapitulate the relations between iSN, SN, liSN, and WN in the non-probabilistic setting. We start with the relation between iSN and SN.

Counterexample 1 (Toyama's Counterexample [44]). The TRS $\mathcal{R}_1$ with the rules $\mathsf{f}(\mathsf{a}, \mathsf{b}, x) \to \mathsf{f}(x, x, x)$, $\mathsf{g} \to \mathsf{a}$, and $\mathsf{g} \to \mathsf{b}$ is not SN since we have $\mathsf{f}(\mathsf{a}, \mathsf{b}, \mathsf{g}) \to_{\mathcal{R}_1} \mathsf{f}(\mathsf{g}, \mathsf{g}, \mathsf{g}) \to_{\mathcal{R}_1} \mathsf{f}(\mathsf{a}, \mathsf{g}, \mathsf{g}) \to_{\mathcal{R}_1} \mathsf{f}(\mathsf{a}, \mathsf{b}, \mathsf{g}) \to_{\mathcal{R}_1} \ldots$ But the only innermost rewrite sequences starting with $\mathsf{f}(\mathsf{a}, \mathsf{b}, \mathsf{g})$ are $\mathsf{f}(\mathsf{a}, \mathsf{b}, \mathsf{g}) \xrightarrow{\mathsf{i}}_{\mathcal{R}_1} \mathsf{f}(\mathsf{a}, \mathsf{b}, \mathsf{a}) \xrightarrow{\mathsf{i}}_{\mathcal{R}_1} \mathsf{f}(\mathsf{a}, \mathsf{a}, \mathsf{a})$ and $\mathsf{f}(\mathsf{a}, \mathsf{b}, \mathsf{g}) \xrightarrow{\mathsf{i}}_{\mathcal{R}_1} \mathsf{f}(\mathsf{a}, \mathsf{b}, \mathsf{b}) \xrightarrow{\mathsf{i}}_{\mathcal{R}_1} \mathsf{f}(\mathsf{b}, \mathsf{b}, \mathsf{b})$, i.e., both reach normal forms in the end. Thus, $\mathcal{R}_1$ is iSN as we have to rewrite the inner g before we can use the f-rule.

The first property known to ensure equivalence of SN and iSN is orthogonality. A TRS is *orthogonal* if it is non-overlapping and left-linear.

Theorem 2 (From iSN to SN (1), [41]). *If a TRS $\mathcal{R}$ is orthogonal, then $\mathcal{R}$ is SN iff $\mathcal{R}$ is iSN.*

Then, in [24] it was shown that one can remove the left-linearity requirement.

Theorem 3 (From iSN to SN (2), [24]). *If a TRS $\mathcal{R}$ is non-overlapping, then $\mathcal{R}$ is SN iff $\mathcal{R}$ is iSN.*

Finally, [24] also refined Thm. 3 further. A TRS $\mathcal{R}$ is an *overlay system* (OS) if its rules may only overlap at the root position, i.e., $\pi = \varepsilon$. For Ex. 1 one can see that the overlaps occur at non-root positions, i.e., $\mathcal{R}_1$ is not an overlay system. Furthermore, a TRS is *locally confluent* (or *weakly Church-Rosser*, abbreviated WCR) if for all terms s, t_1, t_2 such that $t_1 \; {}_{\mathcal{R}}\!\!\leftarrow s \to_{\mathcal{R}} t_2$ the terms t_1 and t_2 are joinable. So $\mathcal{R}_1$ is *not* WCR, as we have $\mathsf{f}(\mathsf{a},\mathsf{b},\mathsf{a}) \; {}_{\mathcal{R}_1}\!\!\leftarrow \mathsf{f}(\mathsf{a},\mathsf{b},\mathsf{g}) \to_{\mathcal{R}_1} \mathsf{f}(\mathsf{a},\mathsf{b},\mathsf{b})$, but $\mathsf{f}(\mathsf{a},\mathsf{b},\mathsf{a}) \not\downarrow_{\mathcal{R}_1} \mathsf{f}(\mathsf{a},\mathsf{b},\mathsf{b})$. If a TRS has both of these properties, then iSN and SN are again equivalent.

Theorem 4 (From iSN to SN (3), [24]). *If a TRS $\mathcal{R}$ is a locally confluent overlay system, then $\mathcal{R}$ is SN iff $\mathcal{R}$ is iSN.*

Thm. 4 is stronger than Thm. 3 as every non-overlapping TRS is a locally confluent overlay system. We recapitulate the relation between WN and SN next.

Counterexample 5. Consider the TRS $\mathcal{R}_2$ with the rules $\mathsf{f}(x) \to \mathsf{b}$ and $\mathsf{a} \to \mathsf{f}(\mathsf{a})$. This TRS is not SN since we can always rewrite the inner a to get $\mathsf{a} \to_{\mathcal{R}_2} \mathsf{f}(\mathsf{a}) \to_{\mathcal{R}_2} \mathsf{f}(\mathsf{f}(\mathsf{a})) \to_{\mathcal{R}_2} \ldots$, but it is WN since we can also rewrite the outer $\mathsf{f}(\ldots)$ before we use the a-rule twice, resulting in the term b, which is a normal form. For the TRS $\mathcal{R}_3$ with the rules $\mathsf{f}(\mathsf{a}) \to \mathsf{b}$ and $\mathsf{a} \to \mathsf{f}(\mathsf{a})$, the situation is similar.

The TRS $\mathcal{R}_2$ from Ex. 5 is erasing and $\mathcal{R}_3$ is overlapping. For TRSs with neither of those two properties, SN and WN are equivalent.

Theorem 6 (From WN to SN [24]). *If a TRS $\mathcal{R}$ is non-overlapping and non-erasing, then $\mathcal{R}$ is SN iff $\mathcal{R}$ is WN.*

Finally, we look at the difference between rewrite strategies that use an ordering for parallel redexes like leftmost innermost rewriting compared to just innermost rewriting. It turns out that such an ordering does not interfere with termination at all.

Theorem 7 (From liSN to iSN [34]). *For all TRSs $\mathcal{R}$ we have that $\mathcal{R}$ is iSN iff $\mathcal{R}$ is liSN.*

The relations between the different properties for non-probabilistic TRSs (given in Thm. 4, 6, and 7) are summarized below.

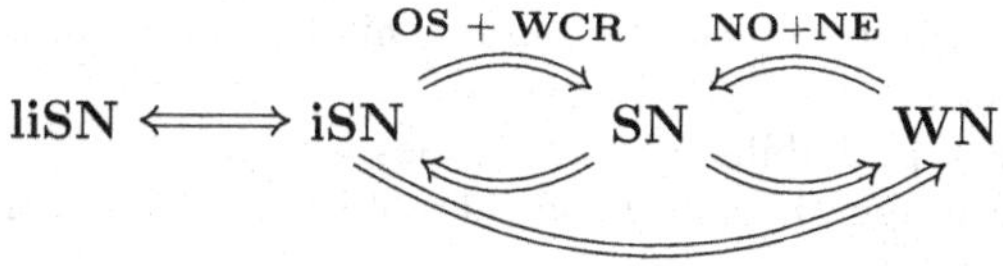

3 Probabilistic Term Rewriting

In this section, we recapitulate *probabilistic TRSs* [4, 10, 29]. In contrast to TRSs, a PTRS has finite multi-distributions[1] on the right-hand sides of its rewrite rules.[2] A finite *multi-distribution* μ on a set $A \neq \varnothing$ is a finite multiset of pairs $(p : a)$, where $0 < p \leq 1$ is a probability and $a \in A$, such that $\sum_{(p:a)\in\mu} p = 1$. $\mathrm{FDist}(A)$ is the set of all finite multi-distributions on A. For $\mu \in \mathrm{FDist}(A)$, its *support* is the multiset $\mathrm{Supp}(\mu) = \{a \mid (p : a) \in \mu \text{ for some } p\}$. A *probabilistic rewrite rule* is a pair $\ell \to \mu \in \mathcal{T}(\Sigma, \mathcal{V}) \times \mathrm{FDist}(\mathcal{T}(\Sigma, \mathcal{V}))$ such that $\ell \notin \mathcal{V}$ and $\mathcal{V}(r) \subseteq \mathcal{V}(\ell)$ for every $r \in \mathrm{Supp}(\mu)$. A *probabilistic TRS* (PTRS) is a (possibly infinite) set $\mathcal{S}$ of probabilistic rewrite rules. Similar to TRSs, the PTRS $\mathcal{S}$ induces a *rewrite relation* $\to_{\mathcal{S}} \subseteq \mathcal{T}(\Sigma, \mathcal{V}) \times \mathrm{FDist}(\mathcal{T}(\Sigma, \mathcal{V}))$ where $s \to_{\mathcal{S}} \{p_1 : t_1, \ldots, p_k : t_k\}$ if there is a position π, a rule $\ell \to \{p_1 : r_1, \ldots, p_k : r_k\} \in \mathcal{S}$, and a substitution σ such that $s|_\pi = \ell\sigma$ and $t_j = s[r_j\sigma]_\pi$ for all $1 \leq j \leq k$. We call $s \to_{\mathcal{S}} \mu$ an *innermost* rewrite step (denoted $s \xrightarrow{\mathrm{i}}_{\mathcal{S}} \mu$) if all proper subterms of the used redex $\ell\sigma$ are in normal form w.r.t. $\mathcal{S}$. We have $s \xrightarrow{\mathrm{li}}_{\mathcal{S}} \mu$ if the rewrite step $s \xrightarrow{\mathrm{i}}_{\mathcal{S}} \mu$ at position π is leftmost (i.e., there is no redex at a position τ with $\tau \prec \pi$). For example, the PTRS $\mathcal{S}_{\mathsf{rw}}$ with the only rule $\mathsf{g} \to \{1/2 : \mathsf{c}(\mathsf{g}, \mathsf{g}),\ 1/2 : \bot\}$ corresponds to a symmetric random walk on the number of g-symbols in a term.

As in [4, 14, 15, 29], we *lift* $\to_{\mathcal{S}}$ to a rewrite relation between multi-distributions in order to track all probabilistic rewrite sequences (up to non-determinism) at once. For any $0 < p \leq 1$ and any $\mu \in \mathrm{FDist}(A)$, let $p \cdot \mu = \{(p \cdot q : a) \mid (q : a) \in \mu\}$.

Definition 8 (Lifting). *The* lifting $\rightrightarrows \subseteq \mathrm{FDist}(\mathcal{T}(\Sigma, \mathcal{V})) \times \mathrm{FDist}(\mathcal{T}(\Sigma, \mathcal{V}))$ *of a relation* $\to \subseteq \mathcal{T}(\Sigma, \mathcal{V}) \times \mathrm{FDist}(\mathcal{T}(\Sigma, \mathcal{V}))$ *is the smallest relation with:*

- *If* $t \in \mathcal{T}(\Sigma, \mathcal{V})$ *is in normal form w.r.t.* $\to$, *then* $\{1 : t\} \rightrightarrows \{1 : t\}$.
- *If* $t \to \mu$, *then* $\{1 : t\} \rightrightarrows \mu$.
- *If for all* $1 \leq j \leq k$ *there are* $\mu_j, \nu_j \in \mathrm{FDist}(\mathcal{T}(\Sigma, \mathcal{V}))$ *with* $\mu_j \rightrightarrows \nu_j$ *and* $0 < p_j \leq 1$ *with* $\sum_{1 \leq j \leq k} p_j = 1$, *then* $\biguplus_{1 \leq j \leq k} p_j \cdot \mu_j \rightrightarrows \biguplus_{1 \leq j \leq k} p_j \cdot \nu_j$.

For a PTRS $\mathcal{S}$, we write $\rightrightarrows_{\mathcal{S}}$, $\xrightarrow{\mathrm{i}}\!\!\!\rightrightarrows_{\mathcal{S}}$, and $\xrightarrow{\mathrm{li}}\!\!\!\rightrightarrows_{\mathcal{S}}$ for the liftings of $\to_{\mathcal{S}}$, $\xrightarrow{\mathrm{i}}_{\mathcal{S}}$, and $\xrightarrow{\mathrm{li}}_{\mathcal{S}}$, respectively.

Example 9. For example, we obtain the following $\rightrightarrows_{\mathcal{S}_{\mathsf{rw}}}$-rewrite sequence (which is also a $\overset{\mathrm{i}}{\rightrightarrows}_{\mathcal{S}_{\mathsf{rw}}}$-sequence, but not a $\overset{\mathrm{li}}{\rightrightarrows}_{\mathcal{S}_{\mathsf{rw}}}$-sequence).

$$
\begin{aligned}
& \{1 : \mathsf{g}\} \\
\rightrightarrows_{\mathcal{S}_{\mathsf{rw}}} & \{1/2 : \mathsf{c}(\mathsf{g}, \mathsf{g}), 1/2 : \bot\} \\
\rightrightarrows_{\mathcal{S}_{\mathsf{rw}}} & \{1/4 : \mathsf{c}(\mathsf{c}(\mathsf{g}, \mathsf{g}), \mathsf{g}), 1/4 : \mathsf{c}(\bot, \mathsf{g}), 1/2 : \bot\} \\
\rightrightarrows_{\mathcal{S}_{\mathsf{rw}}} & \{1/8 : \mathsf{c}(\mathsf{c}(\mathsf{g}, \mathsf{g}), \mathsf{c}(\mathsf{g}, \mathsf{g})), 1/8 : \mathsf{c}(\mathsf{c}(\mathsf{g}, \mathsf{g}), \bot), 1/8 : \mathsf{c}(\bot, \mathsf{c}(\mathsf{g}, \mathsf{g})), 1/8 : \mathsf{c}(\bot, \bot), 1/2 : \bot\}
\end{aligned}
$$

---

[1] The restriction to finite multi-distributions allows us to simplify the handling of PTRSs in the proofs.

[2] A different form of probabilistic rewrite rules was proposed in PMaude [1], where numerical extra variables in right-hand sides of rules are instantiated according to a probability distribution.

To express the concept of almost-sure termination, one has to determine the probability for normal forms in a multi-distribution.

Definition 10 ($|\mu|_{\mathcal{S}}$). *For a PTRS $\mathcal{S}$, $\mathtt{NF}_{\mathcal{S}} \subseteq \mathcal{T}(\Sigma, \mathcal{V})$ denotes the set of all* normal forms *w.r.t. $\mathcal{S}$. For any $\mu \in \mathrm{FDist}(\mathcal{T}(\Sigma, \mathcal{V}))$, let $|\mu|_{\mathcal{S}} = \sum_{(p:t)\in\mu, t\in\mathtt{NF}_{\mathcal{S}}} p$.*

Example 11. Consider $\{1/8 : \mathsf{c}(\mathsf{c}(\mathsf{g},\mathsf{g}),\mathsf{c}(\mathsf{g},\mathsf{g})), 1/8 : \mathsf{c}(\mathsf{c}(\mathsf{g},\mathsf{g}),\bot), 1/8 : \mathsf{c}(\bot,\mathsf{c}(\mathsf{g},\mathsf{g})), 1/8 : \mathsf{c}(\bot,\bot), 1/2 : \bot\} = \mu$ from Ex. 9. Then $|\mu|_{\mathcal{S}_{\mathsf{rw}}} = 1/8 + 1/2 = 5/8$, since $\mathsf{c}(\bot,\bot)$ and $\bot$ are both normal forms w.r.t. $\mathcal{S}_{\mathsf{rw}}$.

Definition 12 (AST). *Let $\mathcal{S}$ be a PTRS and $\vec{\mu} = (\mu_n)_{n\in\mathbb{N}}$ be an infinite $\rightrightarrows_{\mathcal{S}}$-rewrite sequence, i.e., $\mu_n \rightrightarrows_{\mathcal{S}} \mu_{n+1}$ for all $n \in \mathbb{N}$. We say that $\vec{\mu}$* converges *with probability $\lim\limits_{n\to\infty} |\mu_n|_{\mathcal{S}}$. $\mathcal{S}$ is* almost-surely terminating (AST) *(*innermost AST (iAST) */* leftmost innermost AST (liAST)*) if $\lim\limits_{n\to\infty} |\mu_n|_{\mathcal{S}} = 1$ holds for every infinite $\rightrightarrows_{\mathcal{S}}$- ($\overset{\mathsf{i}}{\rightrightarrows}_{\mathcal{S}}$- / $\overset{\mathsf{li}}{\rightrightarrows}_{\mathcal{S}}$-) rewrite sequence $(\mu_n)_{n\in\mathbb{N}}$. To highlight the consideration of AST for* full *(instead of innermost) rewriting, we also speak of* full AST *(fAST) instead of "AST". We say that $\mathcal{S}$ is* weakly AST *(wAST) if for every term t there exists an infinite $\rightrightarrows_{\mathcal{S}}$-rewrite sequence $(\mu_n)_{n\in\mathbb{N}}$ with $\lim\limits_{n\to\infty} |\mu_n|_{\mathcal{S}} = 1$ and $\mu_0 = \{1 : t\}$.*

Example 13. For every infinite extension $(\mu_n)_{n\in\mathbb{N}}$ of the $\rightrightarrows_{\mathcal{S}_{\mathsf{rw}}}$-rewrite sequence in Ex. 9, we have $\lim\limits_{n\to\infty} |\mu_n|_{\mathcal{S}} = 1$. Indeed, $\mathcal{S}_{\mathsf{rw}}$ is fAST and thus also iAST, liAST, and wAST.

Next, we define *positive* almost-sure termination that considers the *expected derivation length* $\mathrm{edl}(\vec{\mu})$ of a rewrite sequence $\vec{\mu}$, i.e., the expected number of steps until one reaches a normal form. For PAST, we require that the expected derivation lengths of all possible rewrite sequences are finite. In the following definition, $(1 - |\mu_n|_{\mathcal{S}})$ is the probability of terms that are *not* in normal form w.r.t. $\mathcal{S}$ after the n-th step.

Definition 14 (edl, **PAST**). *Let $\mathcal{S}$ be a PTRS and $\vec{\mu} = (\mu_n)_{n\in\mathbb{N}}$ be an infinite $\rightrightarrows_{\mathcal{S}}$-rewrite sequence. By $\mathrm{edl}(\vec{\mu}) = \sum_{n=0}^{\infty}(1 - |\mu_n|_{\mathcal{S}})$ we denote the* expected derivation length *of $\vec{\mu}$. $\mathcal{S}$ is* positively almost-surely terminating (PAST) *(*innermost PAST (iPAST) */* leftmost innermost AST (liPAST)*) if $\mathrm{edl}(\vec{\mu})$ is finite for every infinite $\rightrightarrows_{\mathcal{S}}$- ($\overset{\mathsf{i}}{\rightrightarrows}_{\mathcal{S}}$- / $\overset{\mathsf{li}}{\rightrightarrows}_{\mathcal{S}}$-) rewrite sequence $\vec{\mu} = (\mu_n)_{n\in\mathbb{N}}$.[3] Again, we also speak of* full PAST *(fPAST) when considering PAST for the* full *rewrite relation $\rightrightarrows_{\mathcal{S}}$. We say that $\mathcal{S}$ is* weakly PAST *(wPAST) if for every term t there exists an infinite $\rightrightarrows_{\mathcal{S}}$-rewrite sequence $\vec{\mu} = (\mu_n)_{n\in\mathbb{N}}$ such that $\mathrm{edl}(\vec{\mu})$ is finite and $\mu_0 = \{1 : t\}$.*

It is well known that PAST implies AST, but not vice versa.

Example 15. For every infinite extension $\vec{\mu} = (\mu_n)_{n\in\mathbb{N}}$ of the $\rightrightarrows_{\mathcal{S}_{\mathsf{rw}}}$-rewrite sequence in Ex. 9, the expected derivation length $\mathrm{edl}(\vec{\mu})$ is infinite, hence $\mathcal{S}_{\mathsf{rw}}$ is not PAST w.r.t. any of the strategies regarded in this paper.

[3] This definition is from [4], where it is also explained why this definition of PAST is equivalent to the one of, e.g., [10].

In [4, 18], PAST was strengthened further to *bounded* or *strong almost-sure termination* (SAST). Indeed, our results on PAST can also be adapted to SAST (see [30]).

Many properties of TRSs from Sect. 2 can be lifted to PTRSs in a straightforward way: A PTRS $\mathcal{S}$ is right-linear (non-erasing) iff the TRS $\{\ell \to r \mid \ell \to \mu \in \mathcal{S}, r \in \mathrm{Supp}(\mu)\}$ has the respective property. Moreover, all properties that just consider the left-hand sides, e.g., left-linearity, being non-overlapping, orthogonality, and being an overlay system, can be lifted to PTRSs directly as well, since their rules again only have a single left-hand side.

4 Relating Variants of AST

Our goal is to relate AST of full rewriting to restrictions of fAST, i.e., to iAST (Sect. 4.1), wAST (Sect. 4.2), and liAST (Sect. 4.3). More precisely, we want to find properties of PTRSs which are suitable for automated checking and which guarantee that two variants of AST are equivalent. Then for example, we can use existing tools that analyze iAST in order to prove fAST. Clearly, we have to impose at least the same requirements as in the non-probabilistic setting, as every TRS $\mathcal{R}$ can be transformed into a PTRS $\mathcal{S}$ by replacing every rule $\ell \to r$ with $\ell \to \{1 : r\}$. Then $\mathcal{R}$ is SN / iSN / liSN iff $\mathcal{S}$ is fAST / iAST / liAST. While we mostly focus on AST, all results and counterexamples in this section also hold for PAST.

4.1 From iAST to fAST

Again, we start by analyzing the relation between iAST and fAST. The following example shows that Thm. 2 does not carry over to the probabilistic setting, i.e., orthogonality is not sufficient to ensure that iAST implies fAST.

Counterexample 16 (Orthogonality Does Not Suffice). Consider the orthogonal PTRS $\mathcal{S}_1$ with the two rules:

$$\mathsf{g} \to \{3/4 : \mathsf{d}(\mathsf{g}), 1/4 : \bot\} \qquad\qquad \mathsf{d}(x) \to \{1 : \mathsf{c}(x, x)\}$$

This PTRS is not fAST (and thus, also not fPAST), as we have $\{1 : \mathsf{g}\} \rightrightarrows^2_{\mathcal{S}_1} \{3/4 : \mathsf{c}(\mathsf{g}, \mathsf{g}), 1/4 : \bot\}$, which corresponds to a random walk biased towards non-termination (since $\frac{3}{4} > \frac{1}{2}$).

However, the d-rule can only duplicate normal forms in innermost evaluations. To see that $\mathcal{S}_1$ is iPAST (and thus, also iAST), consider the following rewrite sequence $\vec{\mu}$:

$$\{1 : \mathsf{g}\} \overset{\mathrm{i}}{\rightrightarrows}_{\mathcal{S}_1} \{3/4 : \mathsf{d}(\mathsf{g}), 1/4 : \bot\} \overset{\mathrm{i}}{\rightrightarrows}_{\mathcal{S}_1} \{(3/4)^2 : \mathsf{d}(\mathsf{d}(\mathsf{g})), 1/4 \cdot 3/4 : \mathsf{d}(\bot), 1/4 : \bot\} \overset{\mathrm{i}}{\rightrightarrows}_{\mathcal{S}_1} \cdots$$

We can also view this rewrite sequence as a tree:

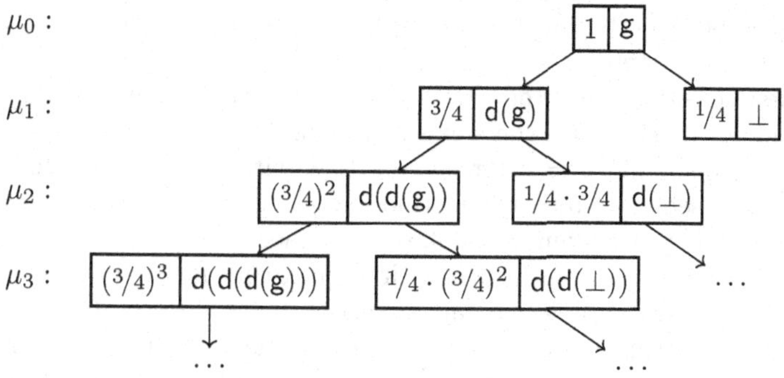

The branch to the right that starts with $\bot$ stops after 0 innermost steps, the branch that starts with $\mathsf{d}(\bot)$ stops after 1 innermost steps, the branch that starts with $\mathsf{d}(\mathsf{d}(\bot))$ stops after 2 innermost steps, and so on. So if we start with the term $\mathsf{d}^n(\bot)$, then we reach a normal form after n steps, and we reach $\mathsf{d}^n(\bot)$ after $n+1$ steps from the initial term g, where $\mathsf{d}^n(\bot) = \underbrace{\mathsf{d}(\ldots(\mathsf{d}}_{n\text{-times}}(\bot))\ldots)$. Hence, for every $k \in \mathbb{N}$ we have $|\mu_{2\cdot k+1}|_{\mathcal{S}_1} = |\mu_{2\cdot k+2}|_{\mathcal{S}_1} = \sum_{n=0}^{k} {}^1\!/\!_4 \cdot ({}^3\!/\!_4)^n$ and thus

$$\begin{aligned} \operatorname{edl}(\vec{\mu}) &= \textstyle\sum_{n=0}^{\infty}(1 - |\mu_n|_{\mathcal{S}_1}) &&= 1 + 2 \cdot \textstyle\sum_{k\in\mathbb{N}}(1 - |\mu_{2\cdot k+1}|_{\mathcal{S}_1}) \\ &= 1 + 2 \cdot \textstyle\sum_{k\in\mathbb{N}}(1 - \sum_{n=0}^{k} {}^1\!/\!_4 \cdot ({}^3\!/\!_4)^n) &&= 1 + 2 \cdot \textstyle\sum_{k\in\mathbb{N}}({}^3\!/\!_4)^{k+1} \\ &= (2 \cdot \textstyle\sum_{k\in\mathbb{N}}({}^3\!/\!_4)^k) - 1 &&= 7 \end{aligned}$$

Analogously, in all other innermost rewrite sequences, the d-rule can also only duplicate normal forms. Thus, all possible innermost rewrite sequences have finite expected derivation length. Therefore, $\mathcal{S}_1$ is iPAST and thus, also iAST. The latter can also be proved automatically by our implementation of the probabilistic DP framework for iAST [29] in AProVE.

To construct a counterexample for AST of $\mathcal{S}_1$, we exploited the fact that $\mathcal{S}_1$ is not right-linear. Indeed, requiring right-linearity yields our desired result. For reasons of space, here we only give a proof sketch. As mentioned, all full proofs can be found in [30].

Theorem 17 (From iAST/iPAST to fAST/fPAST (1)). *If a PTRS $\mathcal{S}$ is orthogonal and right-linear (i.e., non-overlapping and linear), then:*

$$\mathcal{S} \textit{ is fAST} \Longleftrightarrow \mathcal{S} \textit{ is iAST}$$
$$\mathcal{S} \textit{ is fPAST} \Longleftrightarrow \mathcal{S} \textit{ is iPAST}$$

Proof Sketch. We only have to prove the non-trivial direction "$\Longleftarrow$". The proofs for all theorems in this section (for both AST and PAST) follow a similar structure. We always iteratively replace rewrite steps by steps that use the desired strategy and ensure that this does not increase the probability of termination (resp. the expected derivation length). For this replacement, we lift the corresponding construction from the non-probabilistic to the probabilistic setting. However, this cannot be done directly but instead, we have to regard the "limit" of a sequence of transformation steps.

We first consider fAST and iAST. Let $\mathcal{S}$ be a PTRS that is non-overlapping, linear, and not fAST. Thus, there exists an infinite rewrite sequence $\vec{\mu} = (\mu_n)_{n\in\mathbb{N}}$ such that $\lim_{n\to\infty} |\mu_n|_{\mathcal{S}} = c$ for some $c \in \mathbb{R}$ with $0 \le c < 1$. Our goal is to transform this sequence into an innermost sequence that converges at most with probability c. If the sequence is not yet an innermost one, then in $(\mu_n)_{n\in\mathbb{N}}$ at least one rewrite step is performed with a redex that is not an innermost redex. Since $\mathcal{S}$ is non-overlapping, we can replace a first such non-innermost rewrite step with an innermost rewrite step using a similar construction as in the non-probabilistic setting. In this way, we result in a rewrite sequence $\vec{\mu}^{(1)} = (\mu_n^{(1)})_{n\in\mathbb{N}}$ with $\lim_{n\to\infty} |\mu_n^{(1)}|_{\mathcal{S}} = \lim_{n\to\infty} |\mu_n|_{\mathcal{S}} = c$. Here, linearity is needed to ensure that the probability of termination does not increase during this replacement. We can then repeat this replacement for every non-innermost rewrite step, i.e., we again replace a first non-innermost rewrite step in $(\mu_n^{(1)})_{n\in\mathbb{N}}$ to obtain $(\mu_n^{(2)})_{n\in\mathbb{N}}$ with the same termination probability, etc. In the end, the limit of all these rewrite sequences $\lim_{i\to\infty}(\mu_n^{(i)})_{n\in\mathbb{N}}$ is an innermost rewrite sequence that converges with probability at most $c < 1$, and hence, the PTRS $\mathcal{S}$ is not innermost AST.

For fPAST and iPAST, we start with an infinite rewrite sequence $\vec{\mu}$ such that $\mathrm{edl}(\vec{\mu}) = \infty$. Again, we replace the first non-innermost rewrite step with an innermost rewrite step using exactly the same construction as before to obtain $\vec{\mu}^{(1)}$, etc., since $\vec{\mu}^{(1)}$ does not only have the same termination probability as $\vec{\mu}$, but we also have $\mathrm{edl}(\vec{\mu}^{(1)}) \ge \mathrm{edl}(\vec{\mu})$. In the end, the limit of all these rewrite sequences $\lim_{i\to\infty} \vec{\mu}^{(i)}$ is an innermost rewrite sequence such that $\mathrm{edl}(\lim_{i\to\infty} \vec{\mu}^{(i)}) \ge \mathrm{edl}(\vec{\mu}) = \infty$, and hence, the PTRS $\mathcal{S}$ is not innermost PAST. □

One may wonder whether we can remove the left-linearity requirement from Thm. 17, as in the non-probabilistic setting. It turns out that this is not possible.

Counterexample 18 (Left-Linearity Cannot be Removed). Consider the PTRS $\mathcal{S}_2$ with the rules:

$$\mathsf{f}(x,x) \to \{1 : \mathsf{f}(\mathsf{a},\mathsf{a})\} \qquad \mathsf{a} \to \{1/2 : \mathsf{b}, 1/2 : \mathsf{c}\}$$

$\mathcal{S}_2$ is not fAST (hence also not fPAST), since $\{1 : \mathsf{f}(\mathsf{a},\mathsf{a})\} \rightrightarrows_{\mathcal{S}_2} \{1 : \mathsf{f}(\mathsf{a},\mathsf{a})\} \rightrightarrows_{\mathcal{S}_2} \cdots$ is an infinite rewrite sequence that converges with probability 0. However, it is iPAST (and hence, iAST) since the corresponding innermost sequence has the form $\{1 : \mathsf{f}(\mathsf{a},\mathsf{a})\} \overset{\mathrm{i}}{\rightrightarrows}_{\mathcal{S}_2} \{\frac{1}{2} : \mathsf{f}(\mathsf{b},\mathsf{a}), \frac{1}{2} : \mathsf{f}(\mathsf{c},\mathsf{a})\} \overset{\mathrm{i}}{\rightrightarrows}_{\mathcal{S}_2} \{\frac{1}{4} : \mathsf{f}(\mathsf{b},\mathsf{b}), \frac{1}{4} : \mathsf{f}(\mathsf{b},\mathsf{c}), \frac{1}{4} : \mathsf{f}(\mathsf{c},\mathsf{b}), \frac{1}{4} : \mathsf{f}(\mathsf{c},\mathsf{c})\}$. Here, the last distribution contains two normal forms $\mathsf{f}(\mathsf{b},\mathsf{c})$ and $\mathsf{f}(\mathsf{c},\mathsf{b})$ that did not occur in the previous rewrite sequence. Since all innermost rewrite sequences keep on adding such normal forms after a certain number of steps for each start term, they always have finite expected derivation length and thus, converge with probability 1 (again, iAST can be shown automatically by AProVE). Note that adding the requirement of being non-erasing would not help to get rid of the left-linearity either, as shown by the PTRS $\mathcal{S}_3$ which results from $\mathcal{S}_2$ by replacing the f-rule with $\mathsf{f}(x,x) \to \{1 : \mathsf{d}(\mathsf{f}(\mathsf{a},\mathsf{a}),x)\}$.

The problem here is that although we rewrite both occurrences of a with the same rewrite rule, the two a-symbols are replaced by two different terms (each with a probability > 0). This is impossible in the non-probabilistic setting.

Next, one could try to adapt Thm. 4 to the probabilistic setting (when requiring linearity in addition). So one could investigate whether iAST implies fAST for PTRSs that are linear locally confluent overlay systems. A PTRS $\mathcal{S}$ is *locally confluent* if for all multi-distributions μ, μ_1, μ_2 such that $\mu_1 \leftleftarrows_{\mathcal{S}} \mu \rightrightarrows_{\mathcal{S}} \mu_2$, there exists a multi-distribution μ' such that $\mu_1 \rightrightarrows_{\mathcal{S}}^* \mu' \leftleftarrows_{\mathcal{S}}^* \mu_2$, see [14]. Note that in contrast to the probabilistic setting, there are non-overlapping PTRSs that are not locally confluent (e.g., the variant $\mathcal{S}_2'$ of $\mathcal{S}_2$ that consists of the rules $\mathsf{f}(x, x) \to \{1 : \mathsf{d}\}$ and $\mathsf{a} \to \{1/2 : \mathsf{b}, 1/2 : \mathsf{c}\}$, since we have $\{1 : \mathsf{d}\} \leftleftarrows_{\mathcal{S}_2'} \{1 : \mathsf{f}(\mathsf{a}, \mathsf{a})\} \rightrightarrows_{\mathcal{S}_2'} \{1/2 : \mathsf{f}(\mathsf{b}, \mathsf{a}), 1/2 : \mathsf{f}(\mathsf{c}, \mathsf{a})\}$ and the two resulting multi-distributions are not joinable). Thus, such an adaption of Thm. 4 would not subsume Thm. 17.

In contrast to the proof of Thm. 2, the proof of Thm. 4 relies on a minimality requirement for the used redex. In the non-probabilistic setting, whenever a term t starts an infinite rewrite sequence, then there exists a position π of t such that there is an infinite rewrite sequence of t starting with the redex $t|_\pi$, but no infinite rewrite sequence of t starting with a redex at a position $\tau > \pi$ which is strictly below π. In other words, if t starts an infinite rewrite sequence, then there is a "minimal" infinite rewrite sequence starting in t, i.e., as soon as one reduces a proper subterm of one of the redexes in the sequence, then one obtains a term which is terminating. However, such minimal infinite sequences do not always exist in the probabilistic setting.

Example 19 (No Minimal Infinite Rewrite Sequence for AST). Reconsider the PTRS $\mathcal{S}_1$ from Ex. 16, which is not fAST. However, there is no "minimal" rewrite sequence with convergence probability < 1 such that one rewrite step at a proper subterm of a redex would modify the multi-distribution in such a way that now only rewrite sequences with convergence probability 1 are possible. We have $\{1 : \mathsf{g}\} \rightrightarrows_{\mathcal{S}_1} \{3/4 : \mathsf{d}(\mathsf{g}), 1/4 : \bot\}$. In Ex. 16, we now alternated between the d- and the g-rule, resulting in a biased random walk, i.e., we obtained $\{3/4 : \mathsf{d}(\mathsf{g}), 1/4 : \bot\} \rightrightarrows_{\mathcal{S}_1} \{3/4 : \mathsf{c}(\mathsf{g}, \mathsf{g}), 1/4 : \bot\} \rightrightarrows_{\mathcal{S}_1} \{3/4 : \mathsf{c}(\mathsf{d}(\mathsf{g}), \mathsf{g}), 1/4 : \bot\} \rightrightarrows_{\mathcal{S}_1} \ldots$ The steps with the d-rule use redexes that have g as a proper subterm.

However, there does not exist any "minimal" non-fAST sequence. If we rewrite the proper subterm g of a redex $\mathsf{d}(\mathsf{g})$, then this still yields a multi-distribution that is not fAST, i.e., it can still start a rewrite sequence with convergence probability < 1. For example, we have $\{3/4 : \mathsf{d}(\mathsf{g}), 1/4 : \bot\} \rightrightarrows_{\mathcal{S}_1} \{(3/4)^2 : \mathsf{d}(\mathsf{d}(\mathsf{g})), 1/4 \cdot 3/4 : \mathsf{d}(\bot), 1/4 : \bot\}$, but the obtained multi-distribution still contains the subterm g, and thus, one can still continue the rewrite sequence in such a way that its convergence probability is < 1. Again, the same example also shows that there is no "minimal" non-fPAST sequence.

It remains open whether one can also adapt Thm. 4 to the probabilistic setting (e.g., if one can replace non-overlappingness in Thm. 17 by the requirement of locally confluent overlay systems). There are two main difficulties when trying to adapt the proof of this theorem to PTRSs. First, the minimality requirement cannot be imposed in the probabilistic setting, as discussed above. In the non-probabilistic setting, this requirement is needed to ensure that rewriting below a position that was reduced in the original (minimal) infinite rewrite sequence leads to a strongly normalizing rewrite sequence. Second, the original proof of Thm. 4 uses Newman's Lemma [39] which states that local confluence implies confluence for strongly normalizing terms t, and thus it implies that t has a unique normal form. Local confluence and adaptions of the unique normal form property for the probabilistic setting have been studied in [14, 15], which concluded that obtaining an analogous statement to Newman's Lemma for PTRSs that are AST (or PAST) would be very difficult. The reason is that one cannot use well-founded induction on the length of a rewrite sequence of a PTRS that is AST (or PAST), since these rewrite sequences may be infinite.

4.2 From wAST to fAST

Next, we investigate wAST. Since iAST implies wAST, we essentially have the same problems as for innermost AST, i.e., in addition to non-overlappingness, we need linearity, as seen in Ex. 16 and 18, as $\mathcal{S}_1$ and $\mathcal{S}_3$ are iAST (and hence wAST) but not fAST, while they are non-overlapping and non-erasing, but not linear. Furthermore, we need non-erasingness as we did in the non-probabilistic setting for the same reasons, see Ex. 5.

Theorem 20 (From wAST/wPAST to fAST/fPAST). *If a PTRS $\mathcal{S}$ is non-overlapping, linear, and non-erasing, then*

$$\mathcal{S} \text{ is fAST} \iff \mathcal{S} \text{ is wAST}$$
$$\mathcal{S} \text{ is fPAST} \iff \mathcal{S} \text{ is wPAST}$$

4.3 From liAST to fAST

Finally, we look at leftmost-innermost AST as an example for a rewrite strategy that uses an ordering for parallel redexes. In contrast to the non-probabilistic setting, it turns out that liAST and iAST are not equivalent in general. The counterexample is similar to Ex. 18, which illustrated that fAST and iAST are not equivalent without left-linearity.

Counterexample 21. Consider the PTRS $\mathcal{S}_4$ with the five rules:

$$\mathsf{a} \to \{1 : \mathsf{c}_1\} \qquad \mathsf{b} \to \{1/2 : \mathsf{d}_1, 1/2 : \mathsf{d}_2\}$$
$$\mathsf{a} \to \{1 : \mathsf{c}_2\} \qquad \mathsf{f}(\mathsf{c}_1, \mathsf{d}_1) \to \{1 : \mathsf{f}(\mathsf{a}, \mathsf{b})\}$$
$$\mathsf{f}(\mathsf{c}_2, \mathsf{d}_2) \to \{1 : \mathsf{f}(\mathsf{a}, \mathsf{b})\}$$

This PTRS is not iAST (and hence not iPAST) since there exists the infinite rewrite sequence $\{1 : \mathsf{f}(\mathsf{a}, \mathsf{b})\} \overset{\mathsf{i}}{\rightrightarrows}_{\mathcal{S}_4} \{1/2 : \mathsf{f}(\mathsf{a}, \mathsf{d}_1), 1/2 : \mathsf{f}(\mathsf{a}, \mathsf{d}_2)\} \overset{\mathsf{i}}{\rightrightarrows}^2_{\mathcal{S}_4} \{1/2 : \mathsf{f}(\mathsf{c}_1, \mathsf{d}_1), 1/2 : \mathsf{f}(\mathsf{c}_2, \mathsf{d}_2)\} \overset{\mathsf{i}}{\rightrightarrows}^2_{\mathcal{S}_4} \{1/2 : \mathsf{f}(\mathsf{a}, \mathsf{b}), 1/2 : \mathsf{f}(\mathsf{a}, \mathsf{b})\} \overset{\mathsf{i}}{\rightrightarrows}_{\mathcal{S}_4} \ldots$, which converges with probability 0. It first "splits" the term $\mathsf{f}(\mathsf{a}, \mathsf{b})$ with the b-rule, and then applies one of the two different a-rules to each of the resulting terms. In contrast, when applying a leftmost innermost rewrite strategy, we have to decide which a-rule to use. For example, we have $\{1 : \mathsf{f}(\mathsf{a}, \mathsf{b})\} \overset{\mathsf{li}}{\rightrightarrows}_{\mathcal{S}_4} \{1 : \mathsf{f}(\mathsf{c}_1, \mathsf{b})\} \overset{\mathsf{li}}{\rightrightarrows}_{\mathcal{S}_4} \{1/2 : \mathsf{f}(\mathsf{c}_1, \mathsf{d}_1), 1/2 : \mathsf{f}(\mathsf{c}_1, \mathsf{d}_2)\}$. Here, the second term $\mathsf{f}(\mathsf{c}_1, \mathsf{d}_2)$ is a normal form. Since all leftmost innermost rewrite sequences keep on adding such normal forms after a certain number of steps for each start term, the PTRS is liAST (and also liPAST).

The counterexample above can easily be adapted to variants of innermost rewriting that impose different orders on parallel redexes like, e.g., *rightmost* innermost rewriting.

However, liAST and iAST are again equivalent for non-overlapping TRSs. For such TRSs, at most one rule can be used to rewrite at a given position, which prevents the problem illustrated in Ex. 21.

Theorem 22 (From liAST/liPAST to iAST/iPAST). *If a PTRS $\mathcal{S}$ is non-overlapping, then*

$$\mathcal{S} \text{ is iAST} \iff \mathcal{S} \text{ is liAST}$$
$$\mathcal{S} \text{ is iPAST} \iff \mathcal{S} \text{ is liPAST}$$

The relations between the different properties for AST of PTRSs (given in Thm. 17, 20, and 22) are summarized below. An analogous figure also holds for PAST.

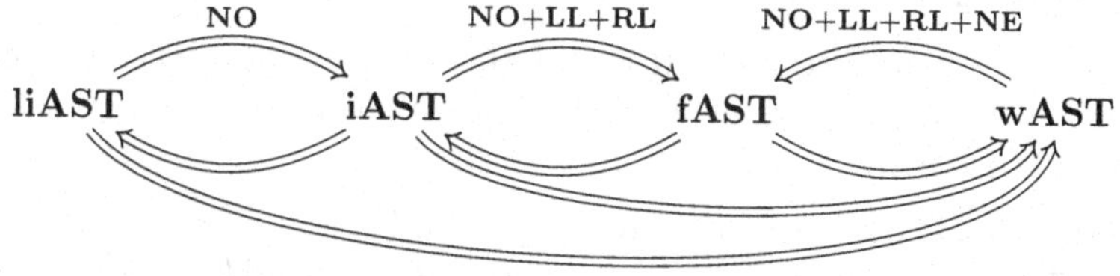

5 Improving Applicability

In this section, we improve the applicability of Thm. 17, which relates fAST and iAST. The results of Sect. 5.1 allow us to remove the requirement of left-linearity by modifying the rewrite relation to *simultaneous rewriting*. Then in Sect. 5.2 we show that the requirement of right-linearity can be weakened to *spareness* if one only considers rewrite sequences that start with *basic terms*.

5.1 Removing Left-Linearity by Simultaneous Rewriting

First, we will see that we do not need to require left-linearity if we allow the simultaneous reduction of several copies of identical redexes. For a PTRS $\mathcal{S}$, this results in the notion of *simultaneous rewriting*, denoted $\rightarrowtail_{\mathcal{S}}$. While $\overset{\mathsf{i}}{\rightarrowtail}_{\mathcal{S}}$ over-approximates $\overset{\mathsf{i}}{\to}_{\mathcal{S}}$, existing techniques for proving iAST [29, 32] (except for the rewriting processor[4]) do not distinguish between both notions of rewriting, i.e., these techniques even prove that every rewrite sequence with the lifting $\overset{\mathsf{i}}{\rightrightarrows}_{\mathcal{S}}$ of $\overset{\mathsf{i}}{\rightarrowtail}_{\mathcal{S}}$ converges with probability 1. So for non-overlapping and right-linear PTRSs, these techniques can be used to prove innermost almost-sure termination w.r.t.

[4] This processor is an optional transformation technique which was added in [32] when improving the DP framework further since it sometimes helps to increase power, but all other (major) DP processors do not distinguish between $\overset{\mathsf{i}}{\to}_{\mathcal{S}}$ and $\overset{\mathsf{i}}{\rightarrowtail}_{\mathcal{S}}$.

$\rightarrowtail_{\mathcal{S}}$, which then implies fAST. The following example illustrates our approach for handling non-left-linear PTRSs by applying the same rewrite rule at parallel positions simultaneously.

Example 23 (Simultaneous Rewriting). Reconsider the PTRS $\mathcal{S}_2$ from Ex. 18 with the rules $\mathsf{f}(x, x) \to \{1 : \mathsf{f}(\mathsf{a}, \mathsf{a})\}$ and $\mathsf{a} \to \{1/2 : \mathsf{b}, 1/2 : \mathsf{c}\}$ which is iAST, but not fAST. Our new rewrite relation $\rightrightarrows_{\mathcal{S}_2}$ allows us to reduce several copies of the same redex simultaneously, so that we get $\{1 : \mathsf{f}(\mathsf{a}, \mathsf{a})\} \overset{\mathsf{i}}{\rightrightarrows}_{\mathcal{S}_2} \{\frac{1}{2} : \mathsf{f}(\mathsf{b}, \mathsf{b}), \frac{1}{2} : \mathsf{f}(\mathsf{c}, \mathsf{c})\} \overset{\mathsf{i}}{\rightrightarrows}^2_{\mathcal{S}_2} \{1/2 : \mathsf{f}(\mathsf{a}, \mathsf{a}), 1/2 : \mathsf{f}(\mathsf{a}, \mathsf{a})\}$, i.e., this $\overset{\mathsf{i}}{\rightrightarrows}_{\mathcal{S}_2}$-sequence converges with probability 0 and thus, $\mathcal{S}_2$ is *not* iAST w.r.t. $\rightarrowtail_{\mathcal{S}_2}$. Note that we simultaneously reduced both occurrences of a in the first step.

Definition 24 (Simultaneous Rewriting). *Let $\mathcal{S}$ be a PTRS. A term s rewrites* simultaneously *to a multi-distribution $\mu = \{p_1 : t_1, \ldots, p_k : t_k\}$ (denoted $s \rightarrowtail_{\mathcal{S}} \mu$) if there is a non-empty set of parallel positions Π, a rule $\ell \to \{p_1 : r_1, \ldots, p_k : r_k\} \in \mathcal{S}$, and a substitution σ such that $s|_\pi = \ell\sigma$ and $t_j = s[r_j\sigma]_\pi$ for every position $\pi \in \Pi$ and for all $1 \leq j \leq k$. We call $s \rightarrowtail_{\mathcal{S}} \mu$ an* innermost simultaneous *rewrite step (denoted $s \overset{\mathsf{i}}{\rightarrowtail}_{\mathcal{S}} \mu$) if all proper subterms of the redex $\ell\sigma$ are in normal form w.r.t. $\mathcal{S}$.*

Clearly, if the set of positions Π from Def. 24 is a singleton, then the resulting simultaneous rewrite step is an "ordinary" probabilistic rewrite step, i.e., $\to_{\mathcal{S}} \subseteq \rightarrowtail_{\mathcal{S}}$ and $\overset{\mathsf{i}}{\to}_{\mathcal{S}} \subseteq \overset{\mathsf{i}}{\rightarrowtail}_{\mathcal{S}}$.

Corollary 25 (From $\rightarrowtail_{\mathcal{S}}$ to $\to_{\mathcal{S}}$). *If $\mathcal{S}$ is fAST (iAST) w.r.t. $\rightarrowtail_{\mathcal{S}}$, i.e., every infinite $\rightrightarrows_{\mathcal{S}}$- (resp. $\overset{\mathsf{i}}{\rightrightarrows}_{\mathcal{S}}$-) rewrite sequence converges with probability 1, then $\mathcal{S}$ is fAST (iAST). Analogously, if $\mathcal{S}$ is fPAST (iPAST) w.r.t. $\rightarrowtail_{\mathcal{S}}$, i.e., every infinite $\rightrightarrows_{\mathcal{S}}$- (resp. $\overset{\mathsf{i}}{\rightrightarrows}_{\mathcal{S}}$-) rewrite sequence has finite expected derivation length, then $\mathcal{S}$ is fPAST (iPAST).*

However, the converse of Cor. 25 does not hold. Ex. 23 shows that $\overset{\mathsf{i}}{\rightarrowtail}_{\mathcal{S}}$ allows for rewrite sequences that are not possible with $\overset{\mathsf{i}}{\to}_{\mathcal{S}}$, and the following example shows the same for $\rightarrowtail_{\mathcal{S}}$ and $\to_{\mathcal{S}}$.

Counterexample 26. Consider the PTRS $\overline{\mathcal{S}_2}$ with the three rules:

$$\mathsf{f}(\mathsf{b}, \mathsf{b}) \to \{1 : \mathsf{f}(\mathsf{a}, \mathsf{a})\} \qquad \mathsf{a} \to \{1/2 : \mathsf{b}, 1/2 : \mathsf{c}\}$$
$$\mathsf{f}(\mathsf{c}, \mathsf{c}) \to \{1 : \mathsf{f}(\mathsf{a}, \mathsf{a})\}$$

This PTRS is fAST. But as in Ex. 23, we have $\{1 : \mathsf{f}(\mathsf{a}, \mathsf{a})\} \overset{\mathsf{i}}{\rightrightarrows}_{\overline{\mathcal{S}_2}} \{\frac{1}{2} : \mathsf{f}(\mathsf{b}, \mathsf{b}), \frac{1}{2} : \mathsf{f}(\mathsf{c}, \mathsf{c})\} \overset{\mathsf{i}}{\rightrightarrows}^2_{\overline{\mathcal{S}_2}} \{1/2 : \mathsf{f}(\mathsf{a}, \mathsf{a}), 1/2 : \mathsf{f}(\mathsf{a}, \mathsf{a})\}$, i.e., there are rewrite sequences with $\overset{\mathsf{i}}{\rightrightarrows}_{\overline{\mathcal{S}_2}}$ and thus, also with $\rightrightarrows_{\overline{\mathcal{S}_2}}$ that converge with probability 0. Hence, $\overline{\mathcal{S}_2}$ is not iAST or fAST w.r.t. $\rightarrowtail_{\overline{\mathcal{S}_2}}$. Again, the same example also shows that fPAST and fPAST w.r.t. simultaneous rewriting are not equivalent either.

Note that this kind of simultaneous rewriting is different from the "ordinary" parallelism used for non-probabilistic rewriting, which is typically denoted by $\to_{||}$. There, one may reduce multiple parallel redexes in a single rewrite step. Here, we do not only allow reducing multiple redexes, but in addition we "merge" the corresponding terms in the multi-distributions that result from rewriting the different redexes. Because of this merging, we only allow the simultaneous reduction of *equal* redexes, whereas "ordinary" parallel rewriting allows the simultaneous reduction of arbitrary parallel redexes. For example, for $\mathcal{S}_2$ from Ex. 18 we have $\{1 : \mathsf{f}(\mathsf{a},\mathsf{a})\} \overset{\mathsf{i}}{\rightarrowtail\!\!\rightrightarrows}_{\mathcal{S}_2} \{\frac{1}{2} : \mathsf{f}(\mathsf{b},\mathsf{b}), \frac{1}{2} : \mathsf{f}(\mathsf{c},\mathsf{c})\}$, whereas using ordinary parallel rewriting we would get $\{1 : \mathsf{f}(\mathsf{a},\mathsf{a})\} \overset{\mathsf{i}}{\rightrightarrows}_{||\mathcal{S}_2} \{\frac{1}{4} : \mathsf{f}(\mathsf{b},\mathsf{b}), \frac{1}{4} : \mathsf{f}(\mathsf{b},\mathsf{c}), \frac{1}{4} : \mathsf{f}(\mathsf{c},\mathsf{b}), \frac{1}{4} : \mathsf{f}(\mathsf{c},\mathsf{c})\}$.

The following theorem shows that indeed, we do not need to require left-linearity when moving from iAST/iPAST w.r.t. $\rightarrowtail_{\mathcal{S}}$ to fAST/fPAST w.r.t. $\to_{\mathcal{S}}$.

Theorem 27 (From iAST/iPAST to fAST/fPAST (2)). *If a PTRS $\mathcal{S}$ is non-overlapping and right-linear, then*

$$\mathcal{S} \textit{ is fAST} \Longleftarrow \mathcal{S} \textit{ is iAST w.r.t. } \rightarrowtail_{\mathcal{S}}$$
$$\mathcal{S} \textit{ is fPAST} \Longleftarrow \mathcal{S} \textit{ is iPAST w.r.t. } \rightarrowtail_{\mathcal{S}}$$

Proof Sketch. We use an analogous construction as for the proof of Thm. 17, but in addition, if we replace a non-innermost rewrite step by an innermost one, then we check whether in the original rewrite sequence, the corresponding innermost redex is "inside" the substitution used for the non-innermost rewrite step. In that case, if this rewrite step applied a non-left-linear rule, then we identify all other (equal) innermost redexes and use $\overset{\mathsf{i}}{\rightarrowtail}_{\mathcal{S}}$ to rewrite them simultaneously (as we did for the innermost redex a in Ex. 23). □

Note that Ex. 26 shows that the direction " $\Longrightarrow$ " does not hold in Thm. 27. The following example shows that right-linearity in Thm. 27 cannot be weakened to the requirement that $\mathcal{S}$ is *non-duplicating* (i.e., that no variable occurs more often in a term on the right-hand side of a rule than on its left-hand side).

Counterexample 28 (Non-Duplicating Does Not Suffice). Let $\mathsf{d}(\mathsf{f}(\mathsf{a},\mathsf{a})^3)$ abbreviate $\mathsf{d}(\mathsf{f}(\mathsf{a},\mathsf{a}),\mathsf{f}(\mathsf{a},\mathsf{a}),\mathsf{f}(\mathsf{a},\mathsf{a}))$. Consider the PTRS $\mathcal{S}_5$ with the four rules:

$$\mathsf{f}(x,x) \to \{1 : \mathsf{g}(x,x)\} \qquad \mathsf{g}(\mathsf{b},\mathsf{c}) \to \{1 : \mathsf{d}(\mathsf{f}(\mathsf{a},\mathsf{a})^3)\}$$
$$\mathsf{a} \to \{1/2 : \mathsf{b}, 1/2 : \mathsf{c}\} \qquad \mathsf{g}(\mathsf{c},\mathsf{b}) \to \{1 : \mathsf{d}(\mathsf{f}(\mathsf{a},\mathsf{a})^3)\}$$

$\mathcal{S}_5$ is not fAST (and thus, also not fPAST), since the infinite rewrite sequence $\{1 : \mathsf{f}(\mathsf{a},\mathsf{a})\} \rightrightarrows_{\mathcal{S}_5} \{1 : \mathsf{g}(\mathsf{a},\mathsf{a})\} \rightrightarrows^2_{\mathcal{S}_5} \{1/4 : \mathsf{g}(\mathsf{b},\mathsf{b}), 1/4 : \mathsf{g}(\mathsf{b},\mathsf{c}), 1/4 : \mathsf{g}(\mathsf{c},\mathsf{b}), 1/4 : \mathsf{g}(\mathsf{c},\mathsf{c})\} \rightrightarrows^2_{\mathcal{S}_5} \{1/4 : \mathsf{g}(\mathsf{b},\mathsf{b}), 1/4 : \mathsf{d}(\mathsf{f}(\mathsf{a},\mathsf{a})^3), 1/4 : \mathsf{d}(\mathsf{f}(\mathsf{a},\mathsf{a})^3), 1/4 : \mathsf{g}(\mathsf{c},\mathsf{c})\}$ can be seen as a biased random walk on the number of $\mathsf{f}(\mathsf{a},\mathsf{a})$-subterms that is not AST. However, for every innermost evaluation with $\overset{\mathsf{i}}{\to}_{\mathcal{S}_5}$ or $\overset{\mathsf{i}}{\rightarrowtail}_{\mathcal{S}_5}$ we have to rewrite the inner a-symbols first. Afterwards, the f-rule can only be used on redexes $\mathsf{f}(t,t)$ where the resulting term $\mathsf{g}(t,t)$ is a normal form. Thus, $\mathcal{S}_5$ is iPAST (and hence, iAST) w.r.t. $\rightarrowtail_{\mathcal{S}_5}$.

Note that for wAST, the direction of the implication in Cor. 25 is reversed, since wAST requires that for each start term, there *exists* an infinite rewrite sequence that is almost-surely terminating, whereas fAST requires that *all* infinite rewrite sequences are almost-surely terminating. Thus, if there exists an infinite $\rightrightarrows_{\mathcal{S}}$-rewrite sequence that converges with probability 1 (showing that $\mathcal{S}$ is wAST), then this is also a valid $\rightarrowtail\!\!\rightrightarrows_{\mathcal{S}}$-rewrite sequence that converges with probability 1 (showing that $\mathcal{S}$ is wAST w.r.t. $\rightarrowtail_{\mathcal{S}}$).

Corollary 29 (From $\to_{\mathcal{S}}$ to $\rightarrowtail_{\mathcal{S}}$ for wAST/wPAST). *If $\mathcal{S}$ is wAST (wPAST), then $\mathcal{S}$ is wAST (wPAST) w.r.t. $\rightarrowtail_{\mathcal{S}}$.*

One may wonder whether simultaneous rewriting could also be used to improve Thm. 20 by removing the requirement of left-linearity, but Ex. 30 shows this is not possible.

Counterexample 30. Consider the non-left-linear PTRS $\mathcal{S}_6$ with the two rules:

$$\mathsf{g} \to \{3/4 : \mathsf{d}(\mathsf{g},\mathsf{g}), 1/4 : \bot\} \qquad\qquad \mathsf{d}(x,x) \to \{1 : x\}$$

This PTRS is not fAST (and thus, also not fPAST), as we have $\{1 : \mathsf{g}\} \rightrightarrows_{\mathcal{S}_6} \{3/4 : \mathsf{d}(\mathsf{g},\mathsf{g}), 1/4 : \bot\}$, which corresponds to a random walk biased towards non-termination if we never use the d-rule (since $\frac{3}{4} > \frac{1}{2}$). However, if we always use the d-rule directly after the g-rule, then we essentially end up with a PTRS whose only rule is $\mathsf{g} \to \{3/4 : \mathsf{c}(\mathsf{g}), 1/4 : \bot\}$, which corresponds to flipping a biased coin until heads comes up. This proves that $\mathcal{S}_6$ is wPAST and hence, also wAST. As $\mathcal{S}_6$ is non-overlapping, right-linear, and non-erasing, this shows that a variant of Thm. 20 without the requirement of left-linearity needs more than just moving to simultaneous rewriting.

5.2 Weakening Right-Linearity to Spareness

To improve our results further, we introduce the notion of *spareness*. The idea of spareness is to require that variables which occur non-linear in right-hand sides may only be instantiated by normal forms. We already used spareness for non-probabilistic TRSs in [17] to find classes of TRSs where innermost and full runtime complexity coincide. For a PTRS $\mathcal{S}$, we decompose its signature $\Sigma = \Sigma_C \uplus \Sigma_D$ such that $f \in \Sigma_D$ iff $f = \mathrm{root}(\ell)$ for some rule $\ell \to \mu \in \mathcal{S}$. The symbols in Σ_C and Σ_D are called *constructors* and *defined symbols*, respectively.

Definition 31 (Spareness). *Let $\ell \to \mu \in \mathcal{S}$. A rewrite step $\ell\sigma \to_{\mathcal{S}} \mu\sigma$ is* spare *if $\sigma(x)$ is in normal form w.r.t. $\mathcal{S}$ for every $x \in \mathcal{V}$ that occurs more than once in some $r \in \mathrm{Supp}(\mu)$. A $\rightrightarrows_{\mathcal{S}}$-sequence is spare if each of its $\to_{\mathcal{S}}$-steps is spare. $\mathcal{S}$ is spare if each $\rightrightarrows_{\mathcal{S}}$-sequence that starts with $\{1 : t\}$ for a basic term t is spare. A term $t \in \mathcal{T}(\Sigma, \mathcal{V})$ is* basic *if $t = f(t_1, \ldots, t_n)$ such that $f \in \Sigma_D$ and $t_i \in \mathcal{T}(\Sigma_C, \mathcal{V})$ for all $1 \leq i \leq n$.*

Example 32. Consider the PTRS $\mathcal{S}_7$ with the two rules:

$$\mathsf{g} \to \{3/4 : \mathsf{d}(\bot), 1/4 : \mathsf{g}\} \qquad\qquad \mathsf{d}(x) \to \{1 : \mathsf{c}(x,x)\}$$

It is similar to the PTRS $\mathcal{S}_1$ from Ex. 16, but we exchanged the symbols g and $\bot$ in the right-hand side of the g-rule. This PTRS is orthogonal but duplicating due to the d-rule. However, in any rewrite sequence that starts with $\{1 : t\}$ for a basic term t we can only duplicate the constructor symbol $\bot$ but no defined symbol. Hence, $\mathcal{S}_7$ is spare.

In general, it is undecidable whether a PTRS is spare, since spareness is already undecidable for non-probabilistic TRSs. However, there exist computable sufficient conditions for spareness, see [17].

If a PTRS is spare, and we start with a basic term, then we will only duplicate normal forms with our duplicating rules. This means that the duplicating rules do not influence the (expected) runtime and, more importantly for AST, the probability of termination. As in [17], which analyzed runtime complexity, we have to restrict ourselves to rewrite sequences that start with basic terms. So we only consider start terms where a single algorithm is applied to data, i.e., we may not have any nested defined symbols in our start terms. This leads to the following theorem, where "*on basic terms*" means that one only considers rewrite sequences that start with $\{1 : t\}$ for a basic term t. It can be proved by an analogous limit construction as in the proof of Thm. 17.

Theorem 33 (From iAST/iPAST to fAST/fPAST (3)). *If a PTRS $\mathcal{S}$ is orthogonal and spare, then*

$$\mathcal{S} \textit{ is fAST on basic terms} \Longleftrightarrow \mathcal{S} \textit{ is iAST on basic terms}$$
$$\mathcal{S} \textit{ is fPAST on basic terms} \Longleftrightarrow \mathcal{S} \textit{ is iPAST on basic terms}$$

While iAST on basic terms is the same as iAST in general, the requirement of basic start terms is real restriction for fAST, i.e., there exists PTRSs that are fAST on basic terms, but not fAST in general.

Counterexample 34. Consider the PTRS $\mathcal{S}_8$ with the two rules:

$$\mathsf{g} \to \{{}^{3}/_{4} : \mathsf{s}(\mathsf{g}), {}^{1}/_{4} : \bot\} \qquad\qquad \mathsf{f}(\mathsf{s}(x)) \to \{1 : \mathsf{c}(\mathsf{f}(x), \mathsf{f}(x))\}$$

This PTRS behaves similarly to $\mathcal{S}_1$ (see Ex. 16). It is not fAST (and thus, also not fPAST), as we have $\{1 : \mathsf{f}(\mathsf{g})\} \rightrightarrows^2_{\mathcal{S}_8} \{{}^{3}/_{4} : \mathsf{c}(\mathsf{f}(\mathsf{g}), \mathsf{f}(\mathsf{g})), {}^{1}/_{4} : \mathsf{f}(\bot)\}$, which corresponds to a random walk biased towards non-termination (since $\frac{3}{4} > \frac{1}{2}$).

However, the only basic terms for this PTRS are g and $\mathsf{f}(t)$ for terms t that do not contain g or f. A sequence starting with g corresponds to flipping a biased coin and a sequence starting with $\mathsf{f}(t)$ will clearly terminate. Hence, $\mathcal{S}_8$ is fAST (and even fPAST) on basic terms. Furthermore, note that $\mathcal{S}_8$ is iPAST (and thus, also iAST) analogous to $\mathcal{S}_1$. This shows that Thm. 33 cannot be extended to fAST or fPAST in general.

One may wonder whether Thm. 33 can nevertheless be used in order to prove fAST of a PTRS $\mathcal{S}$ on all terms by using a suitable transformation from $\mathcal{S}$ to another PTRS $\mathcal{S}'$ such that $\mathcal{S}$ is fAST on all terms iff $\mathcal{S}'$ is fAST on basic terms.

There is an analogous difference in the complexity analysis of non-probabilistic term rewrite systems. There, the concept of *runtime complexity* is restricted to rewrite sequences that start with a basic term, whereas the concept of *derivational complexity* allows arbitrary start terms. In [19], a transformation was presented that extends any (non-probabilistic) TRS $\mathcal{R}$ by so-called generator rules $\mathcal{G}(\mathcal{R})$ such that the derivational complexity of $\mathcal{R}$ is the same as the runtime complexity of $\mathcal{R} \cup \mathcal{G}(\mathcal{R})$, where $\mathcal{G}(\mathcal{R})$ are considered to be *relative* rules whose rewrite steps do not "count" for the complexity. This transformation can indeed be reused to move from fAST *on basic terms* to fAST in general.

Lemma 35. *A PTRS $\mathcal{S}$ is fAST iff $\mathcal{S} \cup \mathcal{G}(\mathcal{S})$ is fAST on basic terms.*

For every defined symbol f, the idea of the transformation is to introduce a new constructor symbol cons_f and for every function symbol f it introduces a new defined symbol enc_f. As an example for $\mathcal{S}_8$ from Ex. 32, then instead of starting with the non-basic term $\mathsf{c}(\mathsf{g}, \mathsf{f}(\mathsf{g}))$, we start with the basic term $\mathsf{enc}_\mathsf{c}(\mathsf{cons}_\mathsf{g}, \mathsf{cons}_\mathsf{f}(\mathsf{cons}_\mathsf{g}))$, its so-called *basic variant.* The new defined symbol enc_c is used to first build the term $\mathsf{c}(\mathsf{g}, \mathsf{f}(\mathsf{g}))$ at the beginning of the rewrite sequence, i.e., it converts all occurrences of cons_f for $f \in \Sigma_D$ back into the defined symbol f, and then we can proceed as if we started with the term $\mathsf{c}(\mathsf{g}, \mathsf{f}(\mathsf{g}))$ directly. For this conversion, we need another new defined symbol argenc that iterates through the term and replaces all new constructors cons_f by the original defined symbol f. Thus, we define the generator rules as in [19] (just with trivial probabilities in the right-hand sides $\ell \to \{1 : r\}$), since we do not need any probabilities during this initial construction of the original start term.

Definition 36 (Generator Rules $\mathcal{G}(\mathcal{S})$). *Let $\mathcal{S}$ be a PTRS over the signature Σ. Its* generator rules *$\mathcal{G}(\mathcal{S})$ are the following set of rules*

$$
\begin{aligned}
&\{\mathsf{enc}_f(x_1, \dots, x_n) \to \{1 : f(\mathsf{argenc}(x_1), \dots, \mathsf{argenc}(x_n))\} \mid f \in \Sigma\} \\
\cup\, &\{\mathsf{argenc}(\mathsf{cons}_f(x_1, \dots, x_n)) \to \{1 : f(\mathsf{argenc}(x_1), \dots, \mathsf{argenc}(x_n))\} \mid f \in \Sigma_D\} \\
\cup\, &\{\mathsf{argenc}(f(x_1, \dots, x_n)) \to \{1 : f(\mathsf{argenc}(x_1), \dots, \mathsf{argenc}(x_n))\} \mid f \in \Sigma_C\},
\end{aligned}
$$

where $x_1, \dots, x_n$ are pairwise different variables and where the function symbols argenc, cons_f, and enc_f are fresh (i.e., they do not occur in $\mathcal{S}$). Moreover, we define $\Sigma_{\mathcal{G}(\mathcal{S})} = \{\mathsf{enc}_f \mid f \in \Sigma\} \cup \{\mathsf{argenc}\} \cup \{\mathsf{cons}_f \mid f \in \Sigma_D\}$.

Example 37. For the PTRS $\mathcal{S}_8$ from Ex. 34, we obtain the following generator rules $\mathcal{G}(\mathcal{S}_8)$:

$$
\begin{aligned}
\mathsf{enc}_\mathsf{g} &\to \{1 : \mathsf{g}\} \\
\mathsf{enc}_\mathsf{f}(x_1) &\to \{1 : \mathsf{f}(\mathsf{argenc}(x_1))\} \\
\mathsf{enc}_\mathsf{c}(x_1, x_2) &\to \{1 : \mathsf{c}(\mathsf{argenc}(x_1), \mathsf{argenc}(x_2))\} \\
\mathsf{enc}_\mathsf{s}(x_1) &\to \{1 : \mathsf{s}(\mathsf{argenc}(x_1))\} \\
\mathsf{enc}_\bot &\to \{1 : \bot\} \\
\mathsf{argenc}(\mathsf{cons}_\mathsf{g}) &\to \{1 : \mathsf{g}\} \\
\mathsf{argenc}(\mathsf{cons}_\mathsf{f}(x_1)) &\to \{1 : \mathsf{f}(\mathsf{argenc}(x_1))\} \\
\mathsf{argenc}(\mathsf{c}(x_1, x_2)) &\to \{1 : \mathsf{c}(\mathsf{argenc}(x_1), \mathsf{argenc}(x_2))\} \\
\mathsf{argenc}(\mathsf{s}(x_1)) &\to \{1 : \mathsf{s}(\mathsf{argenc}(x_1))\} \\
\mathsf{argenc}(\bot) &\to \{1 : \bot\}
\end{aligned}
$$

As mentioned, using the symbols cons_f and enc_f, as in [19] every term over Σ can be transformed into a basic term over $\Sigma \cup \Sigma_{\mathcal{G}(\mathcal{S})}$.

However, even if $\mathcal{S}$ is spare, the PTRS $\mathcal{S} \cup \mathcal{G}(\mathcal{S})$ is not guaranteed to be spare, although the generator rules themselves are right-linear. The problem is that the generator rules include a rule like $\mathsf{enc_f}(x_1) \to \{1 : \mathsf{f}(\mathsf{argenc}(x_1))\}$ where a defined symbol argenc occurs below the duplicating symbol f on the right-hand side. Indeed, while $\mathcal{S}_8$ is spare, $\mathcal{S}_8 \cup \mathcal{G}(\mathcal{S}_8)$ is not. For example, when starting with the basic term $\mathsf{enc_f}(\mathsf{s}(\mathsf{cons_g}))$, we have

$$\begin{aligned} \{1 : \mathsf{enc_f}(\mathsf{s}(\mathsf{cons_g}))\} &\rightrightarrows^2_{\mathcal{G}(\mathcal{S}_8)} \{1 : \mathsf{f}(\mathsf{s}(\mathsf{argenc}(\mathsf{cons_g})))\} \\ &\rightrightarrows_{\mathcal{S}_8} \{1 : \mathsf{c}(\mathsf{f}(\mathsf{argenc}(\mathsf{cons_g})), \mathsf{f}(\mathsf{argenc}(\mathsf{cons_g}))), \end{aligned}$$

where the last step is not spare. In general, $\mathcal{S} \cup \mathcal{G}(\mathcal{S})$ is guaranteed to be spare if $\mathcal{S}$ is right-linear. So we could modify Thm. 33 into a theorem which states that $\mathcal{S}$ is fAST on all terms iff $\mathcal{S} \cup \mathcal{G}(\mathcal{S})$ is iAST on basic terms (and thus, on all terms) for orthogonal and right-linear PTRSs $\mathcal{S}$. However, this theorem would be subsumed by Thm. 17, where we already showed the equivalence of fAST and iAST if $\mathcal{S}$ is orthogonal and right-linear. Indeed, our goal in Thm. 33 was to find a weaker requirement than right-linearity. Hence, such a transformational approach to move from fAST on all start terms to fAST on basic terms does not seem viable for Thm. 33.

Finally, we can also combine our results on simultaneous rewriting and spareness to relax both left- and right-linearity in case of basic start terms. The proof for the following theorem combines the proofs for Thm. 27 and Thm. 33.

Theorem 38 (From iAST/iPAST to fAST/fPAST (4)). *If $\mathcal{S}$ is non-overlapping and spare, then*

$$\mathcal{S} \textit{ is fAST on basic terms} \Longleftarrow \mathcal{S} \textit{ is iAST w.r.t. } \rightarrowtail_{\mathcal{S}} \textit{ on basic terms}$$
$$\mathcal{S} \textit{ is fPAST on basic terms} \Longleftarrow \mathcal{S} \textit{ is iPAST w.r.t. } \rightarrowtail_{\mathcal{S}} \textit{ on basic terms}$$

6 Conclusion and Evaluation

In this paper, we presented numerous new results on the relationship between full and restricted forms of AST, including several criteria for PTRSs such that innermost AST implies full AST. All of our results also hold for PAST, and all of our criteria are suitable for automation (for spareness, there exist sufficient conditions that can be checked automatically).

We implemented our new criteria in our termination prover AProVE [21]. For every PTRS, one can indicate whether one wants to analyze its termination behavior for all start terms or only for basic start terms. Up to now, AProVE's main technique for termination analysis of PTRSs was the probabilistic DP framework from [29, 32] which however can only prove iAST. If one wants to analyze fAST for a PTRS $\mathcal{S}$, then AProVE now first tries to prove that the conditions of Thm. 33 are satisfied if one is restricted to basic start terms, or that the conditions of Thm. 17 hold if one wants to consider arbitrary start terms. If this succeeds, then we can use the full probabilistic DP framework in order to prove iAST, which then implies fAST. Otherwise, we try to prove all conditions of Thm. 38 or Thm. 27, respectively. If this succeeds, then we can use most of the processors from the probabilistic DP framework to prove iAST, which again implies fAST. If none of these theorems can be applied, then AProVE tries to prove fAST using a direct application of polynomial orderings [29]. Note that for AST w.r.t. basic start terms, Thm. 33 generalizes Thm. 17 and Thm. 38 generalizes Thm. 27, since right-linearity implies spareness.

For our evaluation, we compare the *old* AProVE without any of the new theorems (which only uses direct applications of polynomial orderings to prove fAST), to variants of AProVE where we activated each of the theorems individually, and finally to the *new* AProVE strategy explained above. The following diagram shows the theoretical subsumptions of each of these strategies for basic start terms, where an arrow from strategy A to strategy B means that B is strictly better than A.

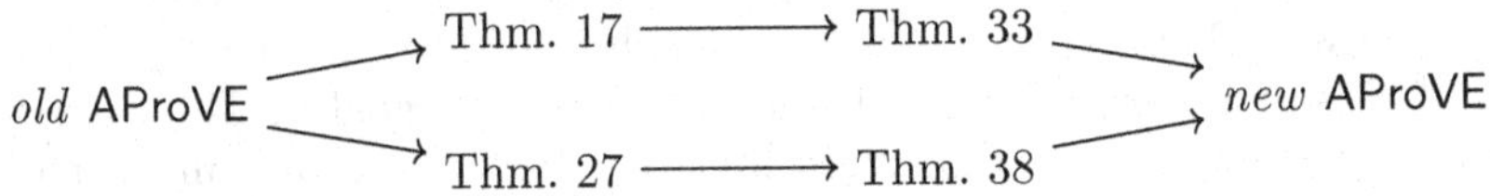

We used the benchmark set of 100 PTRSs from [32], and extended it by 15 new PTRSs that contain all the examples presented in this paper and some additional examples which illustrate the power of each strategy. AProVE can prove iAST for 93 of these 118 PTRSs. The following table shows for how many of these 93 PTRSs the respective strategy allows us to conclude fAST for basic start terms from AProVE's proof of iAST.

old AProVE	Thm. 17	Thm. 27	Thm. 33	Thm. 38	*new* AProVE
36	48	44	58	56	61

From the 61 examples that we can solve by using both Thm. 33 and Thm. 38 in "*new* AProVE", 5 examples (that are all right-linear) can only be solved by Thm. 33, 3 examples (where one is right-linear and the others only spare) can only be solved by Thm. 38, and 53 examples can be solved by both. If one considers arbitrary start terms, then the *new* AProVE can conclude fAST (using only Thm. 17 and Thm. 27) for 49 examples.

Currently, we only use the switch from full to innermost rewriting as a preprocessing step before applying the DP framework. As future work, we want to develop a processor within the DP framework that can perform this switch in a modular way. Then, the criteria of our theorems do not have to be required for the whole PTRS anymore, but just for specific sub-problems within the termination proof. This, however, requires developing a DP framework for fAST directly, which we will investigate in future work.

For details on our experiments, our collection of examples, and for instructions on how to run our implementation in AProVE via its *web interface* or locally, we refer to:

https://aprove-developers.github.io/InnermostToFullAST/

In addition, an artifact is available at [31].

Acknowledgements. We thank Stefan Dollase for pointing us to [19].

References

[1] G. Agha, J. Meseguer, and K. Sen. "PMaude: Rewrite-based Specification Language for Probabilistic Object Systems". In: *Proc. QAPL '05*. ENTCS 153. 2006, pp. 213–239. DOI: 10.1016/j.entcs.2005.10.040.

[2] S. Agrawal, K. Chatterjee, and P. Novotný. "Lexicographic Ranking Supermartingales: An Efficient Approach to Termination of Probabilistic Programs". In: *Proc. ACM Program. Lang.* 2.POPL (2017). DOI: 10.1145/3158122.

[3] T. Arts and J. Giesl. "Termination of Term Rewriting Using Dependency Pairs". In: *Theor. Comput. Sc.* 236.1-2 (2000), pp. 133–178. DOI: 10.1016/S0304-3975(99)00207-8.

[4] M. Avanzini, U. Dal Lago, and A. Yamada. "On Probabilistic Term Rewriting". In: *Sci. Comput. Program.* 185 (2020). DOI: 10.1016/j.scico.2019.102338.

[5] M. Avanzini, G. Moser, and M. Schaper. "A Modular Cost Analysis for Probabilistic Programs". In: *Proc. ACM Program. Lang.* 4.OOPSLA (2020). DOI: 10.1145/3428240.

[6] F. Baader and T. Nipkow. *Term Rewriting and All That*. Cambridge University Press, 1998. DOI: 10.1017/CBO9781139172752.

[7] K. Batz, B. L. Kaminski, J.-P. Katoen, C. Matheja, and L. Verscht. "A Calculus for Amortized Expected Runtimes". In: *Proc. ACM Program. Lang.* 7.POPL (2023). DOI: 10.1145/3571260.

[8] R. Beutner and L. Ong. "On Probabilistic Termination of Functional Programs with Continuous Distributions". In: *Proc. PLDI '21*. 2021, pp. 1312–1326. DOI: 10.1145/3453483.3454111.

[9] O. Bournez and C. Kirchner. "Probabilistic Rewrite Strategies. Applications to ELAN". In: *Proc. RTA '02*. LNCS 2378. 2002, pp. 252–266. DOI: 10.1007/3-540-45610-4_18.

[10] O. Bournez and F. Garnier. "Proving Positive Almost-Sure Termination". In: *Proc. RTA '05*. LNCS 3467. 2005, pp. 323–337. DOI: 10.1007/978-3-540-32033-3_24.

[11] K. Chatterjee, H. Fu, and P. Novotný. "Termination Analysis of Probabilistic Programs with Martingales". In: *Foundations of Probabilistic Programming*. Ed. by G. Barthe, J. Katoen, and A. Silva. Cambridge University Press, 2020, 221–258. DOI: 10.1017/9781108770750.008.

[12] U. Dal Lago and C. Grellois. "Probabilistic Termination by Monadic Affine Sized Typing". In: *Proc. ESOP '17*. LNCS 10201. 2017, pp. 393–419. DOI: 10.1007/978-3-662-54434-1_15.

[13] U. Dal Lago, C. Faggian, and S. R. Della Rocca. "Intersection Types and (Positive) Almost-Sure Termination". In: *Proc. ACM Program. Lang.* 5.POPL (2021). DOI: 10.1145/3434313.

[14] A. Díaz-Caro and G. Martínez. "Confluence in Probabilistic Rewriting". In: *Proc. LSFA '17*. ENTCS 338. 2018, pp. 115–131. DOI: 10.1016/j.entcs.2018.10.008.

[15] C. Faggian. "Probabilistic Rewriting and Asymptotic Behaviour: On Termination and Unique Normal Forms". In: *Log. Methods in Comput. Sci.* 18.2 (2022). DOI: 10.46298/lmcs-18(2:5)2022.

[16] L. M. Ferrer Fioriti and H. Hermanns. "Probabilistic Termination: Soundness, Completeness, and Compositionality". In: *Proc. POPL '15.* 2015, pp. 489–501. DOI: 10.1145/2676726.2677001.

[17] F. Frohn and J. Giesl. "Analyzing Runtime Complexity via Innermost Runtime Complexity". In: *Proc. LPAR '17.* EPiC 46. 2017, pp. 249–228. DOI: 10.29007/1nbh.

[18] H. Fu and K. Chatterjee. "Termination of Nondeterministic Probabilistic Programs". In: *Proc. VMCAI '19.* LNCS 11388. 2019, pp. 468–490. DOI: 10.1007/978-3-030-11245-5_22.

[19] C. Fuhs. "Transforming Derivational Complexity of Term Rewriting to Runtime Complexity". In: *Proc. FroCoS '19.* LNCS 11715. 2019, pp. 348–364. DOI: 10.1007/978-3-030-29007-8_20.

[20] J. Giesl, R. Thiemann, P. Schneider-Kamp, and S. Falke. "Mechanizing and Improving Dependency Pairs". In: *J. Autom. Reason.* 37.3 (2006), pp. 155–203. DOI: 10.1007/s10817-006-9057-7.

[21] J. Giesl, C. Aschermann, M. Brockschmidt, F. Emmes, F. Frohn, C. Fuhs, J. Hensel, C. Otto, M. Plücker, P. Schneider-Kamp, T. Ströder, S. Swiderski, and R. Thiemann. "Analyzing Program Termination and Complexity Automatically with AProVE". In: *J. Autom. Reason.* 58.1 (2017), pp. 3–31. DOI: 10.1007/s10817-016-9388-y.

[22] J. Giesl, P. Giesl, and M. Hark. "Computing Expected Runtimes for Constant Probability Programs". In: *Proc. CADE '19.* LNCS 11716. 2019, pp. 269–286. DOI: 10.1007/978-3-030-29436-6_16.

[23] A. D. Gordon, T. A. Henzinger, A. V. Nori, and S. K. Rajamani. "Probabilistic Programming". In: *Proc. FOSE '14.* 2014, pp. 167–181. DOI: 10.1145/2593882.2593900.

[24] B. Gramlich. "Abstract Relations between Restricted Termination and Confluence Properties of Rewrite Systems". In: *Fundamenta Informaticae* 24 (1995), pp. 2–23. DOI: 10.3233/FI-1995-24121.

[25] R. Gutiérrez and S. Lucas. "MU-TERM: Verify Termination Properties Automatically (System Description)". In: *Proc. IJCAR '20.* LNCS 12167. 2020, pp. 436–447. DOI: 10.1007/978-3-030-51054-1_28.

[26] M. Huang, H. Fu, K. Chatterjee, and A. K. Goharshady. "Modular Verification for Almost-Sure Termination of Probabilistic Programs". In: *Proc. ACM Program. Lang.* 3.OOPSLA (2019). DOI: 10.1145/3360555.

[27] B. L. Kaminski, J.-P. Katoen, C. Matheja, and F. Olmedo. "Weakest Precondition Reasoning for Expected Runtimes of Randomized Algorithms". In: *J. ACM* 65 (2018), pp. 1–68. DOI: 10.1145/3208102.

[28] B. L. Kaminski, J. Katoen, and C. Matheja. "Expected Runtime Analyis by Program Verification". In: *Foundations of Probabilistic Programming.* Ed. by G. Barthe, J. Katoen, and A. Silva. Cambridge University Press, 2020, 185–220. DOI: 10.1017/9781108770750.007.

[29] J.-C. Kassing and J. Giesl. "Proving Almost-Sure Innermost Termination of Probabilistic Term Rewriting Using Dependency Pairs". In: *Proc. CADE '23*. LNCS 14132. 2023, pp. 344–364. DOI: 10.1007/978-3-031-38499-8_20.

[30] J.-C. Kassing, F. Frohn, and J. Giesl. "From Innermost to Full Almost-Sure Termination of Probabilistic Term Rewriting". In: *CoRR* abs/2310.06121 (2023). DOI: 10.48550/arXiv.2310.06121.

[31] J.-C. Kassing, F. Frohn, and J. Giesl. *From Innermost to Full Almost-Sure Termination of Probabilistic Term Rewriting* - AProVE *Artifact*. 2024. DOI: 10.5281/zenodo.10449299.

[32] J.-C. Kassing, S. Dollase, and J. Giesl. "A Complete Dependency Pair Framework for Almost-Sure Innermost Termination of Probabilistic Term Rewriting". In: *Proc. FLOPS '24*. LNCS. To appear. Long version at *CoRR* abs/2309.00344. 2024. DOI: 10.48550/arXiv.2309.00344.

[33] M. Korp, C. Sternagel, H. Zankl, and A. Middeldorp. "Tyrolean Termination Tool 2". In: *Proc. RTA '09*. LNCS 5595. 2009, pp. 295–304. DOI: 10.1007/978-3-642-02348-4_21.

[34] M. R. K. Krishna Rao. "Some Characteristics of Strong Innermost Normalization". In: *Theor. Comput. Sc.* 239 (2000), pp. 141–164. DOI: 10.1016/S0304-3975(99)00215-7.

[35] L. Leutgeb, G. Moser, and F. Zuleger. "Automated Expected Amortised Cost Analysis of Probabilistic Data Structures". In: *Proc. CAV '22*. LNCS 13372. 2022, pp. 70–91. DOI: 10.1007/978-3-031-13188-2_4.

[36] A. McIver, C. Morgan, B. L. Kaminski, and J.-P. Katoen. "A New Proof Rule for Almost-Sure Termination". In: *Proc. ACM Program. Lang.* 2.POPL (2018). DOI: 10.1145/3158121.

[37] F. Meyer, M. Hark, and J. Giesl. "Inferring Expected Runtimes of Probabilistic Integer Programs Using Expected Sizes". In: *Proc. TACAS '21*. LNCS 12651. 2021, pp. 250–269. DOI: 10.1007/978-3-030-72016-2_14.

[38] M. Moosbrugger, E. Bartocci, J. Katoen, and L. Kovács. "Automated Termination Analysis of Polynomial Probabilistic Programs". In: *Proc. ESOP '21*. LNCS 12648. 2021, pp. 491–518. DOI: 10.1007/978-3-030-72019-3_18.

[39] M. H. A. Newman. "On Theories with a Combinatorial Definition of Equivalence". In: *Annals of Mathematics* 43.2 (1942), pp. 223–242. URL: http://www.ens-lyon.fr/LIP/REWRITING/TERMINATION/NEWMAN/Newman.pdf.

[40] V. C. Ngo, Q. Carbonneaux, and J. Hoffmann. "Bounded Expectations: Resource Analysis for Probabilistic Programs". In: *Proc. PLDI '18*. 2018, pp. 496–512. DOI: 10.1145/3192366.3192394.

[41] M. J. O'Donnell. *Computing in Systems Described by Equations*. LNCS 58. 1977. DOI: 10.1007/3-540-08531-9.

[42] N. Saheb-Djahromi. "Probabilistic LCF". In: *Proc. MFCS '78*. LNCS 64. 1978, pp. 442–451. DOI: 10.1007/3-540-08921-7_92.

[43] R. Thiemann and C. Sternagel. "Certification of Termination Proofs Using CeTA". In: *Proc. TPHOLs '09.* LNCS 5674. 2009, pp. 452–468. DOI: 10.1007/978-3-642-03359-9_31.

[44] Y. Toyama. "Counterexamples to the Termination for the Direct Sum of Term Rewriting Systems". In: *Inf. Proc. Lett.* 25 (1987), pp. 141–143. DOI: 10.1016/0020-0190(87)90122-0.

[45] D. Wang, D. M. Kahn, and J. Hoffmann. "Raising Expectations: Automating Expected Cost Analysis with Types". In: *Proc. ACM Program. Lang.* 4.ICFP (2020). DOI: 10.1145/3408992.

[46] A. Yamada, K. Kusakari, and T. Sakabe. "Nagoya Termination Tool". In: *Proc. RTA-TLCA '14.* LNCS 8560. 2014, pp. 466–475. DOI: 10.1007/978-3-319-08918-8_32.

Dimension-Minimality and Primality of Counter Nets***

Shaull Almagor[1(✉)], Guy Avni[2(✉)], Henry Sinclair-Banks[3(✉)], and Asaf Yeshurun[1(✉)]

[1] Technion, Haifa, Israel
shaull@technion.ac.il, asafyeshurun@campus.technion.ac.il
[2] Department of Computer Science, University of Haifa, Haifa, Israel
gavni@cs.haifa.ac.il
[3] Centre for Discrete Mathematics and its Applications (DIMAP) & Department of Computer Science, University of Warwick, Coventry, UK
h.sinclair-banks@warwick.ac.uk

Abstract. A k-Counter Net (k-CN) is a finite-state automaton equipped with k integer counters that are not allowed to become negative, but do not have explicit zero tests. This language-recognition model can be thought of as labelled vector addition systems with states, some of which are accepting. Certain decision problems for k-CNs become easier, or indeed decidable, when the dimension k is small. Yet, little is known about the effect that the dimension k has on the class of languages recognised by k-CNs. Specifically, it would be useful if we could simplify algorithmic reasoning by reducing the dimension of a given CN.

To this end, we introduce the notion of dimension-primality for k-CN, whereby a k-CN is prime if it recognises a language that cannot be decomposed into a finite intersection of languages recognised by d-CNs, for some $d < k$. We show that primality is undecidable. We also study two related notions: dimension-minimality (where we seek a single language-equivalent d-CN of lower dimension) and language regularity. Additionally, we explore the trade-offs in expressiveness between dimension and non-determinism for CN.

1 Introduction

A k-*dimensional Counter Net* (k-CN) is a finite-state automaton equipped with k integer counters that are not allowed to become negative, but do not have explicit zero tests (see Fig. 1a for an example). This language-recognition model can be thought of as an alphabet-labelled Vector Addition System with States (VASS), some of whose states are accepting [7]. A k-CN $\mathcal{A}$ over alphabet Σ

* S. Almagor was supported by the ISRAEL SCIENCE FOUNDATION (grant No. 989/22), G. Avni was supported by the ISRAEL SCIENCE FOUNDATION (grant No. 1679/21), H. Sinclair-Banks was supported by EPSRC Standard Research Studentship (DTP), grant number EP/T5179X/1.

** The full version can be found on https://arxiv.org/abs/2307.14492

N. Kobayashi and J. Worrell (Eds.): FoSSaCS 2024, LNCS 14575, pp. 229–249, 2024.
https://doi.org/10.1007/978-3-031-57231-9_11

accepts a word $w \in \Sigma^*$ if there is a run of $\mathcal{A}$ on w that ends in an accepting state in which the counters stay non-negative. The *language* of $\mathcal{A}$ is the set $\mathcal{L}(\mathcal{A})$ of words accepted by $\mathcal{A}$.

Counter nets are a natural model of concurrency and are closely related — and equivalent, in some senses — to labelled Petri Nets. These models have received significant attention over the years [6,7,13,14,17,19,27], with specific interest in the one-dimensional case, often referred to as one-counter nets [20,21,1,2]. Unfortunately, most decision problems for k-CNs are notoriously difficult and are often undecidable [1,2]. In particular, k-CNs subsume VASS and Petri nets, for which many problems are known to be Ackermann-complete, for example see the recent breakthrough in the complexity of reachability in VASS [11,25].

In many cases, the complexity of decision problems for VASS, sometimes with extensions, depends on the dimension, with low dimensions admitting more tractable solutions. [9,8,10,16]. For example, reachability in dimensions one and two is NP-complete [18] and PSPACE-complete [4], respectively, when counter updates are encoded in binary.

A natural question, therefore, is whether we can *decrease* the dimension of a given a k-CN whilst maintaining its language, to facilitate reasoning about it. More generally, the trade-off between expressiveness and the dimension of Counter Nets is poorly understood. We tackle this question in this work by introducing two approaches. The first is straightforward *dimension-minimality*: given a k-CN, does there exist a d-CN $\mathcal{B}$ recognising the same language for some $d < k$?

The second approach is *primality*: given a k-CN, does there exist some $d < k$ and d-CNs $\mathcal{B}_1, \ldots, \mathcal{B}_n$ such that $L(\mathcal{A}) = \bigcap_{i=1}^n \mathcal{L}(\mathcal{B}_i)$? That is, we ask whether the language of $\mathcal{A}$ can be decomposed as an intersection of languages recognised by several lower-dimension CNs. We also consider *compositeness*, the dual of primality. Intuitively, in a composite k-CN the usage of the counters can be "split" across several lower-dimension CNs, allowing for properties (such as universality) to be checked on each conjunct separately.

Example 1. We illustrate the model and the definition of compositeness. Consider the 2-CN $\mathcal{A}$ depicted in Fig. 1a, and consider a word $w = a^m\#b^n\#c^k$. We have that $\mathcal{A}$ has an accepting run on w iff $m \geq n$ and $m \geq k$. Indeed, if $m < n$, the first counter drops below 0 while cycling in the second state and so the run is "stuck", and similarly if $m < k$. It is not hard to show that there is no 1-CN that recognizes the language of $\mathcal{A}$. However, Fig. 1b shows two 1-CNs $\mathcal{B}_1$ and $\mathcal{B}_2$ such that $\mathcal{L}(\mathcal{B}) = \mathcal{L}(\mathcal{B}_1) \cap \mathcal{L}(\mathcal{B}_2)$. Indeed, a word $w = a^m\#b^n\#c^k \in \mathcal{L}(\mathcal{B}_1)$ iff $m \geq n$, and $w \in \mathcal{L}(\mathcal{B}_2)$ iff $m \geq k$.

Note that the decomposition in Example 1 is obtained by "splitting" the counters between the two 1-CNs. This raises the question of whether such splittings are always possible. As we show in Proposition 1, for deterministic k-CNs (k-DCNs) this is indeed the case. In general, however, it is not hard to find examples where a k-CN cannot simply be split to an intersection by projecting on each counter. This however, does not rule out that other decompositions are

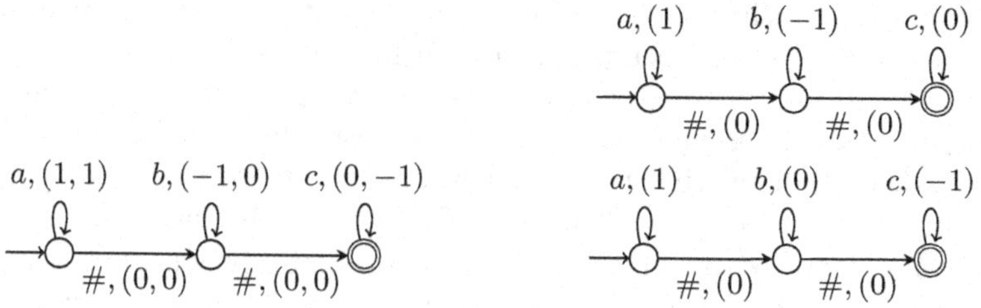

(a) A composite 2-CN. (b) Two 1-CNs showing compositeness of the 2-CN.

Fig. 1: A composite 2-CN whose language is $\{a^m \# b^n \# c^k \mid m \geq n \wedge m \geq k\}$ and its decomposition into two 1-CNs recognising the languages $\{a^m \# b^n \# c^k \mid m \geq n\}$ and $\{a^m \# b^n \# c^k \mid m \geq k\}$.

possible. Our main result, Theorem 1, gives an example of a prime 2-CN. That is, a 2-CN whose language cannot be expressed as an intersection of 1-CNs.

The notion of primality has been studied for regular languages in [24,23,22], the exact complexity of deciding primality is still open. There, an automaton is composite if it can be written as an intersection of finite automata with fewer *states*. In this work we introduce primality for CNs. We focus on *dimension* as a measure of size, a notion which does not exist for regular languages. Thus, unlike regular languages, the differences between prime and composite CNs is not only in succinctness, but actually in expressiveness, as we later demonstrate.

We parameterise primality and compositeness by the dimension d and the number n of lower-dimension factors. Thus, a k-CN $\mathcal{A}$ is (d, n)*-composite* if it can be written as the intersection above. Then, $\mathcal{A}$ is *composite* if it is (d, n)-composite for some $d < k$ and $n \in \mathbb{N}$. Under this view, dimension-minimality is a special case of compositeness, namely $\mathcal{A}$ is dimension-minimal if it is not $(k-1, 1)$-composite. Another particular problem captured by compositeness is *regularity*. Indeed, $\mathcal{L}(\mathcal{A})$ is regular if and only if $\mathcal{A}$ is $(0, 1)$-composite, since 0-CNs are just NFAs. Since regularity is already undecidable for 1-CNs [2,28], it follows that deciding whether a k-CN is (d, n)-composite is undecidable. Moreover, it follows that both primality and dimension-minimality are undecidable for 1-CNs.

The undecidability of the above problems is not surprising, as the huge difference in expressive power between 1-CNs and regular languages is well understood. In contrast, even the expressive power difference between 1-CNs and 2-CNs is poorly understood, let alone what effect the dimension has on the expressive power beyond regular languages. Already, 1-VASS and 2-VASS are known to have *flat* equivalents with respect to reachability [26,4], but the complexity differs greatly.

Our goal in this work is to shed light on these differences. In Section 4, we give a concrete example of a prime 2-CN, which turns out to be technically challenging. This example is the heart of our technical contribution, and we emphasise that we *do not* currently have a proved example of a prime 3-CN, let alone for general k-CN (although we conjecture a candidate for such languages). We consider this an interesting open problem, as it highlights the type of pumping machinery that is currently missing from the VASS/CN reasoning arsenal. The technical intricacy in proving our example suggests that generalising it is highly nontrivial. Indeed, proving this claim would require intricate pumping arguments, which are notoriously difficult even for low-dimensional CNs [9].

Using our example, we obtain in Section 5, the undecidability of primality and of dimension-minimality for 2-CNs. To complement this, we show in Theorem 3, that regularity of k-DCNs is decidable. In Section 6, we explore trade-offs in expressiveness of CNs with increasing dimension and with nondeterminism. In particular, we show that there is a strict hierarchy of expressiveness with respect to the dimension. We conclude with a discussion in Section 7. For brevity, some proofs only appear in the full version of the paper.

2 Preliminaries

We denote the non-negative integers $\{0, 1, \ldots\}$ by $\mathbb{N}$. We write vectors in bold, e.g., $\boldsymbol{e} \in \mathbb{Z}^k$, and $\boldsymbol{e}[i]$ is the i-th coordinate. We use $[k] = \{1, \ldots, k\}$ for $k \geq 1$. We use Σ^* to denote the set of all words over an alphabet Σ, and $|w|$ is the length of $w \in \Sigma^*$.

A *k-dimensional Counter Net* (k-CN) $\mathcal{A}$ is a quintuple $\mathcal{A} = \langle \Sigma, Q, Q_0, \delta, F \rangle$ where Σ is a finite alphabet, Q is a finite set of states, $Q_0 \subseteq Q$ is the set of initial states, $\delta \subseteq Q \times \Sigma \times \mathbb{Z}^k \times Q$ is a set of transitions, and $F \subseteq Q$ are the accepting states. A k-CN is *deterministic*, denoted k-DCN, if $|Q_0| = 1$, and for every $p \in Q$ and $\sigma \in \Sigma$ there is at most one transition of the form $(p, \sigma, \boldsymbol{v}, q) \in \delta$. For a transition $(p, \sigma, \boldsymbol{v}, q) \in \delta$, we refer to $\boldsymbol{v} \in \mathbb{Z}^k$ as its *effect*.

An $\mathbb{N}$-configuration (resp. $\mathbb{Z}$-configuration) of a k-CN $\mathcal{A}$ is a pair $(q, \boldsymbol{v}) \in Q \times \mathbb{N}^k$ (resp. $(q, \boldsymbol{v}) \in Q \times \mathbb{Z}^k$) representing the current state and values of the counters. A transition $(p, \sigma, \boldsymbol{e}, q) \in \delta$ is *valid* from $\mathbb{N}$-configuration $(q, \boldsymbol{v})$ if $\boldsymbol{v}+\boldsymbol{e} \in \mathbb{N}^k$, i.e., if all k counters remain non-negative after the transition. A *$\mathbb{Z}$-run* ρ of $\mathcal{A}$ on w is a sequence of $\mathbb{Z}$-configurations $\rho = (q_0, \boldsymbol{v}_0), (q_1, \boldsymbol{v}_1), \ldots, (q_n, \boldsymbol{v}_n)$ such that $(q_i, \sigma_i, \boldsymbol{v}_{i+1} - \boldsymbol{v}_i, q_{i+1}) \in \delta$ for every $0 \leq i \leq n-1$, we may also say that *ρ reads* $w = \sigma_0\sigma_1 \cdots \sigma_n$. An *$\mathbb{N}$-run* is a $\mathbb{Z}$-run that visits only $\mathbb{N}$-configurations. Note that all the transitions in an $\mathbb{N}$-run are valid. We may omit $\mathbb{N}$ or $\mathbb{Z}$ from the run when it does not matter. For a run $\rho = (q_0, \boldsymbol{v}_0), (q_1, \boldsymbol{v}_1), \ldots, (q_n, \boldsymbol{v}_n)$ of $\mathcal{A}$, we denote $(q_0, \boldsymbol{v}_0) \xrightarrow{\rho} (q_n, \boldsymbol{v}_n)$. We define the *effect* of ρ to be $\mathsf{eff}(\rho) = \boldsymbol{v}_n - \boldsymbol{v}_0$.

An $\mathbb{N}$-run ρ is *accepting* if $q_0 \in Q_0$, $\boldsymbol{v}_0 = \boldsymbol{0}$, and $q_n \in F$. We say that *$\mathcal{A}$ accepts w* if there is an accepting $\mathbb{N}$-run of $\mathcal{A}$ on w. The *language* of $\mathcal{A}$ is $\mathcal{L}(\mathcal{A}) = \{w \in \Sigma^* \mid \mathcal{A} \text{ accepts } w\}$. We say that $\mathcal{A}$ is *unambiguous* if it has at most one accepting run on any given word. Otherwise we say that it is ambiguous.

An infix $\pi = (q_k, \boldsymbol{v}_k), (q_{k+1}, \boldsymbol{v}_{k+1}), \ldots, (q_{k+n}, \boldsymbol{v}_{k+n})$ of a run ρ is a *cycle* if $q_k = q_{k+n}$ and is a *simple cycle* if it does not contain a cycle as a proper infix. When discussing an infix π of a 1-CN – we write that π is > 0, ≥ 0, or < 0 if $\mathsf{eff}(\pi) > 0$, $\mathsf{eff}(\pi) \geq 0$, or $\mathsf{eff}(\pi) < 0$, respectively.

3 Primality and Compositeness

We begin by presenting our main definitions, followed by some introductory properties.

Definition 1 (Compositeness, Primality, and Dimension-Minimality). *Consider a k-CN $\mathcal{A}$, and let $d, n \in \mathbb{N}$. We say that $\mathcal{A}$ is* (d, n)-composite *if there exist d-CNs $\mathcal{B}_1, \ldots, \mathcal{B}_n$ such that $\mathcal{L}(\mathcal{A}) = \bigcap_{i=1}^{n} \mathcal{L}(\mathcal{B}_i)$. If $\mathcal{A}$ is (d, n)-composite for some $d < k$ and $n \in \mathbb{N}$, we say $\mathcal{A}$ is* composite. *Otherwise, $\mathcal{A}$ is* prime. *If $\mathcal{A}$ is not $(k-1, 1)$-composite, we say that $\mathcal{A}$ is* dimension-minimal. *We also extend the definition of primality to languages, and say that a language $\mathcal{L}$ is prime if there is an integer $d > 0$ such that $\mathcal{L} = \mathcal{L}(\mathcal{A})$ for some d-CN $\mathcal{A}$, but there are no $(d-1)$-CNs $\mathcal{B}_1, \ldots \mathcal{B}_n$ such that $\mathcal{L} = \bigcap_{i=1}^{n} \mathcal{L}(\mathcal{B}_i)$.*

Remark 1. Note that the special case where $\mathcal{A}$ is $(0, n)$-composite coincides with the regularity of $\mathcal{L}(\mathcal{A})$, and hence also with being $(0, 1)$-composite.

Observe that in Fig. 1 we in fact show a composite 2-DCN. We now show that every k-DCN is $(1, k)$-composite, by projecting to each of the counters separately. In particular, a k-DCN is prime only when $k = 1$ and it recognises a non-regular language, or when $k = 0$. Formally, consider a k-DCN $\mathcal{D} = \langle \Sigma, Q, Q_0, \delta, F \rangle$ and let $1 \leq i \leq k$. We define the *i-projection* to be the 1-DCN $\mathcal{D}|_i = \langle \Sigma, Q, Q_0, \delta|_i, F \rangle$ where $\delta|_i = \{(p, \sigma, \boldsymbol{v}[i], q) \mid (p, \sigma, \boldsymbol{v}, q) \in \delta\}$.

Proposition 1. *Every k-DCN $\mathcal{D}$ is $(1, k)$-composite, and $\mathcal{L}(\mathcal{D}) = \bigcap_{i=1}^{k} \mathcal{L}(\mathcal{D}|_i)$.*

Proof. Let $w \in \mathcal{L}(\mathcal{D})$ and let ρ be the accepting run of $\mathcal{D}$ on w, then the projection of ρ on counter i induces an accepting run of $\mathcal{D}|_i$ on w, thus $w \in \bigcap_{i=1}^{k} \mathcal{L}(\mathcal{D}|_i)$. Note that this direction does not use the determinism of $\mathcal{D}$.

Conversely, let $w \in \bigcap_{i=1}^{k} \mathcal{L}(\mathcal{D}|_i)$, then each $\mathcal{D}|_i$ has an accepting run ρ_i on w. Since the structure of all the $\mathcal{D}|_i$ is identical to that of $\mathcal{D}$, all the runs ρ_i have identical state sequences, and therefore are also a $\mathbb{Z}$-run of $\mathcal{D}$ on w. Moreover, due to this being a single $\mathbb{N}$-run in each $\mathcal{D}|_i$, it follows that all counter values remain non-negative in the corresponding run of $\mathcal{D}$ on w. Hence, this is an accepting $\mathbb{N}$-run of $\mathcal{D}$ on w, so $w \in \mathcal{L}(\mathcal{D})$. □

Remark 2 (Unambiguous Counter Nets are Composite). The proof of Proposition 1 applies also to *structurally unambiguous* CNs, i.e. CNs whose underlying automaton, disregarding the counters, is unambiguous. Thus, every unambiguous CN is $(1, k)$-composite.

Consider k-CNs $\mathcal{B}_1, \ldots, \mathcal{B}_n$. By taking their product, we can construct a $k \cdot n$-CN $\mathcal{A}$ such that $\mathcal{L}(\mathcal{A}) = \bigcap_{i=1}^{n} \mathcal{L}(\mathcal{B}_i)$. In particular, if each $\mathcal{B}_i$ is a 1-DCN, then $\mathcal{A}$ is an n-DCN. Combining this with Proposition 1, we can deduce the following (proof can be found in the full version).

Proposition 2. *A k-DCN is dimension-minimal if and only if it is not $(1, k-1)$-composite.*

4 A Prime Two-Counter Net

In this section we present our main technical contribution, namely an example of a prime 2-CN. The technical difficulty arises from the need to prove that this example cannot be decomposed as an *intersection* of *nondeterministic* 1-CNs. Since intersection has a "universal flavour", and nondeterminism has an "existential flavour", we have a sort of "quantifier alternation" which is often a source of difficulty.

The importance of this example is threefold. First, it enables us to show that primality is undecidable in Section 5. Second, it offers intuition on what makes a language prime. Third, we suspect that the techniques developed here will be useful in other settings when reasoning about nondeterministic automata, perhaps with counters.

We start by presenting the prime 2-CN, followed by an overview of the proof, before delving into the details.

Example 2. Consider the 2-CN $\mathcal{P}$ over alphabet $\Sigma = \{a, b, c, \#\}$ depicted in Fig. 2. Intuitively, $\mathcal{P}$ starts by reading segments of the form $a^m\#$, where in each segment it nondeterministically chooses whether to increase the first or second counter by m. Then, it reads $b^{m_b}c^{m_c}$ and accepts if the value of the first and second counter is at least m_b and m_c, respectively. Thus, $\mathcal{P}$ accepts a word if its $a^m\#$ segments can be partitioned into two sets I and $\overline{I}$ so that the combined lengths of the segments in I (resp. $\overline{I}$) is at least the length of the b segment (resp. c segment). For example, $a^{10}\#a^{20}\#a^{15}\#b^{15}c^{30} \in \mathcal{L}(\mathcal{P})$, since segments 1 and 2 have length 30, matching c^{30} and segment 3 matches b^{15}. However, $a^{10}\#a^{20}\#a^{15}\#b^{21}c^{21} \notin \mathcal{L}(\mathcal{P})$, since in any partition of $\{10, 20, 15\}$, one set will have sum lower than 21. More precisely, we have the following:

$$\mathcal{L}(\mathcal{P}) = \{a^{m_1}\#a^{m_2}\#\cdots\#a^{m_t}\#b^{m_b}c^{m_c} \mid \exists I \subseteq [t] \text{ s.t. } \sum_{i\in I} m_i \geq m_b \wedge \sum_{i\notin I} m_i \geq m_c\}$$

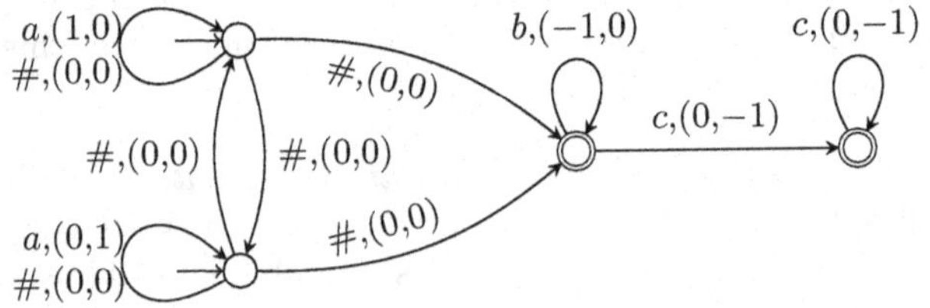

Fig. 2: The prime 2-CN $\mathcal{P}$ for Example 2 and Theorem 1.

Theorem 1. *$\mathcal{P}$ is prime.*

The high-level intuition behind Theorem 1 is that any 1-CN can either guess a subset of segments that covers m_b or m_c, but not both, and in order to make sure the choices between two 1-CNs form a partition, we need to fix the partition in advance. This is only possible if the number of segments is a priori fixed, which is not true (c.f., Remark 3). This intuition, however, is far from a proof.

4.1 Overview of the Proof of Theorem 1

Assume by way of contradiction that $\mathcal{P}$ is not a prime 2-CN. Thus, there exist 1-CNs $\mathcal{V}_1, \dots \mathcal{V}_k$ such that $\mathcal{L}(\mathcal{P}) = \bigcap_{1 \le j \le k} \mathcal{L}(\mathcal{V}_j)$. Throughout the proof, we focus on words of the form $a^{m_1}\#a^{m_2}\#\cdots\#a^{m_{k+1}}\#b^{m_b}c^{m_c}$ for positive integers $\{m_i\}_{i=1}^{k+1}, m_b, m_c$. We index the a^{m_i} segments of these words, so a^{m_i} is the i-th segment. Note that we focus on words with $k+1$ many a segments, one more than the number of $\mathcal{V}_j$ factors in the intersection. It is useful to think about each segment as "paying" for either b or c. Then, a word is accepted if there is a way to choose for each segment whether it pays for b or c, such that there is sufficient budget for both.

Let $i \in [k+1]$ and $j \in [k]$. We say that the i-th segment is *bad* in $\mathcal{V}_j$ if, intuitively, we can pump the length m_i of segment i whilst pumping both m_b and m_c to unbounded lengths, such that the resulting words are accepted by $\mathcal{V}_j$ (see Definition 2 for the formal definition). For example, consider the word $a^{10}\#a^{10}\#a^{10}\#b^{20}c^{10} \in \mathcal{L}(\mathcal{P})$. If the second segment is bad for $\mathcal{V}_j$ then there exist $x, y, z > 0$ such that for every $t, t_b, t_c \in \mathbb{N}$ it holds that the word $a^{10}\#a^{10+tx}\#a^{10}\#b^{20+t_b y}c^{10+t_c z}$ is in $\mathcal{L}(\mathcal{V}_j)$. Observe that such behaviour is undesirable, since for large enough t, t_b, t_c, the resulting word is not in $\mathcal{L}(\mathcal{P})$. Note, however, that the existence of such a bad segment is not a contradiction by itself, since the resulting pumped words might not be accepted by some other 1-CN $\mathcal{V}_{j'}$.

In order to reach a contradiction, we need to show the existence of a segment i that is bad for *every* $\mathcal{V}_j$. Moreover, we must also show that arbitrarily increasing m_i, m_b, m_c can be simultaneously achieved in all the $\mathcal{V}_j$ together (i.e., the above $x, y, z > 0$ are the same for all V_j). This would create a contradiction since all the $\mathcal{V}_j$ accept a word that is not in $\mathcal{L}(\mathcal{P})$. Our goal is therefore to establish a robust and precise definition of a "bad" segment, then find a word w comprising $k+1$ segments where one of the segments is bad for every $\mathcal{V}_j$, and pumping the words in each segment can be done synchronously.

4.2 Pumping Arguments in One-Counter Nets

In this section we establish some pumping results for 1-CN which will be used in the proof of Theorem 1. Throughout this section, we consider a 1-CN $\mathcal{V} = \langle \Sigma, Q, Q_0, \delta, F \rangle$.

Our first lemma states the intuitive fact that without > 0 cycles, the counter value of a run is bounded (proof can be found in the full version).

Lemma 1. *Let (q, n) be a configuration of $\mathcal{V}$, let W be the maximal positive update in $\mathcal{V}$, $\sigma \in \Sigma$, and $N \in \mathbb{N}$. If an $\mathbb{N}$-run ρ of $\mathcal{V}$ on σ^N from configuration (q, n) does not traverse any > 0 cycle, then the maximal possible counter value anywhere along ρ is $n + W|Q|$.*

The next lemma shows that long-enough runs must contain ≥ 0 cycles.

Lemma 2. *Let $\sigma \in \Sigma$ and (q, n) be an $\mathbb{N}$-configuration of $\mathcal{V}$. Then there exists $N \in \mathbb{N}$ such that for all $N' \geq N$, every $\mathbb{N}$-run of $\mathcal{V}$ on $\sigma^{N'}$ from (q, n) traverses a ≥ 0 cycle.*

Proof. Let W be the maximal positive transition update in $\mathcal{V}$, we show that $N = |Q|(n + |Q| \cdot W)$ satisfies the requirements. Assume by way of contradiction that $\mathcal{V}$ can read σ^N via an $\mathbb{N}$-run $\rho = (q_0, n_0 = n) \xrightarrow{\rho} (q_N, n_N)$ that only traverses < 0 cycles.

Since ρ visits $N + 1$ states, then by the Pigeonhole Principle, there exists a state $p \in Q$ that is visited $m \geq (N+1)/|Q| > N/|Q|$ many times in ρ.

Consider all the indices $0 \leq i_1 < i_2 < \ldots < i_m \leq N$ such that $p = q_{i_1} = \ldots = q_{i_m}$. Each run segment $(q_{i_1}, n_{i_1}) \to (q_{i_2}, n_{i_2}), \ldots, (q_{i_{m-1}}, n_{i_{m-1}}) \to (q_{i_m}, n_{i_m})$ is a cycle in ρ, and therefore must have negative effect. Thus $n_{i_1} > n_{i_2} > \ldots > n_{i_m} \geq 0$, so in particular $n_{i_1} \geq n_{i_m} + m - 1 \geq 0$ (as each cycle has effect at most -1). Moreover, $n_{i_1} < n + |Q| \cdot W$ since the prefix $(q_0, n) \to (q_{i_1}, n_{i_1})$ cannot contain a non-negative cycle. However, since $m > N/|Q| = n + |Q| \cdot W$ and $n_{i_1} \geq n_{i_m} + m - 1 \geq n + |Q| \cdot W$, we get $n + |Q| \cdot W < n + |Q| \cdot W$ which is a contradiction. □

Next, we show that runs with ≥ 0 and > 0 cycles have "pumpable" infixes.

Lemma 3. *Let $\sigma \in \Sigma$ and consider a > 0 (resp. ≥ 0) cycle $\pi = (q_0, c_0) \xrightarrow{\sigma} (q_1, c_1) \xrightarrow{\sigma} \ldots (q_n = q_0, c_n)$ on σ^n that induces an $\mathbb{N}$-run. Then, there is a sequence of (not necessarily contiguous) indices $0 \leq i_1 \leq \ldots \leq i_k \leq n$ such that $q_{i_1} \xrightarrow{\sigma} q_{i_2} \xrightarrow{\sigma} \cdots q_{i_k}$ is a simple > 0 (resp. ≥ 0) cycle with some effect $e > 0$ (resp. $e \geq 0$). In addition, this simple cycle is "pumpable" from the first occurrence of q_{i_1} in π; namely, for all $m \in \mathbb{N}$ there is a run π_m obtained from π by traversing the cycle m times so that $\mathsf{eff}(\pi_m) = \mathsf{eff}(\pi) + em$.*

Proof. We prove the ≥ 0 case, the > 0 case can be proved mutatis mutandis.

We define $\pi_m = (q_0, c_0) \xrightarrow{\sigma} \ldots (q_{i_1}, c_{i_1}) \xrightarrow{\sigma} \ldots (q_{i_1}, c_{i_1} + em) \xrightarrow{\sigma} \ldots (q_n, c_n + em)$. The proof is now by induction on the length of π.

The base of the induction is a cyclic $\mathbb{N}$-run of length 2. In this case $\pi = (q_0, c_0) \xrightarrow{\sigma} (q_1 = q_0, c_1)$ is itself a ≥ 0 simple cycle that is infinitely pumpable from (q_0, c_0).

We now assume correctness for length n, and discuss $\pi = (q_0, c_0) \xrightarrow{\sigma} (q_1, c_1) \xrightarrow{\sigma} \ldots (q_n = q_0, c_n)$ of length $n + 1$. Let $0 \leq j_1 < j_2 \leq n$ be indices such that $q_{j_1} = q_{j_2}$, for a maximal j_1. Note that the cycle $\tau = (q_{j_1}, c_{j_1}) \xrightarrow{\sigma} \ldots (q_{j_2}, c_{j_2})$ must be simple. If $j_1 = 0$ and $j_2 = n$, then π itself is a simple ≥ 0 cycle, and the pumping argument is straightforward. Otherwise τ is nested. We now split into two cases, based on whether $\mathsf{eff}(\tau) \geq 0$.

1. τ is ≥ 0: then the induction hypothesis applies on τ. We take the guaranteed constants $j_1 \leq i_1 \leq \ldots \leq i_k \leq j_2$, which apply to π as well.

2. τ is < 0: then we remove τ from π to obtain $\pi' = (q_0, c_0) \xrightarrow{\sigma} \dots (q_{j_1}, c_{j_1}) \xrightarrow{\sigma} (q_{j_2+1}, c'_{j_2+1}) \xrightarrow{\sigma} \dots (q_n, c'_n)$, such that $c'_i \geq c_i$ for all $j_2 + 1 \leq i \leq n$. The induction hypothesis applies on π', so let $i_1, \dots, i_k$ be the guaranteed constants. Note that $i_1 \leq j_1$, since the cycle removed when obtaining π' from π is the last occurrence of a repetition of states in π. We therefore know that $q_{i_1} \xrightarrow{\sigma} q_{i_2} \xrightarrow{\sigma} \cdots q_{i_k}$ is a simple ≥ 0 cycle in π' – which applies to π as well. In addition, it is infinitely pumpable from $\mathbb{N}$-configuration (q_{i_1}, c_{i_1}) in π' for $i_1 \leq j_1$. Indeed, since π and π' coincide up to and including (q_{j_1}, c_{j_1}) between π and π' - this cycle is infinitely pumpable in π as well. □

The simple cycle in Lemma 3 has length $k < |Q|$. By pumping it $\frac{|Q|!}{k}$ times we obtain a pumpable cycle of length $|Q|!$, allowing us to conclude with the following.

Corollary 1. *Let ρ be an $\mathbb{N}$-run of $\mathcal{V}$ on σ^n that traverses a ≥ 0 cycle. For every $m \in \mathbb{N}$, we can construct an $\mathbb{N}$-run ρ' of $\mathcal{V}$ on $\sigma^{n+m|Q|!}$ such that $\mathsf{eff}(\rho') \geq \mathsf{eff}(\rho)$ by pumping a ≥ 0 simple cycle in ρ.*

4.3 Good and Bad Segments

We lift the colour scheme[4] of > 0 and ≥ 0 to words and runs as follows. For a word $w = uv$ and a run ρ, we write e.g., uv to denote that ρ traverses a > 0 cycle when reading u, then a ≥ 0 cycle when reading v. Note that this does not preclude other cycles, e.g., there could also be negative cycles in the u part, etc. That is, the colouring is not unique, but represents elements of the run.

Recall our assumption that $\mathcal{L}(\mathcal{P}) = \bigcap_{1 \leq j \leq k} \mathcal{L}(\mathcal{V}_j)$, and for all $j \in [k]$ denote $\mathcal{V}_j = \langle \Sigma, Q_j, I_j, \delta_j, F_j \rangle$. Let $Q_{\max} = \max\{|Q_j|\}_{j=1}^{k}$ and denote $\alpha = Q_{\max}!$. Further recall that we focus on words of the form $a^{m_1}\#a^{m_2}\#\cdots\#a^{m_{k+1}}\#b^{m_b}c^{m_c}$ for integers $\{m_i\}_{i=1}^{k+1}, m_b, m_c \in \mathbb{N}$, and that we refer to the infix a^{m_i} as the i-th segment, for $1 \leq i \leq k+1$. We proceed to formally define good and bad segments.

Definition 2 (Good and Bad Segments). *The i-th segment is* bad *in $\mathcal{V}_j$ if there exist constants $\{m_i\}_{i=1}^{k+1}, m_b, m_c \in \mathbb{N}$ such that the following hold.*

(a) $\{m_i\}_{i=1}^{k+1}, m_b, m_c$ are multiples of α, and
(b) there is an accepting $\mathbb{N}$-run ρ of $\mathcal{V}_j$ on $w = a^{m_1}\#a^{m_2}\#\cdots\#a^{m_{k+1}}\#b^{m_b}c^{m_c}$ that adheres to one of the three forms:
  (i) $a^{m_1}\#a^{m_2}\#\cdots a^{m_{i-1}}\#a^{m_i}\#a^{m_{i+1}}\#\cdots\#a^{m_{k+1}}\#b^{m_b}c^{m_c}$,
  (ii) $a^{m_1}\#a^{m_2}\#\cdots a^{m_{i-1}}\#a^{m_i}\#a^{m_{i+1}}\#\cdots\#a^{m_{k+1}}\#b^{m_b}c^{m_c}$, or
  (iii) $a^{m_1}\#a^{m_2}\#\cdots a^{m_{i-1}}\#a^{m_i}\#a^{m_{i+1}}\#\cdots\#a^{m_{k+1}}\#b^{m_b}c^{m_c}$.

The i-th segment is good *in $\mathcal{V}_j$ if it is not bad in $\mathcal{V}_j$.*

---

[4] The colours were chosen as accessible for the colourblind. For a greyscale-friendly version, see the full paper.

Lemma 4 formalises the intuition that a bad segment can be pumped simultaneously with both the b and c segments, giving rise to a word accepted by $\mathcal{V}_j$ but rejected by $\mathcal{P}$.

Intuitively, Forms (ii) and (iii) indicate that all segments are bad. Indeed, the i-th segment has a ≥ 0 cycle, so it can be pumped safely, and in Form (ii) both b and c can be pumped using ≥ 0 cycles. Whereas in Form (iii) we can pump b using a > 0 cycle, and can use it to compensate for pumping c, even if the latter requires iterating a negative cycle.

Form (i) is the interesting case, where we use a > 0 cycle in the i-th segment to compensate for pumping both b and c. The requirement that all segments up to the i-th are ≥ 0 is at the core of our proof and is explained in Section 4.4.

Lemma 4. *Suppose the l-th segment is bad in $\mathcal{V}_j$, then there exist $x, y, z \in \mathbb{N}$, that are multiples of α, such that for every $n \in \mathbb{N}$ the following word w is accepted by $\mathcal{V}_j$.*

$$w_n = a^{m_1}\#a^{m_2}\#\cdots\#a^{m_{l-1}}\#a^{m_l+xn}\#a^{m_{l+1}}\#\cdots\#a^{m_{k+1}}\#b^{m_b+yn}c^{m_c+zn}$$

Proof. We can choose $z = \alpha$, then take y to be large enough so that Form (iii) runs can compensate for negative cycles in c^z using > 0 cycles in b^y, whilst not decreasing the counters in Form (ii) runs. We can indeed find such a $y \in \mathbb{N}$ that is a multiple of α, since α is divisible by all lengths of simple cycles. Finally, we choose x so that Form (i) runs can compensate for c^z and b^y using > 0 cycles on a^x in the l-th segment, again whilst not decreasing the counters in Forms (ii) and (iii). □

Recall that our goal is to show that there is a segment $l \in [k+1]$ that is bad in *every* V_j, for $j \in [k]$. In Lemma 5, We show that each V_j has at most one good segment. Therefore, there are at most k good segments in total, leaving at least one segment that is bad in every $\mathcal{V}_j$, as desired.

Lemma 5. *Let $j \in [k]$ and $0 \leq r < s \leq k+1$. Then the r-th or s-th segment is bad in $\mathcal{V}_j$.*

Proof. Since j is fixed, denote $\mathcal{V}_j = \langle \Sigma, Q, Q_0, \delta, F\rangle$. We inductively define constants $\{n_i\}_{i=1}^{k+1}, n_b, n_c \in \mathbb{N}$ as follows. Suppose that n_1 is a large-enough multiple of α so that Lemma 2 guarantees a ≥ 0 cycle in any accepting run of $\mathcal{V}_j$ on a^{n_1} from some $(q_0, 0)$ with $q_0 \in Q_0$. Now, assume that we have defined $n_1, \ldots n_{l-1}$, and consider the word $u = a^{n_1}\#a^{n_2}\#\cdots\#a^{n_{l-1}}\#$. Define $n = |u| \cdot W$ where W is the maximal update of any transition of $\mathcal{V}_j$. Since u consists of $\frac{n}{W}$ letters, $n+1$ is greater than any counter value that can be observed in any run of $\mathcal{V}_j$ on u. We define n_l to be a multiple of α large enough so that Lemma 2 guarantees a ≥ 0 cycle when reading a^{n_l} from any configuration of the form $\{(q, n') \mid q \in Q,\ n' \leq n+1\}$. We set $n_b = n_c = \alpha$, the choice of n_b, n_c is somewhat arbitrary. Finally, we set $w = a^{n_1}\#\cdots\#a^{n_{k+1}}\#b^{n_b}c^{n_c}$.

Now, for every $x \in \mathbb{N}$, we obtain from w a word w_x by pumping $x\alpha$ many a's in the r-th and s-th segments and pumping $x\alpha$ many b's and c's in their segments. That is, let $n'_i = n_i + x\alpha$ for $i \in \{r, s\}$ and $n'_i = n_i$ for $i \notin \{r, s\}$, and let $n'_b = n_b + x\alpha$ and $n'_c = n_c + x\alpha$, then $w_x = a^{n'_1}\#\cdots\#a^{n'_{k+1}}\#b^{n'_b}c^{n'_c}$. Observe that $w_x \in \mathcal{L}(\mathcal{P})$. Indeed, since $n_r \geq n_b = \alpha$ and $n_s \geq n_c = \alpha$ we have that $n_r + x\alpha \geq n_b + x\alpha$ and $n_s + x\alpha \geq n_c + x\alpha$, so the r-th and s-th segments can already pay for the b's and c's, respectively. In particular, $w_x \in \mathcal{L}(\mathcal{V}_j)$ via some accepting $\mathbb{N}$-run ρ_x.

We choose a particular value of x, as follows. Consider x and suppose some accepting $\mathbb{N}$-run ρ_x as above does not traverse a > 0 cycle neither in r-th nor s-th segment. By Lemma 1, the maximal possible counter value of ρ_x after reading

$$a^{n_1}\#\cdots\#a^{n_r+x\alpha}\#\cdots\#a^{n_s+x\alpha}\#\cdots\#a^{n_{k+1}}\#$$

is $M_b = (k+1+\sum_{z\in[k+1]\setminus\{r,s\}} n_z)\cdot W + 2|Q|\cdot W$. Crucially, this value does not depend on x. Further, if there is no > 0 cycle in the segment of b's as well, again the maximal counter value of ρ up to the c segment is bounded by $M_c = (k+2+\sum_{z\in[k+1]\setminus\{r,s\}} n_z)\cdot W + 3|Q|\cdot W$, that is independent of x and M_b. By Lemma 2, we can now choose x large enough to satisfy that for every accepting $\mathbb{N}$-run ρ_x on w_x:

1. If ρ_x does not traverse any > 0 cycle in the r-th or s-th segments, then ρ_x has a ≥ 0 cycle reading $b^{(n_b+x\alpha)}$ from any configuration in $\{(q, M') \mid q \in Q,\ M' \leq M_b\}$.
2. If ρ_x does not traverse any > 0 cycle in the r-th or s-th segment, nor in the b segment, then ρ_x has a ≥ 0 cycle reading $c^{(n_c+x\alpha)}$ from any configuration in $\{(q, M') | q \in Q,\ M' \leq M_c\}$.

Having fixed x, we claim that for the constants of w_x, one of the r-th or s-th segment is bad in $\mathcal{V}_j$. By construction, Lemma 2 guarantees that ρ_x has ≥ 0 cycles in segments $1, \ldots r-1$. If ρ_x has a > 0 cycle in segment r, then ρ_x is of Form (i):

$$a^{n_1}\#a^{n_2}\#\cdots\#a^{n_{r-1}}\#a^{n_r+x\alpha}\#\cdots\#a^{n_s+x\alpha}\#\cdots\#a^{n_{k+1}}\#b^{n_b+x\alpha}c^{n_c+x\alpha}$$

and so the r-th segment must be bad in $\mathcal{V}_j$.

Otherwise, if ρ_x does not have a > 0 cycle in the r-th segment, then the construction in Lemma 2 guarantees ≥ 0 cycles in segments indexed $r, r+1, \ldots, s-1$. Indeed, for the r-th segment, we are guaranteed a ≥ 0 cycle reading a^{n_r}, all the more for $a^{n_r+x\alpha}$. As for segments indexed $r+1, \ldots s-1$, if ρ_x does not have a > 0 cycle in the r-th segment, then the maximal effect of segment r is $|Q|\cdot W$. However, n_{r+1} was constructed to guarantee a ≥ 0 cycle even in case the effect of segment r is $Wn_r \geq W\alpha \geq W|Q|$.

If there is a > 0 cycle in segment s, then ρ_x is again of Form (i):

$$a^{n_1}\#a^{n_2}\#\cdots\#a^{n_{s-1}}\#a^{n_s+x\alpha}\#a^{n_{s+1}}\#\cdots\#a^{n_{k+1}}\#b^{n_b+x\alpha}c^{m_c+x\alpha}$$

and so the s-th segment must be bad in $\mathcal{V}_j$.

Otherwise, using the same arguments as for the r-th segment, we have that segments indexed $s+1, \ldots, k+1$ each contain a ≥ 0 cycle. In this case we are left with the b and c segments. The choice of x guarantees a ≥ 0 cycle in the b segment. If ρ_x traverses a > 0 cycle in the b segment, then w_x is of Form (iii).

$$a^{n_1}\#a^{n_2}\#\cdots\#a^{n_{k+1}}\#b^{n_b+x\alpha}c^{n_c+x\alpha}$$

Finally, if there are no > 0 cycles in the b segment, then the choice of x again guarantees a ≥ 0 cycle in the c segment, so w_x is of Form (ii).

$$a^{n_1}\#a^{n_2}\#\cdots\#a^{n_{k+1}}\#b^{n_b+x\alpha}c^{n_c+x\alpha}$$

In the two latter cases, both the r-th and the s-th segments are bad in $\mathcal{V}_j$. □

4.4 Proof of Theorem 1

Given Lemma 5, we now know that each $\mathcal{V}_j$ has at most one good segment. Therefore, all 1-CNs $\mathcal{V}_1, \ldots, \mathcal{V}_k$ together have at most k good segments. Recall that the words we focus on have $k+1$ segments, and therefore there is at least one segment, say the l-th segment, that is bad in every $\mathcal{V}_j$. Note, however, that this segment may correspond to different constants in each $\mathcal{V}_j$. That is, there exists constants $\{m_i^j, m_b^j, m_c^j \mid i \in [k+1], j \in [k]\}$ witnessing that the l-th segment is bad for each $\mathcal{V}_j$. We group the $\mathcal{V}_j$ according to the form of their accepting runs ρ_j (see Definition 2):

(i) $a^{m_1^j}\#a^{m_2^j}\#\cdots\#a^{m_l^j}\#a^{m_{l+1}^j}\#\cdots\#a^{m_{k+1}^j}\#b^{m_b^j}c^{m_c^j}$,
(ii) $a^{m_1^j}\#a^{m_2^j}\#\cdots\#a^{m_l^j}\#a^{m_{l+1}^j}\#\cdots\#a^{m_{k+1}^j}\#b^{m_b^j}c^{m_c^j}$, or
(iii) $a^{m_1^j}\#a^{m_2^j}\#\cdots\#a^{m_l^j}\#a^{m_{l+1}^j}\#\cdots\#a^{m_{k+1}^j}\#b^{m_b^j}c^{m_c^j}$.

We now find constants resulting in a single word for which the l-th segment is bad in every $\mathcal{V}_j$. First, for $i \in [k+1]\setminus\{l\}$, we define $M_i = \max\{m_i^j \mid j \in [k]\}$, note that these values are still multiples of α. Similarly, we define $M_c = \max\{m_c^j \mid j \in [k]\}$. It remains to fix new constants L and B, which we do in phases in the following. The resulting word is then

$$w = a^{M_1}\#\cdots\#a^{M_{l-1}}\#a^{L}\#a^{M_{l+1}}\#\cdots\#a^{M_{k+1}}\#b^{B}c^{M_c}.$$

Most steps in the analysis below are based on Lemma 3 and Corollary 1. We first, partially, handle Form (iii) runs. For such $\mathcal{V}_j$, there is an accepting $\mathbb{N}$-run ρ_j on

$$a^{m_1^j}\#\cdots\#a^{m_{l-1}^j}\#a^{m_l^j}\#a^{m_{l+1}^j}\#\cdots\#a^{m_{k+1}^j}\#b^{m_b^j}c^{m_c^j}$$

By pumping ≥ 0 cycles as per Corollary 1 in all segments except l we obtain an accepting $\mathbb{N}$-run ρ'_j on

$$a^{M_1}\#\cdots\#a^{M_{l-1}}\#a^{m_l^j}\#a^{M_{l+1}}\#\cdots\#a^{M_{k+1}}\#b^{m_b^j}c^{m_c^j}.$$

We now pump arbitrary cycles in the c segment to construct a $\mathbb{Z}$-run ρ''_j on

$$a^{M_1}\#\cdots\#a^{M_{l-1}}\#a^{m_l^j}\#a^{M_{l+1}}\#\cdots\#a^{M_{k+1}}\#b^{m_b^j}c^{M_c}.$$

Next, we compensate for possible negative cycles in the c segment by pumping a > 0 cycle in the b segment. Thus, we construct an $\mathbb{N}$-run ρ'''_j on

$$a^{M_1}\#\cdots\#a^{M_{l-1}}\#a^{m_l^j}\#a^{M_{l+1}}\#\cdots\#a^{M_{k+1}}\#b^{B}c^{M_c},$$

where B is chosen to be large enough such that ρ_j''' is an $\mathbb{N}$-run for all $\mathcal{V}_j$, $j \in [k]$. Note that it remains to fix L.

We now turn to Form (i) with a similar process we start with an accepting $\mathbb{N}$-run ρ_j on

$$a^{m_1^j}\#\cdots\#a^{m_{l-1}^j}\#a^{m_l^j}\#a^{m_{l+1}^j}\#\cdots\#a^{m_{k+1}^j}\#b^{m_b^j}c^{m_c^j}.$$

Pump ≥ 0 cycles in segments indexed $1,\ldots,l-1$ to obtain an accepting $\mathbb{N}$-run ρ_j' on

$$a^{M_1}\#\cdots\#a^{M_{l-1}}\#a^{m_l^j}\#a^{m_{l+1}^j}\#\cdots\#a^{m_{k+1}^j}\#b^{m_b^j}c^{m_c^j}.$$

Now, obtain a $\mathbb{Z}$-run ρ_j'' by pumping arbitrary cycles in the remaining segments, including the b segment.

$$a^{M_1}\#\cdots\#a^{M_{l-1}}\#a^{m_l^j}\#a^{M_{l+1}}\#\cdots\#a^{M_{k+1}}\#b^{B}c^{M_c}$$

Again, compensate for negative cycles by taking L large enough so that pumping > 0 cycles in the l-th segment yields an accepting $\mathbb{N}$-run ρ_j''' on

$$a^{M_1}\#\cdots\#a^{M_{l-1}}\#a^{L}\#a^{M_{l+1}}\#\cdots\#a^{M_{k+1}}\#b^{B}c^{M_c}.$$

We now return to Form (iii) and fix the l-th segment by pumping ≥ 0 cycles to construct an accepting $\mathbb{N}$-run on

$$a^{M_1}\#\cdots\#a^{M_{l-1}}\#a^{L}\#a^{M_{l+1}}\#\cdots\#a^{M_{k+1}}\#b^{B}c^{M_c}.$$

We are left with Form (ii), which are the most straightforward to handle. We simply pump ≥ 0 cycles in all segments to construct an accepting $\mathbb{N}$-run ρ_j' on

$$a^{M_1}\#\cdots\#a^{M_{l-1}}\#a^{L}\#a^{M_{l+1}}\#\cdots\#a^{M_{k+1}}\#b^{B}c^{M_c}.$$

Note that the requirement for all segments before the l-th to be ≥ 0 is crucial, otherwise we won't be able to pump all the cycles in all forms simultaneously.

We now have that w is accepted by every $\mathcal{V}_j$, and the l-th segment is bad for all $\mathcal{V}_j$. By applying Lemma 4 for each of the $\mathcal{V}_j$ and taking global constants to be the products of the respective constants $x, y, z > 0$ for each $\mathcal{V}_j$, we now obtain $X, Y, Z \in \mathbb{N}$, multiples of α, such that for every $n \in \mathbb{N}$ the word

$$w_n = a^{M_1}\#\cdots\#a^{M_{l-1}}\#a^{L+Xn}\#a^{M_{l+1}}\#\cdots\#a^{M_{k+1}}\#b^{B+Yn}c^{M_c+Zn} \in \mathcal{L}(\mathcal{V}_j)$$

is accepted by every $\mathcal{V}_j$, for every $j \in [k]$.

Finally, we choose n large enough to satisfy $\sum_{i\in[k+1]\setminus\{l\}} M_i < \min\{B + Yn, M_c + Zn\}$, so that $w_n \notin \mathcal{L}(\mathcal{P})$. This is possible because, w.l.o.g, the l-th segment can only pay for b, and the remaining segments $[k+1]\setminus\{l\}$ cannot pay for c. This contradicts the assumption that $\mathcal{L}(\mathcal{P}) = \bigcap_{j\in[k]} \mathcal{L}(\mathcal{V}_j)$, concluding the proof of Theorem 1. □

Remark 3 (Unbounded Compositeness). The proof of Theorem 1 shows that if words with $k+1$ segments are allowed, then the language is not $(1, k)$-composite, we use this to establish primality. By intersecting $\mathcal{L}(\mathcal{P})$ with words that allow at most $k+1$ segments, we obtain a language that is not $(1, k)$-composite, but it is not hard to show that it is $(1, 2^{k+1})$-composite. This demonstrates that a 2-CN can be composite, but may require unboundedly many factors.

The intuition behind Theorem 1 is that separate counters are needed to keep track of the elements that "cover" b^{m_b} and c^{m_c}. Extending this idea to k-CN, we require that the a segments are partitioned to k different sets that cover k "targets".

Conjecture 1. The following language is the language of a prime k-CN:

$$\begin{aligned} L_k = &\{a^{m_1}\#a^{m_2}\#\cdots\#a^{m_t}\#b_1^{n_1}\#b_2^{n_2}\cdots\#b_k^{n_k} \mid \\ &\exists I_1,\ldots,I_k \subseteq [t]\ \forall i \in [k],\ \sum_{j\in I_i} m_j \geq n_i\ \wedge \forall i \neq j,\ I_i \cap I_j = \emptyset\} \end{aligned}$$

While constructing a k-CN for L_k is a simple extension of Example 2, proving that it is indeed prime does not seem to succumb to our techniques, and we leave it as an important open problem (see Section 7).

5 Primality of Counter Nets is Undecidable

In this section we consider the *primality* and *dimension-minimality* decision problems: given a k-CN $\mathcal{A}$, decide whether $\mathcal{A}$ is prime and whether $\mathcal{A}$ is dimension-minimal, respectively.

We use our prime 2-CN from Example 2 and the results of Section 4 to show that both problems are undecidable. Our proof is by reduction from the containment problem[5] for 1-CN: given two 1-CN $\mathcal{A},\mathcal{B}$ over alphabet Σ, decide whether $\mathcal{L}(\mathcal{A}) \subseteq \mathcal{L}(\mathcal{B})$. This problem was shown to be undecidable in [20].

We begin by describing the reduction that applies to both problems. Consider an instance of 1-CN containment with two 1-CNs $\mathcal{A}$ and $\mathcal{B}$ over the alphabet Σ. We construct a 2-CN $\mathcal{C}$ as follows. Let Λ be the alphabet of the 2-CN from Example 2 and Theorem 1, and let $\$ \notin \Sigma \cup \Lambda$ be a fresh symbol. Intuitively, $\mathcal{C}$ accepts words of the form $u\$v$ when either $u \in \mathcal{L}(\mathcal{A})$ and v is accepted by $\mathcal{P}$ starting from the maximal counter $\mathcal{A}$ ended with on u, or when $u \in \mathcal{L}(\mathcal{B})$ and $v \in \Lambda^*$.

Formally, we convert $\mathcal{A}$ and $\mathcal{B}$ to 2-CNs $\mathcal{A}'$ and $\mathcal{B}'$ by adding a counter and never modifying its value, so a transition (p,σ,v,q) in $\mathcal{A}$ becomes $(p,\sigma,(v,0),q))$ in $\mathcal{A}'$, for example. We construct a 2-CN $\mathcal{C}$ as follows (see Fig. 3). We take $\mathcal{A}'$, $\mathcal{B}'$, and $\mathcal{P}$, and for every accepting state q of $\mathcal{A}'$ we introduce a transition $(q,\$,\mathbf{0},p_0)$ where p_0 is an initial state of $\mathcal{P}$. We then add a new accepting state $q_\top$ and add the transitions $(q_\top,\lambda,\mathbf{0},q_\top)$ for every letter $\lambda \in \Lambda$, in other words $q_\top$ is an accepting sink for Λ. We also add transitions $(s,\$,\mathbf{0},q_\top)$ from every accepting state s of $\mathcal{B}'$. The initial states are those of $\mathcal{A}'$ and $\mathcal{B}'$, and the accepting states are those of $\mathcal{P}$ and $q_\top$.

Theorem 2. *Primality and dimension-minimality are undecidable, already for 2-CN.*

Proof. We prove the theorem by establishing that $\mathcal{C}$ is not prime if and only if $\mathcal{L}(\mathcal{A}) \subseteq \mathcal{L}(\mathcal{B})$, and $\mathcal{C}$ is not dimension-minimal if and only if $\mathcal{L}(\mathcal{A}) \subseteq \mathcal{L}(\mathcal{B})$.

[5] Actually, the complement thereof.

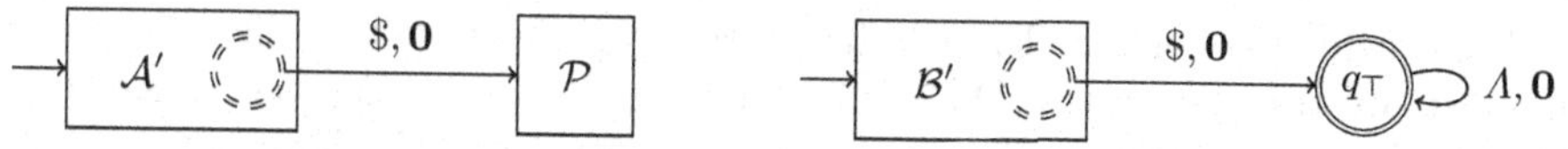

Fig. 3: The reduction from 1-CN non-containment to 2-CN primality and dimension-minimality. The dashed accepting states are those of $\mathcal{A}'$ and $\mathcal{B}'$, and are not accepting in the resulting construction.

Assume that $\mathcal{L}(\mathcal{A}) \subseteq \mathcal{L}(\mathcal{B})$, then the component of $\mathcal{C}$ containing $\mathcal{A}'$ and $\mathcal{P}$ (Fig. 3 left) becomes redundant. Since the component containing $\mathcal{B}'$ and $q_\top$ only makes use of one counter, $\mathcal{C}$ is composite. Formally, we claim that $\mathcal{L}(\mathcal{C}) = \{u\$v \mid u \in \mathcal{L}(\mathcal{B}) \wedge v \in \Lambda^*\}$. Indeed, if $w \in \mathcal{L}(\mathcal{C})$ then $w = u\$v$ so either $u \in \mathcal{L}(\mathcal{A}') = \mathcal{L}(\mathcal{A})$ or $u \in \mathcal{L}(\mathcal{B})$, but since $\mathcal{L}(\mathcal{A}) \subseteq \mathcal{L}(\mathcal{B})$, this is equivalent to $u \in \mathcal{L}(\mathcal{B})$, and in this case there is simply no condition on $v \in \Lambda^*$. Since the second counter is not used in component containing $\mathcal{B}'$ and $q_\top$ (Fig. 3 right), we can construct a 1-CN equivalent to $\mathcal{C}$ by projecting on the first counter and just deleting the component containing $\mathcal{A}'$ and $\mathcal{P}$ completely. It follows that in this case $\mathcal{C}$ is not dimension-minimal, and therefore is not prime either.

For the converse, assume that $\mathcal{L}(\mathcal{A}) \not\subseteq \mathcal{L}(\mathcal{B})$, and let $u \in \mathcal{L}(\mathcal{A}) \setminus \mathcal{L}(\mathcal{B})$. Denote $m = \max\{\mathsf{eff}(\rho) \mid \rho \text{ is an accepting run of } \mathcal{A} \text{ on } u\}$. Thus, for a word $v \in \Lambda^*$ we have that $u\$v \in \mathcal{L}(\mathcal{C})$ if and only if v is accepted in $\mathcal{P}$ with initial counter m. Assume by way of contradiction that $\mathcal{C}$ is not prime, then we can write $\mathcal{L}(\mathcal{C})$ as an intersection of languages of 1-CNs. Loosely speaking, this will create a contradiction as we will be able to argue that $\mathcal{P}$ is not prime. More precisely, take $v = a^{m_1}\#a^{m_2}\#\cdots\#a^{m_{k+1}}\#b^{m_b}c^{m_c}$ for integers $\{m_i\}_{i=1}^{k+1}, m_b, m_c \in \mathbb{N}$ and consider words of the form $u\$v$. Our analysis from Section 4—specifically the arguments used in the proof Lemma 5—on $u\$v$ can show, mutatis mutandis, that the language of $\mathcal{P}$ is not composite regardless of any fixed initial counter value (an analogue of Theorem 1).

We thus have that $\mathcal{C}$ is prime, and in particular $\mathcal{C}$ is dimension-minimal, concluding the correctness of the reduction. □

To contrast the undecidability of primality in nondeterministic CNs, we turn our attention to a decidable fragment of primality, for which we focus on deterministic CNs. Recall that by Proposition 1, a k-DCN is dimension minimal if and only if it is not $(1, k-1)$-composite. Thus, dimension-minimality "captures" primality. We show that regularity, which is equivalent to being $(0, 1)$-composite, is decidable for k-DCNs for every dimension k.

For dimension one, regularity is already known to be decidable in EXPSPACE, even for history-deterministic 1-CNs [5, Theorem 19]. History-determinism is a restricted form of nondeterminism; history-deterministic CNs are less expressive than nondeterministic CNs but more expressive than DCNs. However, already for $k \geq 2$, regularity is undecidable for history-deterministic k-CNs [5, Theorem 20].

Theorem 3. *Regularity of k-DCN is decidable and is in* EXPSPACE.

We provide further details, including a proof of Theorem 3, in the full version. In short, we translate our k-DCN into a regularity preserving Vector Addition System (VAS) and use results on VAS regularity from [3, Theorem 4.5]. We remark that an alternative approach may be taken by adapting the results of [12] on regularity of VASS, although this seems more technically challenging because CNs have accepting states.

6 Expressiveness Trade-Offs between Dimensions and Nondeterminism

Theorem 1 implies that 2-CNs are more expressive than 1-CNs, and that nondeterministic models are more expressive than deterministic ones. In particular, a k-DCN can be decomposed by projection (Proposition 1), and have decidable regularity (Theorem 3). It is therefore interesting to study the interplay between increasing the dimension and introducing nondeterminism. In this section we present two results: first, we show that dimension and nondeterminism, are incomparable notions, in a sense. Second, we show that increasing the dimension strictly increases expressiveness, for both CNs and DCNs. We remark that the latter may seem like an intuitive and simple claim. However, to the best of our knowledge it has never been proved, and moreover, it requires a nontrivial approach to pumping with several counters.

We start by showing that nondeterminism can sometimes compensate for low dimension. Let $k \in \mathbb{N}$ and $\Sigma = \{a_1, \ldots, a_k, b_1, \ldots, b_k, c\}$; consider the language $L_k = \{a_1^{n_1} a_2^{n_2} \cdots a_k^{n_k} b_i c^m \mid i \in [k] \wedge n_i \geq m\}$. It is easy to construct a k-DCN as well as a 1-CN for L_k, as depicted by Figs. 4 and 5 for $k = 3$. To construct a 1-CN we guess which b_i will be later read, and verify the guess using the single counter in the $a_i^{n_i}$ part.

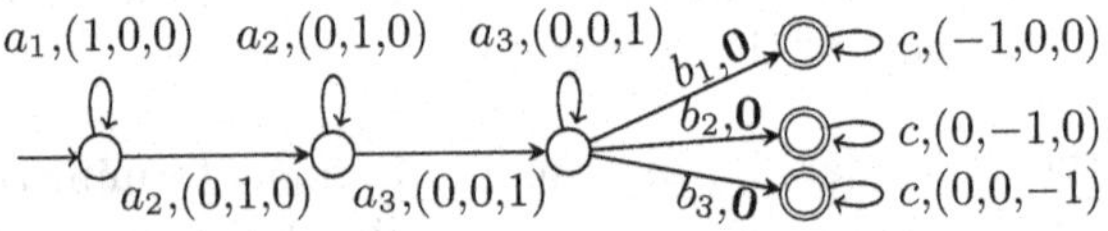

Fig. 4: A 3-DCN for $L_3 = \{a_1^{n_1} a_2^{n_2} a_3^{n_3} b_i c^m \mid i \in [3] \wedge n_i \geq m\}$. Intuitively, the 3-DCN counts the number of occurrences of each letter, and decreases the appropriate counter once the letter b_i selects it.

We now show that $\mathcal{L}_k$'s dimension cannot be minimised whilst maintaining determinism.

Theorem 4. *L_k is not recognisable by a $(k-1)$-DCN.*

Proof. Assume by way of contradiction that there exists a $(k-1)$-DCN $\mathcal{D} = \langle \Sigma, Q, Q_0, \delta, F \rangle$ such that $\mathcal{L}(\mathcal{D}) = L_k$. Let $n > |Q|$ and for every $i \in [k]$ consider

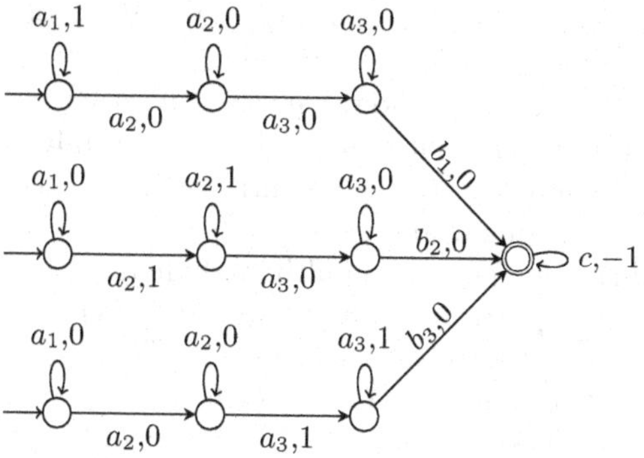

Fig. 5: A 1-CN for $L_3 = \{a_1^{n_1} a_2^{n_2} a_3^{n_3} b_i c^m \mid i \in [3] \wedge n_i \geq m\}$. Intuitively, the CN guesses which b_i will be seen, and counts the respective occurrences of the letter a_i. Then, once b_i is seen, the counter is decreased on c.

the word $w_i = a_1^n a_2^n \cdots a_k^n b_i c^n \in L_k$. Since $\mathcal{D}$ is deterministic and $n > |Q|$, all of the accepting runs on the w_i coincide up to the b_i part and have cycles in each a_i^n segment as well as in the c^n segment (the latter may differ according to i). Let M be the product of the lengths of all these cycles.

First, observe that the cycles in all of the a_i^n segments cannot decrease any counter. Indeed, otherwise by pumping such a cycle for large enough $t > 0$ times, there would not exist an $\mathbb{N}$-run on words with the prefix $a_1^n \cdots a_{i-1}^n a_i^{n+tM}$. This creates a contradiction since, with an appropriate suffix, such words can be accepted.

Thus, all a_i cycles have non-negative effects for all counters. Indeed, for each counter i – associate with i the minimal segment index whose cycle strictly increases i. Since there are $k-1$ counters and k segments this map is not surjective, in other words, there is a segment (without loss of generality, the a_k segment) such that every counter that is increased in the a_k cycle is also increased in a previous segment. Therefore, there exist $s, t > 0$ such that the word

$$a_1^{n+sM} a_2^{n+sM} \cdots a_{k-1}^{s+sM} a_k^n b_k c^{n+tM} \notin L_k$$

is accepted by $\mathcal{D}$, which is a contradiction.

We now turn to show that conversely, dimension can sometimes compensate for nondeterminism. Moreover, we show that there is a strict hierarchy of expressiveness with respect to dimension. Specifically, for $k \in \mathbb{N}$ consider the language $H_k = \{a_1^{m_1} a_2^{m_2} \cdots a_k^{m_k} b_1^{n_1} b_2^{n_2} \cdots b_k^{n_k} \mid \forall 1 \leq i \leq k,\ m_i \geq n_i\}$.

Theorem 5. *H_k is recognisable by a k-DCN, but not by a $(k-1)$-CN.*

Proof (sketch). Constructing a k-DCN for H_k is straightforward, by using the i-th counter to check that $m_i \geq n_i$, for each $i \in [k]$.

We turn to argue that H_k is not recognisable by a $(k-1)$-CN (See the full version for a complete proof). Assume by way of contradiction that $\mathcal{A} =$

$\langle \Sigma, Q, Q_0, \delta, F\rangle$ is a $(k-1)$-CN with $L(\mathcal{A}) = H_k$. We first observe that there exists $m_1 \in \mathbb{N}$ large enough so that every run of $\mathcal{A}$ on $a_1^{m_1}$ must traverse a non-negative cycle, i.e., a cycle whose overall effect is $\boldsymbol{u_1} \in \mathbb{Z}^{k-1}$ such that $\boldsymbol{u_1}[i] \geq 0$ for all $i \in [k-1]$. Indeed, this is immediate by a "uniformly bounded" version of Dickson's lemma [15]; any long-enough "controlled" sequence of vectors in $\mathbb{N}^{k-1}$ must contain an r-increasing chain, for any $r \in \mathbb{N}$.

By repeating this argument we can ultimately find $m_1, \ldots, m_k$ such that any run of $\mathcal{A}$ on $a_1^{m_1} a_2^{m_2} \cdots a_k^{m_k}$ traverses a non-negative cycle in each a_j segment for $j \in [k]$. Consider now the word $w = a_1^{m_1} a_2^{m_2} \cdots a_k^{m_k} b_1^{m_1} b_2^{m_2} \cdots b_k^{m_k} \in H_k$, then there exists an accepting run ρ of $\mathcal{A}$ on w such that for each $j \in [k]$, the run ρ traverses a non-negative cycle in segment a_j, with effect $\boldsymbol{u_j} \in \mathbb{N}^{k-1}$.

Consider the vectors $\boldsymbol{u_1}, \ldots, \boldsymbol{u_k}$. We claim that there exists $\ell \in [k]$ such that the support of $\boldsymbol{u_\ell}$ is *covered* by $\boldsymbol{u_1}, \ldots, \boldsymbol{u_{\ell-1}}$ in the following sense: for every counter $i \in [k-1]$, if $\boldsymbol{u_\ell}[i] > 0$, then there exists $j < \ell$ such that $\boldsymbol{u_j}[i] > 0$. Indeed, this holds since otherwise every $\boldsymbol{u_j}$ must contribute a fresh positive coordinate to the union of supports of the previous vectors, but there are k vectors and only $k-1$ coordinates.

Next, observe that since each $\boldsymbol{u_j}$ is a non-negative cycle taken in ρ, then it can be pumped without decreasing any following counters, and hence induce an accepting run on a pumped word. Intuitively, we now proceed by pumping all the $\boldsymbol{u_j}$ cycles for $j < \ell$ for some large-enough number of times M, which enables us to remove one iteration of the cycle with effect $\boldsymbol{u_\ell}$ while maintaining an accepting run on a word of the form:

$$w' = a_1^{m_1+Md_1} a_2^{m_2+Md_2} \cdots a_{\ell-1}^{m_{\ell-1}+Md_{\ell-1}} a_\ell^{m_\ell - d_\ell} a_{\ell+1}^{m_{\ell+1}} \cdots a_k^{m_k} b_1^{m_1} b_2^{m_2} \cdots b_k^{m_k}.$$

Since $m_\ell > m_\ell - d_\ell$, the b_ℓ segment is longer than the a_ℓ segment. Thus $w' \notin H_k$, this yields a contradiction. □

Apart from showing that nondeterminism cannot always compensate for increased dimension, Theorem 5 also shows that for every dimension k, there are languages recognisable by a $(k+1)$-DCN (and in particular by a $(k+1)$-CN), but not by any k-CN (and in particular not by any k-DCN). Thus, we obtain the following hierarchy.

Corollary 2. *For every $k \in \mathbb{N}$, k-CNs (resp. k-DCNs) are strictly less expressive than $(k+1)$-CNs (resp. $(k+1)$-DCNs).*

7 Discussion

Broadly, this work explores the interplay between the dimension of a CN and its expressive power. This is done by studying the *dimension-minimality* problem, where we ask whether the dimension of a given CN can be decreased while preserving its language, and by the more involved *primality* problem, which allows a decomposition to multiple CNs of lower dimension. We show that both primality and dimension-minimality are undecidable. Moreover, they remain undecidable even when we discard the degenerate dimension 0 case, which corresponds to finite memory, i.e., regular languages. On the other hand, this degenerate case is one where we can show decidability for DCNs.

This work also highlights a technical shortcoming of current understanding of high-dimensional CNs: pumping arguments in the presence of k dimensions and nondeterminism are very involved, and are (to our best efforts) insufficient to prove Conjecture 1. To this end, we present novel pumping arguments in the proof of Theorem 1 and to some extent in the proof of Theorem 5, which make progress towards pumping in the presence of k dimensions and nondeterminism.

References

1. Shaull Almagor, Udi Boker, Piotr Hofman, and Patrick Totzke. Parametrized universality problems for one-counter nets. In *31st International Conference on Concurrency Theory (CONCUR 2020)*. Schloss Dagstuhl-Leibniz-Zentrum für Informatik, 2020.
2. Shaull Almagor and Asaf Yeshurun. Determinization of one-counter nets. In *33rd International Conference on Concurrency Theory (CONCUR 2022)*. Schloss Dagstuhl-Leibniz-Zentrum für Informatik, 2022.
3. Michel Blockelet and Sylvain Schmitz. Model checking coverability graphs of vector addition systems. In Filip Murlak and Piotr Sankowski, editors, *Mathematical Foundations of Computer Science 2011 - 36th International Symposium, MFCS 2011, Warsaw, Poland, August 22-26, 2011. Proceedings*, volume 6907 of *Lecture Notes in Computer Science*, pages 108–119. Springer, 2011. doi: 10.1007/978-3-642-22993-0_13.
4. Michael Blondin, Matthias Englert, Alain Finkel, Stefan Göller, Christoph Haase, Ranko Lazić, Pierre McKenzie, and Patrick Totzke. The reachability problem for two-dimensional vector addition systems with states. *Journal of the ACM (JACM)*, 68(5):1–43, 2021.
5. Sougata Bose, David Purser, and Patrick Totzke. History-deterministic vector addition systems. *arXiv preprint arXiv:2305.01981*, 2023.
6. Maria Paola Cabasino, Alessandro Giua, and Carla Seatzu. Diagnosability of discrete-event systems using labeled Petri nets. *IEEE Transactions on Automation Science and Engineering*, 11(1):144–153, 2013.
7. Wojciech Czerwiński, Diego Figueira, and Piotr Hofman. Universality problem for unambiguous vass. In *31st International Conference on Concurrency Theory (CONCUR 2020)*. Schloss Dagstuhl-Leibniz-Zentrum für Informatik, 2020.
8. Wojciech Czerwiński, Sławomir Lasota, Ranko Lazić, Jérôme Leroux, and Filip Mazowiecki. Reachability in fixed dimension vector addition systems with states. In *31st International Conference on Concurrency Theory (CONCUR 2020)*. Schloss Dagstuhl-Leibniz-Zentrum für Informatik, 2020.
9. Wojciech Czerwiński, Slawomir Lasota, Christof Löding, and Radoslaw Piórkowski. New pumping technique for 2-dimensional VASS. In *44th International Symposium on Mathematical Foundations of Computer Science (MFCS 2019)*. Schloss Dagstuhl-Leibniz-Zentrum fuer Informatik, 2019.
10. Wojciech Czerwiński, Sławomir Lasota, and Łukasz Orlikowski. Improved lower bounds for reachability in vector addition systems. In *48th International Colloquium on Automata, Languages, and Programming (ICALP 2021)*. Schloss Dagstuhl-Leibniz-Zentrum für Informatik, 2021.

11. Wojciech Czerwiński and Łukasz Orlikowski. Reachability in vector addition systems is Ackermann-complete. In *2021 IEEE 62nd Annual Symposium on Foundations of Computer Science (FOCS)*, pages 1229–1240. IEEE, 2022.
12. Stéphane Demri. On selective unboundedness of VASS. *J. Comput. Syst. Sci.*, 79(5):689–713, 2013. doi:10.1016/j.jcss.2013.01.014.
13. Javier Esparza. Decidability and complexity of petri net problems—an introduction. *Lectures on Petri Nets I: Basic Models: Advances in Petri Nets*, pages 374–428, 2005.
14. Diego Figueira. Co-finiteness of VASS coverability languages. working paper or preprint, July 2019. URL: https://hal.science/hal-02193089.
15. Diego Figueira, Santiago Figueira, Sylvain Schmitz, and Philippe Schnoebelen. Ackermannian and primitive-recursive bounds with dickson's lemma. In *2011 IEEE 26th Annual Symposium on Logic in Computer Science*, pages 269–278. IEEE, 2011.
16. Alain Finkel, Jérôme Leroux, and Grégoire Sutre. Reachability for two-counter machines with one test and one reset. In *FSTTCS 2018-38th IARCS Annual Conference on Foundations of Software Technology and Theoretical Computer Science*, volume 122, pages 31–1. Schloss Dagstuhl-Leibniz-Zentrum fuer Informatik, 2018.
17. Sheila A. Greibach. Remarks on blind and partially blind one-way multicounter machines. *Theoretical Computer Science*, 7(3):311–324, 1978.
18. Christoph Haase, Stephan Kreutzer, Joël Ouaknine, and James Worrell. Reachability in succinct and parametric one-counter automata. In *CONCUR 2009-Concurrency Theory: 20th International Conference, CONCUR 2009, Bologna, Italy, September 1-4, 2009. Proceedings 20*, pages 369–383. Springer, 2009.
19. Michel Henri Theódore Hack. Petri net language. *Computation Structures Group Memo 124*, 1976. URL: http://publications.csail.mit.edu/lcs/pubs/pdf/MIT-LCS-TR-159.pdf.
20. Piotr Hofman, Richard Mayr, and Patrick Totzke. Decidability of weak simulation on one-counter nets. In *2013 28th Annual ACM/IEEE Symposium on Logic in Computer Science*, pages 203–212. IEEE, 2013.
21. Piotr Hofman and Patrick Totzke. Trace inclusion for one-counter nets revisited. In *Reachability Problems: 8th International Workshop, RP 2014, Oxford, UK, September 22-24, 2014. Proceedings 8*, pages 151–162. Springer, 2014.
22. I. Jecker, N. Mazzocchi, and P. Wolf. Decomposing permutation automata. In *Proc. 32nd CONCUR*, volume 203 of *LIPIcs*, pages 18:1–18:19. Schloss Dagstuhl - Leibniz-Zentrum für Informatik, 2021.
23. Ismael R Jecker, Orna Kupferman, and Nicolas Mazzocchi. Unary prime languages. In *45th International Symposium on Mathematical Foundations of Computer Science*, volume 170, 2020.
24. Orna Kupferman and Jonathan Mosheiff. Prime languages. *Information and Computation*, 240:90–107, 2015.
25. Jérôme Leroux. The reachability problem for petri nets is not primitive recursive. In *2021 IEEE 62nd Annual Symposium on Foundations of Computer Science (FOCS)*, pages 1241–1252. IEEE, 2022.
26. Jérôme Leroux and Grégoire Sutre. On flatness for 2-dimensional vector addition systems with states. In *CONCUR*, volume 4, pages 402–416. Springer, 2004.
27. Elaine Render and Mark Kambites. Rational subsets of polycyclic monoids and valence automata. *Information and Computation*, 207(11):1329–1339, 2009.
28. Rüdiger Valk and Guy Vidal-Naquet. Petri nets and regular languages. *Journal of Computer and system Sciences*, 23(3):299–325, 1981.

Parameterized Broadcast Networks with Registers: from NP to the Frontiers of Decidability*

Lucie Guillou[1], Corto Mascle[2], and Nicolas Waldburger[3(✉)]

[1] IRIF, CNRS, Université Paris Cité, Paris, France
guillou@irif.fr
[2] LaBRI, Université de Bordeaux, Bordeaux, France
corto.mascle@labri.fr
[3] IRISA, Université de Rennes, Rennes, France
nicolas.waldburger@irisa.fr

Abstract. We consider the parameterized verification of networks of agents which communicate through unreliable broadcasts. In this model, agents have local registers whose values are unordered and initially distinct and may therefore be thought of as identifiers. When an agent broadcasts a message, it appends to the message the value stored in one of its registers. Upon reception, an agent can store the received value or test it for equality against one of its own registers. We consider the coverability problem, where one asks whether a given state of the system may be reached by at least one agent. We establish that this problem is decidable, although non-primitive recursive. We contrast this with the undecidability of the closely related target problem where all agents must synchronize on a given state. On the other hand, we show that the coverability problem is NP-complete when each agent only has one register.

Keywords: Parameterized verification · Well quasi-orders · Distributed systems

1 Introduction

We consider Broadcast Networks of Register Automata (BNRA), a model for networks of agents communicating by broadcasts. These systems are composed of an arbitrary number of agents whose behavior is specified with a finite automaton. This automaton is equipped with a finite set of private registers that contain values from an infinite unordered set. Initially, registers all contain distinct values, so these values can be used as identifiers. A broadcast message is composed of a symbol from a finite alphabet along with the value of one of the sender's registers. When an agent broadcasts a message, any subset of agents may receive it; this models unreliable systems with unexpected crashes and disconnections. Upon reception, an agent may store the received value or test it for equality with one of its register values. For example, an agent can check that several received messages have the same value.

* Partly supported by ANR project PaVeDyS (ANR-23-CE48-0005).

N. Kobayashi and J. Worrell (Eds.): FoSSaCS 2024, LNCS 14575, pp. 250–270, 2024.
https://doi.org/10.1007/978-3-031-57231-9_12

This model was introduced in [10], as a natural extension of Reconfigurable Broadcast Networks [12]. In [10], the authors established that coverability is undecidable if the agents are allowed to send two values per message. They moreover claimed that, with one value per message, coverability was decidable and PSPACE-complete; however, the proof turned out to be incorrect [22]. As we will see, the complexity of that problem is in fact much higher.

In this paper we establish the decidability of the coverability problem and its completeness for the hyper-Ackermannian complexity class $\mathbf{F}_{\omega^\omega}$, showing that the problem has nonprimitive recursive complexity. The lower bound comes from lossy channel systems, which consist (in their simplest version) of a finite automaton that uses an unreliable FIFO memory from which any letter may be erased at any time [3, 8, 26]. We further establish that our model lies at the frontier of decidability by showing undecidability of the target problem (where all agents must synchronize in a given state). We contrast these results with the NP-completeness of the coverability problem if each agent has only one register.

Related work Broadcast protocols are a widely studied class of systems in which processes are represented by nodes of a graph and can send messages to their neighbors in the graph. There are many versions depending on how one models processes, the communication graph, the shape of messages... A model with a fully connected communication graph and messages ranging over a finite alphabet was presented in [13]. When working with parameterized questions over this model (*i.e.*, working with systems of arbitrary size), many basic problems are undecidable [14]; similar negative results were found for Ad Hoc Networks where the communication graph is fixed but arbitrary [12]. This lead the community to consider Reconfigurable Broadcast Networks (RBN) where a broadcast can be received by an arbitrary subset of agents [12].

Parameterized verification problems over RBN have been the subject of extensive study in recent years, concerning for instance reachability questions [5, 11], liveness [9] or alternative communication assumptions [4]; however, RBN have weak expressivity, in particular because agents are anonymous. In [10], RBN were extended to BNRA, the model studied in this article, by the addition of registers allowing processes to exchange identifiers.

Other approaches exist to define parameterized models with registers [6], such as dynamic register automata in which processes are allowed to spawn other processes with new identifiers and communicate integers values [1]. While basic problems on these models are in general undecidable, some restrictions on communications allow to obtain decidability [2, 20].

Parameterized verification problems often relate to the theory of well quasi-orders and the associated high complexities obtained from bounds on the length of sequences with no increasing pair (see for example [25]). In particular, our model is linked to data nets, a classical model connected to well-quasi-orders. Data nets are Petri nets in which tokens are labeled with natural numbers and can exchange and compare their labels using inequality tests [18]; in this model, the coverability problem is $\mathbf{F}_{\omega^{\omega^\omega}}$-complete [15]. When one restricts data nets to only equality tests, the coverability problem becomes $\mathbf{F}_{\omega^\omega}$-complete [21]. Data nets with equality tests do not subsume BNRA. Indeed, in data nets, each process can only carry one integer at a time, and problems on models of data nets where tokens carry tuples of integers are typically undecidable [17].

Overview We start with the model definition and some preliminary results in Section 2. As our decidability proof is quite technical, we start by proving decidability of the coverability problem in a subcase called *signature protocols* in Section 3. We then rely on the intuitions built in that subcase to generalize the proof to the general case in Section 4. We also show the undecidability of the closely-related target problem. Finally, we prove the NP-completeness of the coverability problem for protocols with one register in Section 5. Due to space constraints, a lot of proofs, as well as some technical definitions, are only sketched in this version. Detailed proofs can be found in the full version, available here.

In this document, each notion is linked to its *definition* using the knowledge package. On electronic devices, clicking on words or symbols allows to access their definitions.

2 Preliminaries

2.1 Definitions of the Model

A *Broadcast Network of Register Automata* (BNRA) [10] is a model describing broadcast networks of agents with local registers. A finite transition system describes the behavior of an agent; an agent can broadcast and receive messages with integer values, store them in local registers and perform (dis)equality tests. There are arbitrarily many agents. When an agent broadcasts a message, every other agent may receive it, but does not have to do so.

Definition 1. *A* protocol *with r registers is a tuple $\mathcal{P} = (Q, \mathcal{M}, \Delta, q_0)$ with Q a finite set of states, $q_0 \in Q$ an initial state, $\mathcal{M}$ a finite set of* message types *and $\Delta \subseteq Q \times \mathsf{Op} \times Q$ a finite set of transitions, with operations $\mathsf{Op} =$*

$$\{\boldsymbol{br}(m,i), \boldsymbol{rec}(m,i,*), \boldsymbol{rec}(m,i,\downarrow), \boldsymbol{rec}(m,i,=), \boldsymbol{rec}(m,i,\neq) \mid m \in \mathcal{M}, 1 \leq i \leq r\}.$$

Label **br** *stands for* broadcasts *and* **rec** *for* receptions. *In a reception $\boldsymbol{rec}(m,i,\alpha)$, α is its* action. *The set of actions is* $\mathsf{Actions} := \{=, \neq, \downarrow, *\}$, *where '$=$' is an* equality test, *'$\neq$' is a* disequality test, *'$\downarrow$' is a* store action *and '$*$' is a* dummy action *with no effect. The* size *of $\mathcal{P}$ is $|\mathcal{P}| := |Q| + |\mathcal{M}| + |\Delta| + r$.*

We now define the semantics of those systems. Essentially, we have a finite set of agents with r registers each; all registers initially contain distinct values. A step consists of an agent broadcasting a message that other agents may receive.

Definition 2 (Semantics). *Let $(Q, \mathcal{M}, \Delta, q_0)$ be a protocol with r registers, and $\mathbb{A}$ a finite non-empty set of* agents. *A* configuration *over $\mathbb{A}$ is a function*

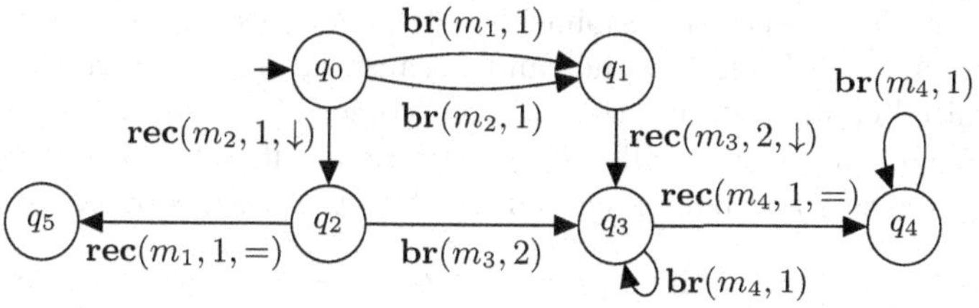

Fig. 1: Example of a protocol.

$\gamma : \mathbb{A} \to Q \times \mathbb{N}^r$ mapping each agent to its state and its register values. We write $\mathsf{st}(\gamma)$ *for the state component of* γ *and* $\mathsf{data}(\gamma)$ *for its register component.*

An initial configuration *γ is one where for all $a \in \mathbb{A}$, $\mathsf{st}(\gamma)(a) = q_0$ and $\mathsf{data}(\gamma)(a, i) \neq \mathsf{data}(\gamma)(a', i')$ for all $(a, i) \neq (a', i')$.*

Given a finite non-empty set of agents $\mathbb{A}$ and two configurations γ, γ' over $\mathbb{A}$, a step *$\gamma \to \gamma'$ is defined when there exist $m \in \mathcal{M}$, $a_0 \in \mathbb{A}$ and $i \in [1, r]$ such that $(\mathsf{st}(\gamma)(a_0), \textbf{br}(m, i), \mathsf{st}(\gamma')(a_0)) \in \Delta$, $\mathsf{data}(\gamma)(a_0) = \mathsf{data}(\gamma')(a_0)$ and, for all $a \neq a_0$, either $\gamma'(a) = \gamma(a)$ or there exists $(\mathsf{st}(\gamma)(a), \textbf{rec}(m, j, \alpha), \mathsf{st}(\gamma')(a)) \in \Delta$ s.t. $\mathsf{data}(\gamma')(a, j') = \mathsf{data}(\gamma)(a, j')$ for $j' \neq j$ and:*

- *if $\alpha =$ '$*$' then $\mathsf{data}(\gamma')(a, j) = \mathsf{data}(\gamma)(a, j)$,*
- *if $\alpha =$ '$\downarrow$' then $\mathsf{data}(\gamma')(a, j) = \mathsf{data}(\gamma)(a_0, i)$,*
- *if $\alpha =$ '$=$' then $\mathsf{data}(\gamma')(a, j) = \mathsf{data}(\gamma)(a, j) = \mathsf{data}(\gamma)(a_0, i)$,*
- *if $\alpha =$ '$\neq$' then $\mathsf{data}(\gamma')(a, j) = \mathsf{data}(\gamma)(a, j) \neq \mathsf{data}(\gamma)(a_0, i)$.*

A run *over $\mathbb{A}$ is a sequence of steps $\rho : \gamma_0 \to \gamma_1 \to \cdots \to \gamma_k$ with $\gamma_0, \ldots, \gamma_k$ configurations over $\mathbb{A}$. We write $\gamma_0 \xrightarrow{*} \gamma_k$ when there exists such a run. A run is* initial *when γ_0 is an initial configuration.*

Remark 3. In our model, agents may only send one value per message. Indeed, coverability is undecidable if agents can broadcast several values at once [10].

Example 4. Figure 1 shows a protocol with 2 registers. Let $\mathbb{A} = \{a_1, a_2\}$. We denote by $\langle \mathsf{st}(\gamma)(a_1), \mathsf{data}(\gamma)(a_1), \mathsf{st}(\gamma)(a_2), \mathsf{data}(\gamma)(a_2) \rangle$ a configuration γ over $\mathbb{A}$. The following sequence is an initial run:

$$\langle q_0, (1,2), q_0, (3,4) \rangle \to \langle q_1, (1,2), q_2, (1,4) \rangle \to \langle q_3, (1,4), q_3, (1,4) \rangle \\ \to \langle q_4, (1,4), q_3, (1,4) \rangle \to \langle q_4, (1,4), q_4, (1,4) \rangle$$

The broadcast messages are, in this order: $(m_2, 1)$ by a_1, $(m_3, 4)$ by a_2, $(m_4, 1)$ by a_2 and $(m_4, 1)$ by a_1. In this run, each broadcast message is received by the other agent; in general, however, this does not have to be true. □

Remark 5. From a run $\rho : \gamma_0 \xrightarrow{*} \gamma$, we can build a larger run ρ' in which, for each agent a of ρ, there are arbitrarily many extra agents in ρ' that end in the same state as a, all with distinct register values. To obtain this, ρ' make many copies of ρ run in parallel on disjoint sets of agents. Because all these copies of ρ do not interact with one another and because all agents start with distinct values in initial configurations, the different copies of ρ have no register values in common. This property is called *copycat principle*: if state q is coverable, then for all n there exists an augmented run which puts n agents on q.

Definition 6. *The* coverability problem COVER *asks, given a protocol $\mathcal{P}$ and a state q_f, whether there is a finite non-empty set of agents $\mathbb{A}$, an initial run $\gamma_0 \xrightarrow{*} \gamma_f$ over $\mathbb{A}$ that* covers q_f, i.e., *there is $a \in \mathbb{A}$ such that* $\mathsf{st}(\gamma_f)(a) = q_f$.

The target problem TARGET *asks, given a protocol $\mathcal{P}$ and a state q_f, whether there is there is a finite non-empty set of agents $\mathbb{A}$ and an initial run $\gamma_0 \xrightarrow{*} \gamma_f$ over $\mathbb{A}$ such that, for every $a \in \mathbb{A}$,* $\mathsf{st}(\gamma_f)(a) = q_f$, i.e., *all agents end on q_f.*

Example 7. Let $\mathcal{P}$ the protocol of Figure 1. As proven in Example 4, $(\mathcal{P}, q_4)$ is a positive instance of COVER and TARGET. However, let $\mathcal{P}'$ the protocol obtained from $\mathcal{P}$ by removing the loop on q_4; $(\mathcal{P}', q_4)$ becomes a negative instance of TARGET. Indeed, there must be an agent staying on q_3 to broadcast m_4. Also, $(\mathcal{P}, q_5)$ is a negative instance of COVER: we would need to be able to have one agent on q_2 and one agent on q_0 with the same value in their first registers. However, an agent in q_0 has performed no transition so it cannot share register values with other agents. □

Remark 8. In [10], the authors consider the *query problem* where one looks for a run reaching a configuration satisfying some *queries*. In fact, this problem exponentially reduces to COVER hence our complexity result of $\mathbf{F}_{\omega^\omega}$ also holds for the query problem. In the case with one register, one can even find a polynomial-time reduction hence our NP result also holds with queries.

We finally introduce *signature BNRA*, an interesting restriction of our model where register 1 is *broadcast-only* and all other registers are *reception-only*. Said otherwise, the first register acts as a permanent identifier with which agents sign their messages. An example of such a protocol is displayed in Fig. 2. Under this restriction, a message is composed of a message type along with the identifier of the sender. This restriction is relevant for pedagogical purposes: we will see that it falls into the same complexity class as the general case but makes the decidability procedure simpler.

Definition 9 (Signature protocols). *A* signature protocol *with r registers is a protocol $\mathcal{P} = (Q, \mathcal{M}, \Delta, q_0)$ where register* 1 *appears only in broadcasts in Δ and registers $i \geq 2$ appear only in receptions in Δ.*

2.2 Classical Definitions

Fast-growing hierarchy For α an ordinal in Cantor normal form, we denote by $\mathscr{F}_\alpha$ the class of functions corresponding to level α in the Fast-Growing Hierarchy. We denote by $\mathbf{F}_\alpha$ the associated complexity class and use the notion of $\mathbf{F}_\alpha$-completeness. All these notions are defined in [23]. We will specifically work with complexity class $\mathbf{F}_{\omega^\omega}$. For readers unfamiliar with these notions, $\mathbf{F}_{\omega^\omega}$-complete problems are decidable but with very high complexity (non-primitive recursive, and even much higher than the Ackermann class $\mathbf{F}_{\omega}$).

We highlight that our main result is the decidability of the problem. We show that the problem lies in $\mathbf{F}_{\omega^\omega}$ because it does not complicate our decidability proof significantly; also, it fits nicely into the landscape of high-complexity problems arising from well quasi-orders.

Well-quasi orders For our decidability result, we rely on the theory of well quasi-orders in the context of subword ordering. Let Σ be a finite alphabet, $w_1, w_2 \in \Sigma^*$, w_1 is a *subword* of w_2, denoted $w_1 \preceq w_2$, when w_1 can be obtained from w_2 by erasing some letters. A sequence of words $w_0, w_1, \dots$ is *good* if there exist $i < j$ such that $w_i \preceq w_j$, and *bad* otherwise. Higman's lemma [16] states that every bad sequence of words over a finite alphabet is finite, but there is no uniform bound. In order to bound the length of all bad sequences, one must bound the growth of the sequence of words. We will use the following result, known as the Length function theorem [24]:

Theorem 10 (*Length function theorem* [24]). *Let Σ a finite alphabet and $g : \mathbb{N} \to \mathbb{N}$ a primitive recursive function. There exists a function $f \in \mathscr{F}_{\omega^{|\Sigma|-1}}$ such that, for all $n \in \mathbb{N}$, every bad sequence $w_1, w_2, \dots$ such that $|w_i| \leq g^{(i)}(n)$ for all i has at most $f(n)$ terms (where $g^{(i)}$ denotes g applied i times).*

2.3 A Complexity Lower Bound for COVER Using LCS

Lossy channel systems (LCS) are systems where finite-state processes communicate by sending messages from a finite alphabet through lossy FIFO channels. Unlike in the non-lossy case [7], reachability of a state is decidable for lossy channel systems [3], but has non-primitive recursive complexity [26] and is in fact $\mathbf{F}_{\omega^\omega}$-complete [8]. By simulating LCS using BNRA, we obtain our $\mathbf{F}_{\omega^\omega}$ lower bound for the coverability problem:

Proposition 11. COVER *for signature BNRA is $\mathbf{F}_{\omega^\omega}$-hard.*

Proof sketch. Given an LCS $\mathcal{L}$, we build a signature protocol $\mathcal{P}$ with two registers. Each agent starts by receiving a foreign identifier and storing it in its second register; using equality tests, it then only accepts messages with this identifier. Each agent has at most one predecessor, so the communication graph is a forest where messages propagate from roots to leaves. Each branch simulates an execution of $\mathcal{L}$. Each agent of the branch simulates a step of the execution: it receives from its predecessor a configuration of $\mathcal{L}$, chooses the next configuration of $\mathcal{L}$ and broadcasts it, sending first the location of $\mathcal{L}$ and then, letter by letter, the content of the channel. It could be that some messages are not received, hence the lossiness. □

3 Coverability Decidability for Signature Protocols

This section and the next one are dedicated to the proof of our main result:

Theorem 12. COVER *for BNRA is decidable and* $\mathbf{F}_{\omega^\omega}$*-complete.*

For the sake of clarity, in this section, we will first focus on the case of signature BNRA. As a preliminary, we start by defining a notion of local run meant to represent the projection of a run onto a given agent.

3.1 Local runs

A *local configuration* is a pair $(q, \nu) \in Q \times \mathbb{N}^r$. An *internal step* from (q, ν) to (q', ν') with transition $\delta \in \Delta$, denoted $(q, \nu) \xrightarrow{\mathsf{int}(\delta)} (q', \nu')$, is defined when $\nu = \nu'$ and $\delta = (q, \mathbf{br}(m, i), q')$ is a broadcast. A *reception step* from (q, ν) to (q', ν') with transition $\delta \in \Delta$ and value $v \in \mathbb{N}$, denoted $(q, \nu) \xrightarrow{\mathsf{ext}(\delta, v)} (q', \nu')$, is defined when δ is of the form $(q, \mathbf{rec}(m, j, \alpha), q')$ with $\nu(j') = \nu'(j')$ for all $j' \neq j$ and:

- if $\alpha =$ '$*$' then $\nu(j) = \nu'(j)$,
- if $\alpha =$ '$\downarrow$' then $\nu'(j) = v$,
- if $\alpha =$ '$=$' then $\nu(j) = \nu'(j) = v$,
- if $\alpha =$ '$\neq$' then $\nu(j) = \nu'(j) \neq v$.

Such a reception step corresponds to receiving message (m, v); in a local run, one does not specify the origin of a received message. A *local step* $(q, \nu) \rightarrow (q', \nu')$ is either a reception step or an internal step. A *local run* u is a sequence of local steps denoted $(q_0, \nu_0) \xrightarrow{*} (q, \nu)$. Its *length* $|u|$ is its number of steps.

A value $v \in \mathbb{N}$ appearing in u is *initial* if it appears in ν_0 and *non-initial* otherwise. For $v \in \mathbb{N}$, the *v-input* $\mathsf{In}_v(u)$ (resp. *v-output* $\mathsf{Out}_v(u)$) is the sequence $m_0 \cdots m_\ell \in \mathcal{M}^*$ of message types received (resp. broadcast) with value v in u.

3.2 Unfolding Trees

We first prove decidability of COVER for signature BNRA. Note that, in signature protocols, the initial values of reception-only registers are not relevant as they can never be shared with other agents. We deduce from this idea the following informal observation:

Observation 13 *In signature BNRA, when some agent receives a message, it can compare the value of the message only with the ones of previously received messages,* i.e., *check whether the sender is the same.*

If we want to turn a local run u of an agent a into an actual run, we must match a's receptions with broadcasts. Because of Observation 13, what matters is not the actual values of the receptions in u but which ones are equal to which. Therefore, for a value v received in u, if $m_1 \dots m_k \in \mathcal{M}^*$ are the message types received in u with value v in this order, it means that to execute u, a need another agent a' to broadcast messages types m_1 to m_k, all with the same value. We describe what an agent needs from other agents as a set of specifications which are words of $\mathcal{M}^*$.

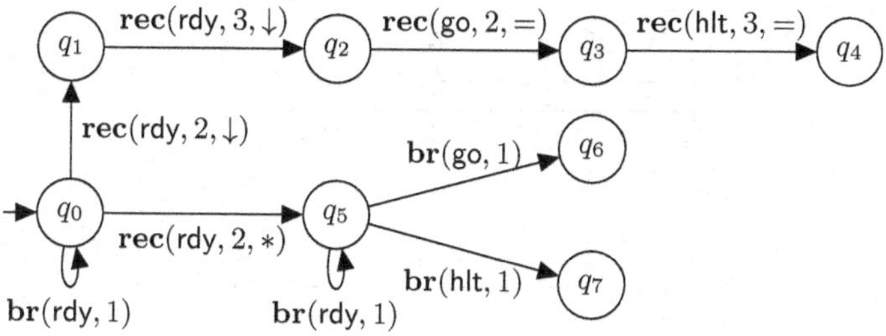

Fig. 2: Example of a signature protocol.

To represent runs, we consider unfolding trees that abstract runs by representing such specifications, dependencies between them and how they are carried out. In this tree, each node is assigned a local run and the specification that it carries out. Because of copycat arguments, we will in fact be able to duplicate agents so that each agent only accomplishes one task, hence the tree structure.

Definition 14. *An* unfolding tree τ *over* $\mathcal{P}$ *is a finite tree where nodes* μ *have three labels:*

- *a local run of* $\mathcal{P}$*, written* $\mathbf{lr}(\mu)$*;*
- *a value in* $\mathbb{N}$*, written* $\mathbf{val}(\mu)$*;*
- *a* specification $\mathbf{spec}(\mu) \in \mathcal{M}^*$.

Moreover, all nodes μ *in* τ *must satisfy the three following conditions:*

(i) Initial values of $\mathbf{lr}(\mu)$ *are never received in* $\mathbf{lr}(\mu)$*,*
(ii) $\mathbf{spec}(\mu) \preceq \mathsf{Out}_{\mathbf{val}(\mu)}(\mathbf{lr}(\mu))$, (recall that $\preceq$ denotes the subword relation)
(iii) For each value v *received in* $\mathbf{lr}(\mu)$*,* μ *has a child* μ' *s.t.* $\mathsf{In}_v(\mathbf{lr}(\mu)) \preceq \mathbf{spec}(\mu')$.

Lastly, given τ *an unfolding tree, we define its* size *by* $|\tau| := \sum_{\mu\in\tau} |\mu|$ *where* $|\mu| := |\mathbf{lr}(\mu)| + |\mathbf{spec}(\mu)|$*. Note that the size of* τ *takes into account the size of its nodes, so that a tree* τ *can be stored in space polynomial in* $|\tau|$ *(renaming the values appearing in* τ *if needed).*

We explain this definition. Condition (i) enforces that the local run cannot cheat by receiving its initial values. Condition (ii) expresses that $\mathbf{lr}(\mu)$ broadcasts (at least) the messages of $\mathbf{spec}(\mu)$. We can use the subword relation $\preceq$ (instead of equality) because messages do not have to be received. Condition (iii) expresses that, for each value v received in the local run $\mathbf{lr}(\mu)$, μ has a child who is able to broadcast the sequence of messages that $\mathbf{lr}(\mu)$ receives with value v.

Example 15. Figure 2 provides an example of a signature protocol. Let $\mathbb{A} = \{a_1, a_2, a_3\}$. We denote a configuration γ by $\langle \mathsf{st}(\gamma)(a_1), (\mathsf{data}(\gamma)(a_1)),$ $\mathsf{st}(\gamma)(a_2), (\mathsf{data}(\gamma)(a_2)), \mathsf{st}(\gamma)(a_3), (\mathsf{data}(\gamma)(a_3))\rangle$. Irrelevant register values are denoted by _. Let ρ be the run over $\mathbb{A}$ of initial configuration $\langle q_0, (1, \_, \_), q_0, (2, \_, \_), q_0, (3, \_, \_)\rangle$ where the following occurs:

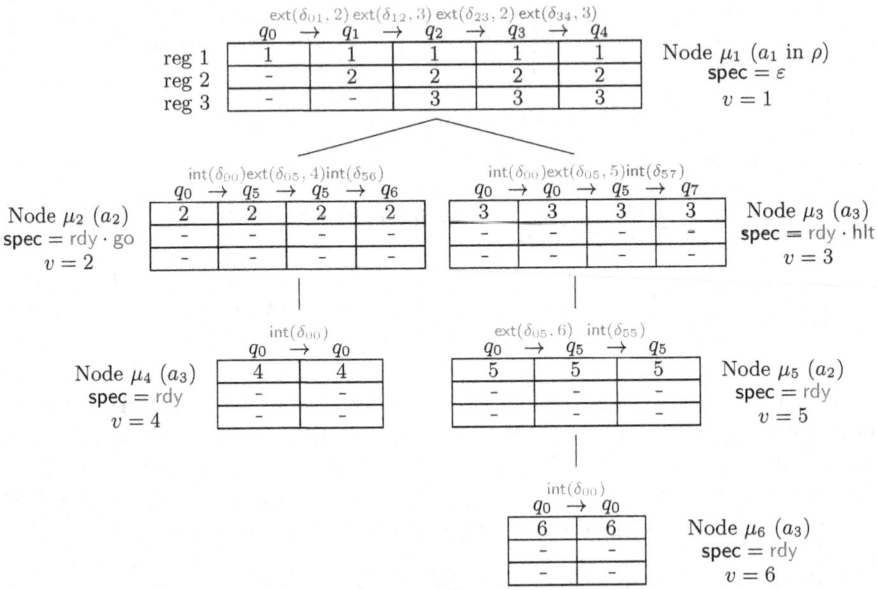

Fig. 3: Example of an unfolding tree derived from ρ. Grids correspond to local runs, a column of a grid is a local configuration. Transition δ_{ij} is the transition between state q_i and state q_j, for example $\delta_{01} = (q_0, \mathbf{rec}(\mathsf{rdy}, 2, \downarrow), q_1)$. If δ is a reception of $m \in \mathcal{M}$, $\mathsf{ext}(\delta, v)$ corresponds to receiving message (m, v); if δ is a broadcast of $m \in \mathcal{M}$, $\mathsf{int}(\delta)$ corresponds to broadcasting (m, id) where id is the value in the first register of the agent. Initial values of reception-only registers are irrevelant and written as '_'. Colors correspond to message types.

- a_2 broadcasts rdy, a_1 receives: $\langle q_1, (1, 2, \_), q_0, (2, \_, \_), q_0, (3, \_, \_)\rangle$,
- a_3 broadcasts rdy, a_1 and a_2 receive: $\langle q_2, (1, 2, 3), q_5, (2, \_, \_), q_0, (3, \_, \_)\rangle$,
- a_2 broadcasts rdy, a_3 receives: $\langle q_2, (1, 2, 3), q_5, (2, \_, \_), q_5, (3, \_, \_)\rangle$,
- a_2 broadcasts go, a_1 receives: $\langle q_3, (1, 2, 3), q_6, (2, \_, \_), q_5, (3, \_, \_)\rangle$,
- a_3 broadcasts hlt, a_1 receives: $\langle q_4, (1, 2, 3), q_6, (2, \_, \_), q_7, (3, \_, \_)\rangle$.

Figure 3 provides an unfolding tree derived from ρ by applying a procedure introduced later. Because agents a_2 and a_3 broadcast to several other agents, they each correspond to several nodes of the tree.

We explain why this tree is an unfolding tree. Condition (i) is trivially satisfied. Condition (ii) holds at every node because the local run of each node exactly broadcasts the specification of the node. Condition (iii) is satisfied at μ_1: $\mathsf{In}_2(\mathbf{lr}(\mu_1)) = \mathsf{rdy} \cdot \mathsf{go} = \mathbf{spec}(\mu_2)$ and $\mathsf{In}_3(\mathbf{lr}(\mu_1)) = \mathsf{rdy} \cdot \mathsf{hlt} = \mathbf{spec}(\mu_3)$. It is also satisfied at μ_2, μ_3 and μ_5 because their local runs only receive rdy and they each have a child with specification rdy. It is trivially satisfied at μ_4 and μ_6 as their local runs have no reception. $\square$

Lemma 16. *Given a signature protocol $\mathcal{P}$ with a state q_f, q_f is coverable in $\mathcal{P}$ if and only if there exists an unfolding tree whose root is labelled by a local run covering q_f. We call such an unfolding tree a* coverability witness.

Proof. Given a run ρ, agent a *satisfies a specification* $w \in \mathcal{M}^*$ in ρ if the sequence of message types broadcast by a admits w as subword.

Let τ be a coverability witness. We prove the following property by strong induction on the depth of μ: for every μ in τ, there exists a run ρ with an agent a whose local run in ρ is $\mathbf{lr}(\mu)$ and who satisfies specification $\mathbf{spec}(\mu)$. This is trivially true for leaves of τ because their local runs have no reception (by condition (iii)) hence are actual runs by themselves. Let μ a node of τ, $u := \mathbf{lr}(\mu)$ and $v_1, \ldots, v_c$ the values received in u. These values are non-initial thanks to condition (i); applying condition (iii) gives the existence of corresponding children $\mu_1, \ldots, \mu_c$ in τ. We apply the induction hypothesis on the subtrees rooted in $\mu_1, \ldots, \mu_c$ to obtain runs $\rho_1, \ldots, \rho_c$ satisfying the specifications of the children of μ. Up to renaming agents, we can assume the set of agents of these runs are disjoint; up to renaming values, we can assume that $v_j = \mathbf{val}(\mu_j)$ for all j and that all agents start with distinct values. We build an initial run ρ whose agents is the union of the agents of the c runs along with a fresh agent a. In ρ, we make ρ_1 to ρ_c progress in parallel and make a follow the local run u, matching each reception with value v_j in u with a broadcast in ρ_j. This is possible because, for all j, $\mathsf{In}_{v_j}(u) \preceq \mathbf{spec}(\mu_j) \preceq \mathsf{Out}_{v_j}(\rho_j)$ (by (ii)).

Conversely, we prove the following by induction on the length of ρ: for every initial run ρ, for every agent a in ρ and for every $v \in \mathbb{N}$, there exists an unfolding tree whose root has as local run the projection of ρ onto a and as specification the v-output of a in ρ. If ρ is the empty run, consider the unfolding tree with a single node whose local run and specification are empty. Suppose now that ρ has non-zero length, let a an agent in ρ, $v \in \mathbb{N}$ and let ρ_p the prefix run of ρ of length $|\rho| - 1$. Let τ_1 the unfolding tree obtained by applying the induction hypothesis to ρ_p, a and v, and consider τ_2 obtained by simply appending the last step of a in ρ to the local run at the root of τ_1. If this last step is a broadcast, we obtain an unfolding tree; if the broadcast value is v, we append the broadcast message type to the specification at the root of τ_2 and we are done. Suppose that, in the last step of ρ, a performs a reception $(q, \mathbf{rec}(m, i, \alpha), q')$ of a message (m, v'). We might need to adapt τ_2 to respect condition (iii) at the root. Let a' the agent broadcasting in the last step of ρ. Let τ_3 the unfolding tree obtained by applying the induction to ρ_p, a' and v'. Let τ_4 the unfolding tree obtained by appending the last broadcast to the local run at the root of τ_3 and the corresponding message type to the specification at the root of τ_3. Attaching τ_4 below the root of τ_2 gives an unfolding tree satisfying the desired properties. □

The unfolding tree τ of Figure 3 is built from ρ of Example 15 using the previous procedure. Observe that the unfolding tree τ is a coverability witness for q_4. However, one can find a smaller coverability witness. Indeed, in the right branch of τ, μ_5 and μ_6 have the same specification, therefore μ_5 can be deleted and replaced with μ_6. More generally, we would have also been able to shorten the tree if we had $\mathbf{spec}(\mu_5) \preceq \mathbf{spec}(\mu_6)$.

Remark 17. With the previous notion of coverability witness, the root has to cover q_f but may have an empty specification. However, we will later need the length of the specification of a node to be equal to the number of tasks that it must carry out. For this reason, we will, in the rest of this paper, consider that the roots of coverability witnesses have a specification of length 1. This can be formally achieved by introducing a new message type m_f that may only be broadcast from q_f and require that, at the root, $\mathsf{spec} = m_f$.

3.3 Bounding the Size of a Coverability Witness

In all the following, we fix a positive instance $(\mathcal{P}, q_f)$ of COVER with $r+1$ registers (*i.e.*, r registers used for reception) and a coverability witness τ of minimal size. We turn the observation above into an argument that will be useful towards bounding the length of branches of a coverability witness:

Lemma 18. *If a coverability witness τ for $(\mathcal{P}, q_f)$ of minimal size has two nodes μ, μ' with μ a strict ancestor of μ' then* $\mathbf{spec}(\mu)$ *cannot be a subword of* $\mathbf{spec}(\mu')$.

Proof. Otherwise, replacing the subtree rooted in μ with the one rooted in μ' would contradict minimality of τ. □

We would now like to use the Length function theorem to bound the height of τ, using the previous lemma. To do so, we need a bound on the size of a node with respect to its depth. The following lemma bounds the number of steps of a local run between two local configurations: we argue that if the local run is long enough we can replace it with a shorter one that can be executed using the same input. This will in turn bound the length of a local run of a node with respect to the size of its specification, which is the first step towards our goal.

Lemma 19. *There exists a primitive recursive function ψ so that, for every local run $u : (q, \nu) \xrightarrow{*} (q', \nu')$, there exists $u' : (q, \nu) \xrightarrow{*} (q', \nu')$ with $|u'| < \psi(|\mathcal{P}|, r)$ and for all value $v' \in \mathbb{N}$, there exists $v \in \mathbb{N}$ such that $\mathsf{In}_{v'}(u') \preceq \mathsf{In}_v(u)$.*

Proof. Let $\psi(n, 0) = n + 1$ and $\psi(n, k + 1) = 2\,\psi(n, k) \cdot (|\Delta|^{2\,\psi(n,k)} + 1) + 1$ for all k. Observe that $\psi(n, k)$ is a tower of exponentials of height k, which is primitive-recursive although non-elementary. A register $i \geq 2$ is *active* in a local run u if u has some '$\downarrow$' action on register i. Let u a local run, k the number of active registers in u, $n := |\mathcal{P}|$ and $M := \psi(n, k)$. We prove by induction on the number k of active registers in u that if $|u| \geq \psi(n, k)$ then u can be shortened.

If $k = 0$, any state repetition can be removed. Suppose that $|u| > \psi(n, k+1)$ and that the set I of active registers of u is such that $|I| = k + 1$. If there exists an infix run of u of length M with only k active registers, we shorten u using the induction hypothesis. Otherwise, every sequence of M steps in u has a '$\downarrow$' on every register of I. Because $|u| > 2M\,(|\Delta|^{2M} + 1)$, u contains at least $|\Delta|^{2M} + 1$ disjoint sequences of length $2M$ and some $s \in \Delta^{2M}$ appears twice: in infix run u_1 first, then in infix run u_2. We build a shorter run u' by removing all steps between u_1 and u_2 and merging u_1 and u_2 (see Fig. 4). We need suitable values

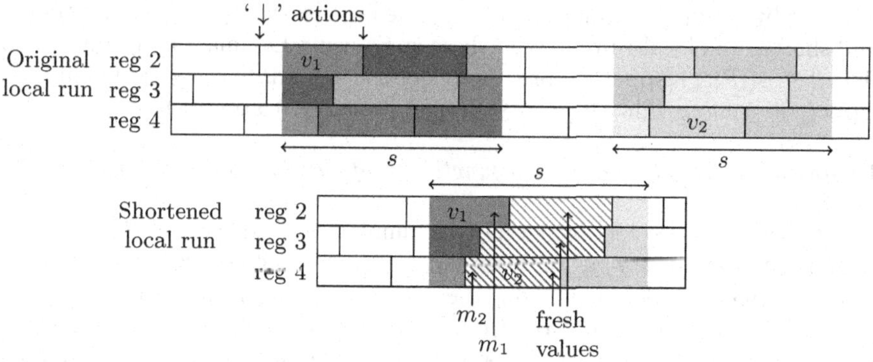

Fig. 4: Illustration of the proof of Lemma 19.

for the reception steps in s in the shortened run u'. For a given register $i \in I$, we would like to pick a ' $\downarrow$ ' step on register i in s, use values from u_1 before that step and values from u_2 after that step. This would guarantee that all equality and disequality tests still pass. However, there is an issue if a value v appears in several registers in u. For example, if $v_1 = v_2 = v$ in Figure 4, we might interleave receptions of v on registers 2 and 4: if we had a $\mathsf{ext}(\mathbf{rec}(m_1, 2, =), v)$ in u_1 and a $\mathsf{ext}(\mathbf{rec}(m_2, 4, =), v)$ in u_2, we could have m_1 before m_2 in $\mathsf{In}_v(u)$ but m_1 after m_2 in $\mathsf{In}_v(u')$, so that we do not have $\mathsf{In}_v(u') \preceq \mathsf{In}_v(u)$. We solve this issue by introducing fresh values between values of u_1 and values of u_2; because $|s| = 2M$, there is a ' $\downarrow$ ' for each register in I in each half of s. In the shortened run u', before the *first* ' $\downarrow$ ' on register i (excluded), we use values of u_1, and after the *last* ' $\downarrow$ ' on register i (included), we use values of u_2. For every value v appearing in register i between these two steps in u_1, we select a fresh value v_f (*i.e.*, a value that does not appear anywhere in the run) and consistently replace v with v_f (hatched blocks in Fig. 4). With this technique, receptions with values from u_1 and receptions with values from u_2 cannot get interleaved in u'. Therefore, for every value that appeared in u, we have $\mathsf{In}_v(u') \preceq \mathsf{In}_v(u)$. Also, for every fresh value v' there is a value v such that $\mathsf{In}_{v'}(u') \preceq \mathsf{In}_v(u)$. Moreover, u' is shorter than u; we conclude by iterating this shortening procedure. □

Using the previous lemma, we will bound the size of a node in τ with respect to its specification therefore with respect to its parent's size. By induction, we will then obtain a bound depending on the depth, and apply the Length function theorem to bound the height of the tree.

Lemma 20. *For all nodes μ, μ' in τ:*

1. $|\mathbf{lr}(\mu)| \leq \psi(|\mathcal{P}|, r)\, |\mathbf{spec}(\mu)|$,
2. *if μ is the child of μ',* $|\mathbf{spec}(\mu)| \leq \psi(|\mathcal{P}|, r)\, |\mathbf{spec}(\mu')|$.

Proof. Thanks to Remark 17, we assume that the specification at the root is of length 1. For the first item, by minimality of τ, $\mathbf{lr}(\mu)$ ends with the last broadcast required by $\mathbf{spec}(\mu)$; we identify in $\mathbf{lr}(\mu)$ the broadcast steps witnessing $\mathbf{spec}(\mu)$ and shorten the local run between these steps using Lemma 19. We thus obtain $|\mathbf{lr}(\mu)| \leq \psi(|\mathcal{P}|, r)\, |\mathbf{spec}(\mu)|$, proving 1. For the second item, by minimality of τ, $|\mathbf{spec}(\mu)| \leq \max_{v \in \mathbb{N}} |\mathsf{In}_v(\mathbf{lr}(\mu'))| \leq |\mathbf{lr}(\mu')| \leq \psi(|\mathcal{P}|, r)\, |\mathbf{spec}(\mu')|$. □

Proposition 21. *There exists a function f of class $\mathscr{F}_{\omega^{|\mathcal{M}|-1}}$ s.t. $|\tau| \leq f(|\mathcal{P}|)$.*

Proof. Let $n := |\mathcal{P}|$, let $r+1$ be the number of registers in $\mathcal{P}$. Thanks to Lemma 18, for all $\mu \neq \mu'$ in τ with μ ancestor of μ', $\mathbf{spec}(\mu)$ is not a subword of $\mathbf{spec}(\mu')$. Let $\mu_1, \ldots, \mu_m$ the node appearing in a branch of τ, from root to leaf. The sequence $\mathbf{spec}(\mu_1), \ldots, \mathbf{spec}(\mu_m)$ is a bad sequence. For all $i \in [1, m]$, $|\mathbf{spec}(\mu_{i+1})| \leq \psi(n, r)\, |\mathbf{spec}(\mu_i)|$ by Lemma 20. By direct induction, $|\mathbf{spec}(\mu_i)|$ is bounded by $g^{(i)}(n)$ where $g : n \mapsto n\, \psi(n, n)$ is a primitive recursive function. Let h of class $\mathscr{F}_{\omega^{|\mathcal{M}|-1}}$ the function obtained when applying the Length function theorem on g and $\mathcal{M}$; we have $m \leq h(n)$.

By immediate induction, thanks to Lemma 20.2, for every node μ at depth d, $|\mathbf{spec}(\mu)| \leq \psi(n, r)^{d+1}$ which, by Lemma 20.1 and because $d \leq h(n)$, bounds the size of every node by $h'(n) = \psi(n, n)^{h(n)+2}$. By minimality of τ, the number of children of a node is bounded by the number of values appearing in its local run hence by $h'(n)$, so the total number of nodes in τ is bounded by $h'(n)^{h(n)+1}$ and the size of τ by $f(n) := h'(n)^{h(n)+2}$. Because $\mathscr{F}_{\omega^{|\mathcal{M}|-1}}$ is closed under composition with primitive-recursive functions, f is in $\mathscr{F}_{\omega^{|\mathcal{M}|-1}}$. □

The previous argument shows that Cover for signature protocols is decidable and lies in complexity class $\mathbf{F}_{\omega^\omega}$. Because the hardness from Proposition 11 holds for signature protocols, Cover is in fact complete for this complexity class.

We now extend this method to the general case.

4 Coverability Decidability in the General Case

4.1 Generalizing Unfolding Trees

In the general case, a new phenomenon appears: an agent may broadcast a value that it did not initially have but that it has received and stored. In particular, an agent starting with value v could broadcast v then require someone else to make a broadcast with value v as well. For example, in the run described in Example 4, 1 is initially a value of a_1 that a_2 receives and rebroadcasts to a_1.

We now have two types of specifications. *Boss specifications* describe the task of broadcasting with one of its own initial values; this is the specification we had in signature protocols and, as before, it consists of a word $\mathsf{bw} \in \mathcal{M}^*$ describing a sequence of message types that should be all broadcast with the same value. *Follower specifications* describe the task of broadcasting with a non-initial value received previously. More precisely, a follower specification is a pair $(\mathsf{fw}, \mathsf{fm}) \in \mathcal{M}^* \times \mathcal{M}$ asking to broadcast a message (fm, v) under the condition of previously receiving the sequence of message types fw with value v.

A key idea is that, if an agent that had v initially receives some message (m, v), then intuitively we can isolate a subset of agents that did not have v initially but that are able to broadcast (m, v) after receiving a sequence of messages with that value. We can then copy them many times in the spirit of the copycat principle. Each copy receives the necessary sequence of messages in parallel, and they then provide us with an unbounded supply of messages (m, v). In short, if an agent broadcasts (m, v) while not having v as an initial value, then we can consider that we have an unlimited supply of messages (m, v).

Example 22. Assume that $\mathbb{A} = \{a_1, a_2, a_3\}$ and let v be initial for a_1. Consider an execution where the broadcasts with value v are: a_1 broadcasts $\mathsf{a} \cdot \mathsf{b}$, then a_2 broadcasts c, then a_1 broadcasts a^3 then a_3 broadcasts b. The follower specification of a_2's task would be of the form (w, c) where $w \preceq \mathsf{a} \cdot \mathsf{b}$: a_2 must be able to broadcast (c, v) once $\mathsf{a} \cdot \mathsf{b}$ has been broadcast with value v. By contrast, a_3's follower specification would be of the form $(w \cdot w', \mathsf{c})$ where $w \preceq \mathsf{a} \cdot \mathsf{b}$ and $w' \in \{\mathsf{a}, \mathsf{c}\}^*$ is a subword of a^3 enriched with as many c as desired, because a_2 may be cloned at will. For example, one could have $w = \mathsf{b}$ and $w' = \mathsf{c} \cdot \mathsf{a} \cdot \mathsf{c}^4 \cdot \mathsf{a} \cdot \mathsf{c}^2$. This idea is formalized in the full version of the paper with the notion of *decomposition*. Using this notion, the previous condition becomes: $w \cdot w'$ *admits decomposition* $(\mathsf{a} \cdot \mathsf{b}, \mathsf{c}, \mathsf{a}^3)$. □

In our new *unfolding trees*, a node is either a *boss node* or a *follower node*, depending on its type of specification. A *boss node* with a boss specification bw must broadcast that sequence of message types with one of its initial values. A *follower node* μ with follower specification $(\mathsf{fw}, \mathsf{fm})$ is allowed to receive sequence of messages fw with value $\mathbf{val}(\mu)$ (which must be non-initial) without it being broadcast by its children. Other conditions are similar to the ones for signature protocols: if μ is a node and $v \neq \mathbf{val}(\mu)$ a non-initial value received in its local run, μ must have a boss child broadcasting this word. Moreover, for each (m, v) received where v is an initial value of the local run, μ must have a follower child that is able to broadcast (m, v) after receiving messages sent previously with value v; the formal statement is more technical because it takes into account the observation of Example 22. The formal definition of *unfolding tree* is given in the full version.

Example 23. Figure 5 depicts the unfolding tree associated to a_1 in the run of Example 4. Follower node μ_3 can have a m_2 reception that is not matched by its children because m_2 is in $\mathsf{fw}(\mu_3)$. μ_1 broadcasts $(m_2, 1)$ before receiving $(m_4, 1)$ hence the follower specification of μ_3 witnesses broadcast of $(m_4, 1)$. □

A *coverability witness* is again an unfolding tree whose root covers q_f (or broadcasts a message m_f, see Remark 17), with the extra condition that the root is a boss node (a follower node implicitly relies on its parent's ability to broadcast).

Proposition 24. *An instance of* COVER $(\mathcal{P}, q_f)$ *is positive if and only if there exists a coverability witness for that instance.*

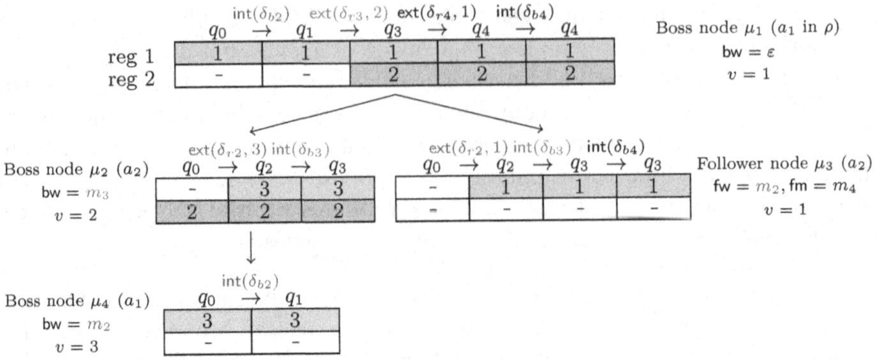

Fig. 5: Example of an unfolding tree. δ_{ri} (resp. δ_{bi}) denotes the reception (resp. broadcast) transition of message m_i in the protocol described in Fig. 1. Values that are never broadcast are omitted and written as '_'.

Proof sketch. The proof is quite similar to the one of Lemma 16, but is made more technical by the addition of follower nodes. When translating an unfolding tree to a run, if the root of the tree is a follower node μ of specification (fw, fm), then we actually obtain a *partial run*, *i.e.*, a run except that the receptions from fw are not matched by broadcasts in the run. We then combine this partial run with the run corresponding to the parent of μ and with the runs of other children of μ so that every reception is matched with a broadcast. For the translation from run to tree, we inductively construct the tree by extracting from the run the agents and values responsible for satisfying the specifications of each node and analyzing the messages they receive to determine their set of children (as in Example 22). □

Bounding the Size of the Unfolding Tree. Our aim is again to bound the size of a minimal coverability witness. In the following, we fix an instance $(\mathcal{P}, q_f)$ with r registers and a coverability witness of minimal size. We start by providing new conditions under which a branch can be shortened; for boss specifications, it is the condition of Lemma 18 but for follower specifications, the subword relation goes the opposite direction because the shorter the requirement fw, the better.

Lemma 25. *Let $\mu \neq \mu'$ be two nodes of τ such that μ is an ancestor of μ'. If one of those conditions holds, then τ can be shortened (contradicting its minimality):*

- *μ and μ' are boss nodes with boss specifications respectively* bw *and* bw′, *and* bw $\preceq$ bw′;
- *μ and μ' are follower nodes with follower specifications respectively* (fw, fm) *and* (fw′, fm′), *and* fw′ $\preceq$ fw *and* fm′ = fm.

We can generalize Lemma 19 to bound the size of a node by the number of messages that it must broadcast times a primitive-recursive function $\psi(|\mathcal{P}|, r)$. The proof is more technical than the one of Lemma 19 but the idea is essentially

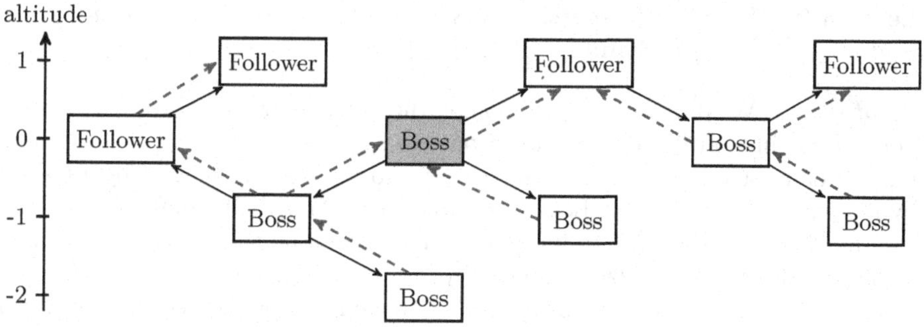

Fig. 6: Rearrangement of a tree. The root is in red, black solid arrows connect parents to children, blue dashed arrows highlight that long words of messages are sent upwards.

the same. The formal statement is given below. One can therefore bound the size of a node with respect to the size of the nodes that it must broadcast to.

Lemma 26. *There exists a primitive recursive function* ψ *such that, for every protocol* $\mathcal{P}$ *with* r *registers, for all local runs* $u_0 : (q_0, \nu_0) \xrightarrow{*} (q, \nu)$, $u : (q, \nu) \xrightarrow{*} (q', \nu')$, $u_f : (q', \nu') \xrightarrow{*} (q_f, \nu_f)$, *there exists a local run* $u' : (q, \nu) \xrightarrow{*} (q', \nu')$ *with* $|u'| \leq \psi(|\mathcal{P}|, r)$ *and for all* $v' \in \mathbb{N}$*:*

1. *if* v' *appears in* u_0, u, *or* u_f, $\mathsf{In}_{v'}(u') \preceq \mathsf{In}_{v'}(u)$,
2. *otherwise, there exists* $v \in \mathbb{N}$, *not initial in* u_0, *such that* $\mathsf{In}_{v'}(u') \preceq \mathsf{In}_v(u)$.

It is however now much harder than in the signature case to bound the size of the coverability witness. Indeed, the broadcasts no longer go only from children to parents in the unfolding tree. If μ_p is the parent of μ_c, then μ_c broadcasts to μ_p if μ_c is a boss node, but μ_p broadcasts to μ_c if μ_c is a follower node, in which case μ_c only broadcasts one message to μ_p. Therefore, we cannot in general bound $|\mu_p|$ with respect to $|\mu_c|$ nor $|\mu_c|$ with respect to $|\mu_p|$, making us unable to apply the Length function theorem immediately.

This leads us to arrange the unfolding tree so that long broadcast sequences are sent upwards, using the notion of altitude depicted in Figure 6, formally defined as follows. The *altitude* of the root is 0, the altitude of a boss node is the altitude of its parent minus one, and the altitude of a follower node is the altitude of its parent plus one. We denote the altitude of μ by $\mathbf{alt}(\mu)$. This way the nodes of maximal altitude are the ones that do not need to send long sequences of messages. We will bound the size of nodes with respect to their altitude, from the highest to the lowest, and then use the Length function theorem to bound the maximal and minimal altitudes. We present here a sketch of the proof.

Let $\mathbf{altmax} \geq 0$ (resp. $\mathbf{altmin} \leq 0$) denote the maximum (resp. minimum) altitude in τ. We first bound the size of a node with respect to the difference between its altitude and $\mathbf{altmax}$.

Lemma 27. *There is a primitive recursive function f_0 such that, for every node μ of τ, $|\mu| \leq f_0(|\mathcal{P}| + \mathbf{altmax} - \mathbf{alt}(\mu))$.*

Proof sketch. We proceed by induction on the altitude, from highest to lowest. A node of maximal altitude has at most one message to broadcast (a follower node must broadcast one message to its parent), so its size is bounded by $\psi(|\mathcal{P}|, r)$ by Lemma 26 (applying the Lemma to its local run minus its final step, *i.e.*, the step making the broadcast to its parent). Let μ be a node of τ whose neighbors of higher altitude have size bounded by K. We claim that $|\mu| \leq (\psi(|\mathcal{P}|, r) + 2)\,(|\mathcal{M}|\, r\, K + K)$, with ψ the primitive-recursive function defined in Lemma 26. The idea is similar to the one for Lemma 20. The neighbors of higher altitude are the nodes which require sequences of messages from μ. Their size bounds the number of messages that μ needs to send; we then apply Lemma 26 to bound the size of the local run of μ. We finally obtain f_0 by iteratively applying the inequality above. □

We now bound **altmax** and **altmin**:

Lemma 28. **altmax** *and* |**altmin**| *are bounded by a function of class $\mathscr{F}_{\omega^{|\mathcal{M}|}}$.*

Proof sketch. We first bound **altmax**. Consider a branch of τ that has a node at altitude **altmax**. We follow this branch from the root to a node of altitude **altmax**: for every $j \in [1, \mathbf{altmax}]$, let μ_j be the first node of the branch that has altitude j. All such nodes are necessarily follower nodes as they are above their parent. Sequence $\mu_{\mathbf{altmax}}, \ldots, \mu_2, \mu_1$ is so that the ith term is at altitude $\mathbf{altmax} - i$ hence its size is bounded by $f_0(|\mathcal{P}| + i)$ (Lemma 27). With the observation of Lemma 25, we retrieve from the follower specifications of this sequence of nodes a bad sequence and we apply the Length function theorem to bound **altmax**. This yields in turn a bound on the size of the root of τ. In order to bound **altmin**, we proceed similarly, using boss nodes this time. We follow a branch from the root to a node of altitude **altmin**. The sequence of nodes that are lower than all previous ones yields a sequence of boss specifications, which is a bad sequence by Lemma 25, and whose growth can be bounded using Lemma 27 and the bound on **altmax**. We apply the Length function theorem to bound |**altmin**|. □

Once we have bounded **altmax** and **altmin**, we can infer a bound on the size of all nodes (Lemma 27), and then on the length of branches: by minimality, a branch cannot have two nodes with the same specification. The bound on the size of the tree then follows from the observation that bounding the size of nodes of τ also allows to bound their number of children.

We obtain a computable bound (of the class $\mathscr{F}_{\omega^\omega}$) on the size of a minimal coverability witness if it exists. Our decidability procedure computes that bound, enumerates all trees of size below the bound and checks for each of them whether it is coverability witness. This yields the main result of this paper:

Theorem 12. COVER *for BNRA is decidable and* $\mathbf{F}_{\omega^\omega}$*-complete.*

4.2 Undecidability of the target problem

A natural next problem, after COVER, is the target problem (TARGET). Our COVER procedure heavily relies on the ability to add agents at no cost. For TARGET we need to guarantee that those agents can then reach the target state, which makes the problem harder. In fact, TARGET is undecidable, which indicates that our model lies at the frontier of decidability.

Proposition 29. *TARGET is undecidable for BNRA, even with two registers.*

Proof sketch. We simulate a Minsky machine with two counters. As in Proposition 11, each agent starts by storing some other agent's identifier, called its "predecessor". It then only accepts messages from its predecessor. As there are finitely many agents, there is a cycle in the predecessor graph.

In a cycle, we use the fact that *all* agents must reach state q_f to simulate faithfully a run of the machine: agents alternate between receptions and broadcasts so that, in the end, they have received and sent the same number of messages, implying that no message has been lost along the cycle. We then simulate the machine by having an agent (the leader) choose transitions and the other ones simulate the counter values by memorizing a counter (1 or 2) and a binary value (0 or 1). For instance, an increment of counter 1 takes the form of a message propagated in the cycle from the leader until it finds an agent simulating counter 1 and having bit 0. This agent switches to 1 and sends an acknowledgment that propagates back to the leader. □

5 Cover in 1-BNRA

In this section, we establish the NP-completeness of the restriction of COVER to BNRA with one register per agent, called 1-BNRA. Here we simply sketch the key observations that allow us to abstract runs into short witnesses, leading to an NP algorithm for the problem.

In 1-BNRA, thanks to the copycat principle, any message can be broadcast with a fresh value, therefore one can always circumvent '$\neq$' tests. In the end, our main challenge for 1-BNRA is '$=$' tests upon reception. For this reason, we look at clusters of agents that share the value in their registers.

Consider a run in which some agent a reaches some state q,; we can duplicate a many times to have an unlimited supply of agents in state q. Now assume that, at some point in the run, agent a stored a received value. Consider the last storing action performed by a: a was in a state q_1 and performed transition $(q_1, \mathbf{rec}(m, 1, \downarrow), q_2)$ upon reception of a message (m, v). Because we can assume that we have an unlimited supply of agents in q_1 thanks to the copycat principle, we can make as many agents as we want take transition $(q_1, \mathbf{rec}(m, 1, \downarrow), q_2)$ at the same time as a by receiving the same message (m, v). These new agents end up in q_2 with value v, and then follow a along every transition until they all reach q, still with value v. In summary, because a has stored a value in the run, we can have an unlimited supply of agents in state q with the same value as a.

Following those observations, we define an abstract semantics with abstract configurations of the form (S, b, K) with $S, K \subseteq Q$ and $b \in Q \cup \{\perp\}$. The first component S is a set of states that we know we can cover (hence we can assume that there are arbitrarily many agents in all these states). We start with $S = \{q_0\}$ and try to increase it. To do so, we use the two other components (the *gang*) to keep track of the set of agents sharing a value v: b (the *boss*) is the state of the agent which had that value at the start, K (the *clique*) is the set of states covered by other agents with that value. As mentioned above, we may assume that every state of K is filled with as many agents with value v as we need. We will thus define abstract steps which allow to simulate steps of the agents with the value we are following. When they cover states outside of S, we may add those to S and reset b to q_0 and K to $\emptyset$, to then start following another value. We can bound the length of relevant abstract runs, and thus use them as witnesses for our NP upper bound.

The NP lower bound follows from a reduction from 3SAT. An agent a sends a sequence of messages representing a valuation, with its identifier, to other agents who play the role of an external memory by broadcasting back the valuation. This then allows a to check the satisfaction of a 3SAT formula.

Theorem 30. *The coverability problem for 1-BNRA is* NP*-complete.*

6 Conclusion

We established the decidability (and $\mathbf{F}_{\omega^\omega}$-completeness) of the coverability problem for BNRA, as well as the NP-completeness of the problem for 1-BNRA. Concerning future work, one may want to push decidability further, for instance by enriching our protocols with inequality tests, as done in classical models such as data nets [15]. Reductions of other distributed models to this one are also being studied.

Acknowledgements. We are grateful to Arnaud Sangnier for encouraging us to work on BNRA, for the discussions about his work in [10] and for his valuable advice. We also thank Philippe Schnoebelen for the interesting discussion and Sylvain Schmitz for the exchange on complexity class $\mathbf{F}_{\omega^\omega}$ and related topics.

References

1. Abdulla, P.A., Atig, M.F., Kara, A., Rezine, O.: Verification of dynamic register automata. In: 34th International Conference on Foundation of Software Technology and Theoretical Computer Science, FSTTCS 2014. LIPIcs, vol. 29, pp. 653–665. Schloss Dagstuhl - Leibniz-Zentrum für Informatik (2014). `https://doi.org/10.4230/LIPIcs.FSTTCS.2014.653`
2. Abdulla, P.A., Atig, M.F., Kara, A., Rezine, O.: Verification of buffered dynamic register automata. In: Networked Systems, NETYS 2015. Lecture Notes in Computer Science, vol. 9466, pp. 15–31. Springer (2015). `https://doi.org/10.1007/978-3-319-26850-7_2`

3. Abdulla, P.A., Jonsson, B.: Verifying programs with unreliable channels. Information and Computation **127**(2), 91–101 (1996). https://doi.org/10.1006/inco.1996.0053
4. Balasubramanian, A.R., Bertrand, N., Markey, N.: Parameterized verification of synchronization in constrained reconfigurable broadcast networks. In: Tools and Algorithms for the Construction and Analysis of Systems, TACAS 2018. Lecture Notes in Computer Science, vol. 10806, pp. 38–54. Springer (2018). https://doi.org/10.1007/978-3-319-89963-3_3
5. Balasubramanian, A.R., Guillou, L., Weil-Kennedy, C.: Parameterized analysis of reconfigurable broadcast networks. In: Foundations of Software Science and Computation Structures, FoSSaCS 2022. Lecture Notes in Computer Science, vol. 13242, pp. 61–80. Springer (2022). https://doi.org/10.1007/978-3-030-99253-8_4
6. Bollig, B., Ryabinin, F., Sangnier, A.: Reachability in distributed memory automata. In: Annual Conference on Computer Science Logic, CSL 2021. LIPIcs, vol. 183, pp. 13:1–13:16. Schloss Dagstuhl - Leibniz-Zentrum für Informatik (2021). https://doi.org/10.4230/LIPIcs.CSL.2021.13
7. Brand, D., Zafiropulo, P.: On communicating finite-state machines. Journal of the ACM **30**(2), 323–342 (1983). https://doi.org/10.1145/322374.322380
8. Chambart, P., Schnoebelen, P.: The ordinal recursive complexity of lossy channel systems. In: Annual IEEE Symposium on Logic in Computer Science, LICS 2008. pp. 205–216. IEEE Computer Society (2008). https://doi.org/10.1109/LICS.2008.47
9. Chini, P., Meyer, R., Saivasan, P.: Liveness in broadcast networks. Computing **104**(10), 2203–2223 (2022). https://doi.org/10.1007/s00607-021-00986-y
10. Delzanno, G., Sangnier, A., Traverso, R.: Parameterized verification of broadcast networks of register automata. In: Reachability Problems , RP 2013. Lecture Notes in Computer Science, vol. 8169, pp. 109–121. Springer (2013). https://doi.org/10.1007/978-3-642-41036-9_11
11. Delzanno, G., Sangnier, A., Traverso, R., Zavattaro, G.: On the complexity of parameterized reachability in reconfigurable broadcast networks. In: IARCS Annual Conference on Foundations of Software Technology and Theoretical Computer Science, FSTTCS 2012. LIPIcs, vol. 18, pp. 289–300. Schloss Dagstuhl - Leibniz-Zentrum für Informatik (2012). https://doi.org/10.4230/LIPIcs.FSTTCS.2012.289
12. Delzanno, G., Sangnier, A., Zavattaro, G.: Parameterized verification of ad hoc networks. In: CONCUR 2010. Lecture Notes in Computer Science, vol. 6269, pp. 313–327. Springer (2010). https://doi.org/10.1007/978-3-642-15375-4_22
13. Emerson, E.A., Namjoshi, K.S.: On model checking for non-deterministic infinite-state systems. In: Annual IEEE Symposium on Logic in Computer Science, LICS 1998. pp. 70–80. IEEE Computer Society (1998). https://doi.org/10.1109/LICS.1998.705644
14. Esparza, J., Finkel, A., Mayr, R.: On the verification of broadcast protocols. In: 14th Annual IEEE Symposium on Logic in Computer Science, Trento, Italy, July 2-5, 1999. pp. 352–359. IEEE Computer Society (1999). https://doi.org/10.1109/LICS.1999.782630
15. Haddad, S., Schmitz, S., Schnoebelen, P.: The ordinal-recursive complexity of timed-arc petri nets, data nets, and other enriched nets. In: Proceedings of the 27th Annual IEEE Symposium on Logic in Computer Science, LICS 2012, Dubrovnik, Croatia, June 25-28, 2012. pp. 355–364. IEEE Computer Society (2012). https://doi.org/10.1109/LICS.2012.46

16. Higman, G.: Ordering by divisibility in abstract algebras. Proceedings of the London Mathematical Society **s3-2**(1), 326–336 (1952). https://doi.org/10.1112/plms/s3-2.1.326
17. Lasota, S.: Decidability border for petri nets with data: WQO dichotomy conjecture. In: Kordon, F., Moldt, D. (eds.) Application and Theory of Petri Nets and Concurrency - 37th International Conference, PETRI NETS 2016, Toruń, Poland, June 19-24, 2016. Proceedings. Lecture Notes in Computer Science, vol. 9698, pp. 20–36. Springer (2016). https://doi.org/10.1007/978-3-319-39086-4_3, https://doi.org/10.1007/978-3-319-39086-4_3
18. Lazic, R., Newcomb, T.C., Ouaknine, J., Roscoe, A.W., Worrell, J.: Nets with tokens which carry data. Fundam. Informaticae **88**(3), 251–274 (2008). https://doi.org/10.1007/978-3-540-73094-1_19
19. Minsky, M.L.: Computation: Finite and Infinite Machines. Prentice-Hall, Inc., USA (1967)
20. Rezine, O.: Verification of networks of communicating processes: Reachability problems and decidability issues. Ph.D. thesis, Uppsala University, Sweden (2017)
21. Rosa-Velardo, F.: Ordinal recursive complexity of unordered data nets. Information and Computation **254**, 41–58 (2017). https://doi.org/10.1016/j.ic.2017.02.002
22. Sangnier, A.: Erratum to parameterized verification of broadcast networks of register automata (2023), https://www.irif.fr/~sangnier/publications.html
23. Schmitz, S.: Complexity hierarchies beyond elementary. ACM Transactions on Computation Theory **8**(1), 3:1–3:36 (2016). https://doi.org/10.1145/2858784
24. Schmitz, S., Schnoebelen, P.: Multiply-recursive upper bounds with Higman's lemma. In: International Colloquium on Automata, Languages and Programming, ICALP 2011. Lecture Notes in Computer Science, vol. 6756, pp. 441–452. Springer (2011). https://doi.org/10.1007/978-3-642-22012-8_35
25. Schmitz, S., Schnoebelen, P.: The power of well-structured systems. In: D'Argenio, P.R., Melgratti, H.C. (eds.) CONCUR 2013 - Concurrency Theory - 24th International Conference, CONCUR 2013, Buenos Aires, Argentina, August 27-30, 2013. Proceedings. Lecture Notes in Computer Science, vol. 8052, pp. 5–24. Springer (2013). https://doi.org/10.1007/978-3-642-40184-8_2
26. Schnoebelen, P.: Verifying lossy channel systems has nonprimitive recursive complexity. Information Processing Letters **83**(5), 251–261 (2002). https://doi.org/10.1016/S0020-0190(01)00337-4

Author Index

N. Kobayashi and J. Worrell (Eds.): FoSSaCS 2024, LNCS 14575, pp. 271–272, 2024.
https://doi.org/10.1007/978-3-031-57231-9

Printed in the United States
by Baker & Taylor Publisher Services